Flammability testing of materials used in construction, transport and mining

Related titles:

Corrosion in reinforced concrete structures
(ISBN-13: 978-1-85573-768-6; ISBN-10: 1-85573-768-X)
In this authoritative new book the fundamental aspects of corrosion in concrete are analysed in detail. An overview of current monitoring techniques together with a discussion of practical applications and current numerical methods that simulate the corrosion process provides the civil and structural engineer with an invaluable guide to best practice when it comes to design aimed at minimising the effects of corrosion in concrete. The corrosion protective properties of concrete and modified cements are also discussed. The most frequently used stainless steels are examined together with an analysis of their reinforcement properties. Special attention is given to their handling and their welding requirements and the economics of their use. A comprehensive overview of surface treatments and corrosion inhibitors is presented alongside their practical applications as well as detailed coverage of electrochemical protection and maintenance techniques.

The deformation and processing of structural materials
(ISBN-13: 978-1-85573-738-9; ISBN-10: 1-85573-738-8)
This new study focuses on the latest research in the performance of a wide range of materials used in the construction of structures particularly structural steels. It considers each material's processing and its deformation behaviour in structural applications. This book will help engineers by providing them with a better understanding of the performance of the major structural materials (especially metals) under different conditions in order to select the right type of material for a job and for setting design specifications. It also shows how the microstructural composition of materials is affected by processing and what influence this has on its subsequent *in situ* performance. This book will be the first to give such comprehensive coverage to the deformation and processing of all types of structural materials and will be a valuable resource for researchers in mechanical, civil and structural engineering.

Engineering catastrophes (3rd edn)
(ISBN-13: 978-1-84569-016-8; ISBN-10: 1-84569-016-8)
This new edition of a well received and popular book contains a general update of historical data, more material concerning road and rail accidents and, most importantly, a new chapter on the human factor. The author provides a broad survey of the accidents to which engineering structures and vehicles may be subject. Historical records are analysed to determine how loss and fatality rates vary with time and these results are displayed in numerous graphs and tables. Notable catastrophes such as the sinking of the *Titanic* and the *Estonia* ferry disaster are described. Natural disasters are considered generally, with more detail in this edition on earthquake-resistant buildings.

Details of these and other Woodhead Publishing materials books and journals, as well as materials books from Maney Publishing, can be obtained by:

- visiting our web site at www.woodheadpublishing.com
- contacting Customer Services (e-mail: sales@woodhead-publishing.com; fax: +44 (0) 1223 893694; tel.: +44 (0) 1223 891358 ext. 30; address: Woodhead Publishing Limited, Abington Hall, Abington, Cambridge CB1 6AH, England)

If you would like to receive information on forthcoming titles, please send your address details to: Francis Dodds (address, tel. and fax as above; email: francisd@woodhead-publishing.com). Please confirm which subject areas you are interested in.

Maney currently publishes 16 peer-reviewed materials science and engineering journals. For further information visit www.maney.co.uk/journals.

Flammability testing of materials used in construction, transport and mining

Edited by
Vivek B. Apte

TH
1091
F52
2006
WEB

Woodhead Publishing and Maney Publishing
on behalf of
The Institute of Materials, Minerals & Mining

CRC Press
Boca Raton Boston New York Washington, DC

WOODHEAD PUBLISHING LIMITED
Cambridge England

Woodhead Publishing Limited and Maney Publishing Limited on behalf of
The Institute of Materials, Minerals & Mining

Published by Woodhead Publishing Limited, Abington Hall, Abington,
Cambridge CB1 6AH, England
www.woodheadpublishing.com

Published in North America by CRC Press LLC, 6000 Broken Sound Parkway, NW,
Suite 300, Boca Raton, FL 33487, USA

First published 2006, Woodhead Publishing Limited and CRC Press LLC
© Woodhead Publishing Limited, 2006
The authors have asserted their moral rights.

British Library Cataloguing in Publication Data
A catalogue record for this book is available from the British Library.

Library of Congress Cataloging in Publication Data
A catalog record for this book is available from the Library of Congress.

Woodhead Publishing Limited ISBN-13: 978-1-85573-935-2 (book)
Woodhead Publishing Limited ISBN-10: 1-85573-935-6 (book)
Woodhead Publishing Limited ISBN-13: 978-1-84569-104-2 (e-book)
Woodhead Publishing Limited ISBN-10: 1-84569-104-0 (e-book)
CRC Press ISBN-10: 0-8493-3468-3
CRC Press order number: WP3468

The publishers' policy is to use permanent paper from mills that operate a sustainable
forestry policy, and which has been manufactured from pulp which is processed using
acid-free and elementary chlorine-free practices. Furthermore, the publishers ensure
that the text paper and cover board used have met acceptable environmental
accreditation standards.

Project managed by Macfarlane Production Services, Dunstable, Bedfordshire
(macfarl@aol.com)
Typeset by Godiva Publishing Services Ltd, Coventry, West Midlands
Printed & bound in Great Britain by TJ International Ltd, Padstow, Cornwall

Contents

13 Flammability testing in the mining sector   302

H C VERAKIS, US Department of Labor, USA

14 Flammability tests for railway passenger cars   336

R D PEACOCK and R W BUKOWSKI, NIST Building and Fire Research Laboratory, USA

15 Ships and submarines   361

B Y LATTIMER, Hughes Association, Inc., USA

(* = main contact)

Chapter 1
Mr Chris Lautenberger
60A Hesse Hall
Department of Mechanical
   Engineering
University of California, Berkeley
Berkeley, CA 94720
USA

Tel: 510-643-5282
Fax: 510-642-1850
E-mail: clauten@me.berkeley.edu

Professor Jose Torero
School of Engineering and
   Electronics
The University of Edinburgh
The King's Buildings
Edinburgh EH9 3JL
UK

Tel: 44-131-650-5723
Fax: 44-131-650-5736
E-mail: jltorero@staffmail.ed.ac.uk

Professor Carlos Fernandez-Pello*
6105A Etcheverry Hall
Department of Mechanical
   Engineering

University of California, Berkeley
Berkeley, CA 94720
USA

Tel: (510) 642-6554
Fax: (510) 642-1850
E-mail: ferpello@me.berkeley.edu

Chapter 2
Dr Marc Janssens
Department of Fire Technology
Southwest Research Institute
6220 Culebra Road
San Antonio, TX 78238-5166
USA

Tel: +1-210-522-6655
Fax: +1-210-522-3377
E-mail: mjanssens@swri.org

Chapter 3
Dr Birgit Östman
SP Trätek / Wood Technology
Box 5609
SE-114 86 Stockholm
Sweden

Tel: +46 8 762 1800 (dir 1871)
Fax: +46 8 762 1801
E-mail: Birgit.Ostman@sp.se

Chapter 4
Mr Edmond Soja* and
    Ms Colleen Wade
BRANZ Ltd
Private Bag 50908
Porirua City
New Zealand

Tel: +64 4 237 1170
E-mail: edsoja@branz.co.nz

Chapter 5
Ms Colleen Wade
BRANZ Ltd
Private Bag 50908
Porirua City
New Zealand

Tel: +64 4 2371178
E-mail: colleenwade@branz.co.nz

Chapter 6
Dr Patrick Van Hees
SP Swedish National Testing and
    Research Institute
Fire Technology
P O Box 857
S-501 15 Borås
Sweden

E-mail: patrick.van.hees@sp.se

Chapter 7
Dr Charles Fleischmann
Dept of Civil Engineering
University of Canterbury
Private Bag 4800
Christchurch
New Zealand

Tel: 364-2399
E-mail: charles.fleischmann@
    canterbury.ac.nz

Chapter 8
Dr Bjorn Sundstrom
SP Swedish National Testing and
    Research Institute
Fire Technology
P O Box 857
S-501 15 Borås
Sweden

E-mail: bjorn.sundstrom@sp.se

Chapter 9
Dr Margaret Simonson
SP Swedish National Testing and
    Research Institute
Fire Technology
P O Box 857
S-501 15 Borås
Sweden

E-mail: margaret.simonson@sp.se

Chapter 10
Kuma Sumathipala
American Forest & Paper
    Association
1111 Nineteenth Street, NW
Suite 800
Washington, DC 20036
USA

E-mail:
    Kuma_Sumathipala@afandpa.org

Robert H. White
Forest Products Laboratory
USDA Forest Service
1 Gifford Pinchot Drive
Madison, WI 53726-2398
USA

E-mail: rhwhite@fs.fed.us;
    rhwhyte@wisc.edu

Chapter 11
Dr Haukur Ingason
SP Swedish National Testing and
    Research Institute
Fire Technology
P O Box 857
S-501 15 Borås
Sweden

Tel: +4633165197
Fax: +463341775
E-mail: haukur.ingason@sp.se

Chapter 12
Dr James M. Peterson
Boeing Commercial Airplanes
PO Box 3707 M/C 73-48
Seattle, WA 98124-2207
USA

Tel: +1-425-237-8243
Fax: +1-425-237-0052
E-mail:
    james.m.peterson@boeing.com

Chapter 13
Mr Harry C. Verakis
Mine Safety and Health
    Administration (MSHA)
Industrial Park Road
Box 251, Building 1
Triadelphia, WV 26059
USA

E-mail: verakis-harry@dol.gov

Chapter 14
Richard D. Peacock* and
    Richard W. Bukowski
National Institute of Standards and
    Technology

100 Bureau Drive
Stop 8664
Gaithersburg, MD 20899-8664
USA

Tel: (301) 975-6664
E-mail: richard.peacock@nist.gov

Chapter 15
Brian Y. Lattimer
Hugh Associates, Inc.
3610 Commerce Drive
Suite 817
Baltimore, MD 21227-1652
USA

E-mail: blattimer@haifire.com

Chapter 16
Haihui Wang
Laboratory of Fire Science and
    Technology
CSIRO Manufacturing and
    Infrastructure Technology
PO Box 310
North Ryde, NSW 1670
Australia

Bogdan Z. Dlugogorski* and
    Eric M. Kennedy
Process Safety and Environment
    Protection Research Group
School of Engineering
The University of Newcastle
Callaghan, NSW 2308
Australia

Fax: +61 2 4921 6920
E-mail: Bogdan.Dlugogorski@
    newcastle.edu.au

Introduction

V A P T E , formerly CSIRO, Australia

The editor and the authors are pleased to present this book on flammability testing of materials and products used in the construction, transportation and mining sectors. A vast array of flammability tests exists for different materials and products for a variety of end uses. However, it is not the intent of this book to present encyclopaedic information on all flammability tests available internationally.

The objective of the book is to reflect current thinking and trends on flammability testing and allied topics, with applications in construction, transportation and mining industries. The book gives an overview of the subject under discussion, and the readers are encouraged to obtain details from the references cited at the end of each chapter. With this viewpoint, the book is designed as a collection of papers written by international experts in fire science and engineering. The book would serve as a good source of information for those who wish to know what flammability is, and what test methods are used to assess the flammability of materials and products for various applications.

Current generation flammability tests, which are the outcomes of the latest advances in fire research are summarised. Traditional *ad-hoc* tests, still practised in many industries are also discussed. Compliance criteria for some applications are presented wherever available and possible.

The current trend in fire research is to minimise the use of *ad-hoc* flammability tests, and instead, use tests which measure 'fundamental' flammability properties, such as heat release rate, mass loss rate, heat of combustion, heat of vapourisation, critical heat flux for ignition, thermal inertia, ignition temperature, radiant fraction, smoke extinction area and combustion species yields under a variety of externally imposed test conditions. These 'fundamental' properties are then input into mathematical fire models to predict and assess the behaviour of large and real-scale fires. A limited number of large-scale tests may be conducted to validate the predictions of the fire models, as such tests are very expensive.

Chapter 1 presents the fundamentals of material flammability including the definition of flammability and its key aspects – ignitability, burning and heat release rate, flame spread and smoke production. Flammability testing involves measurements of temperature, heat flux, mass loss rate during burning, pressure, velocity, flow, smoke density using light attenuation, and species concentration in the fire gases using gas analysers. Chapter 2 gives an overview of the techniques to measure these parameters, and the limitations of these methods.

Chapters 3 to 10 cover flammability testing of building materials and components. Chapter 3 is on the fire performance and classification of wood based products with reference to the recently introduced European classification for building products. External façade forms an important and an expensive part of most modern high-rise commercial buildings. Chapter 4 summarises fire tests for external façades, regulations in Japan, the UK, Canada, USA, Australia and New Zealand, and case studies of some major façade fires. Chapter 5 presents a modelling approach to quantify the fire hazard of combustible wall and ceiling linings in a room. A mathematical model BRANZFIRE developed for this purpose is summarised, and the model predictions are compared with some room fire tests. Chapter 6 gives an overview of fire testing of sandwich panels used in constructing warehouses, cold storage, etc. Sandwich panels present a unique fire safety design challenge, due to flammable materials such as polyurethane and polystyrene foams encased within metallic sheets. Upholstered furniture and mattresses represent a large proportion of flammable materials in buildings, and Chapter 7 covers flammability testing of these materials. Fire testing of cables and the proposed classification system in Europe is summarised in Chapter 8. Cables can pose a major fire safety problem in the service ducts of large buildings, mines, power cabinets, cable trays concealed in ceilings and under the floor and so on. Chapter 9 discusses fire tests for electrical appliances such as TVs, computers, fridges and washing machines. Chapter 10 reviews the regulatory classification of building materials in the USA, Canada, Europe, Japan and China with reference to flammability.

Fire safety in road and rail tunnels is a very important issue with many tunnels being built throughout the world. Chapter 11 summarises the large-scale tunnel fire tests carried over the last 40 years or so. These data can be used for tunnel fire safety design – ventilation, egress routes, tunnel lining and so on. Chapter 12 reviews the flammability requirements for commercial transport airplanes. Bunsen burner type tests are still widely used in this industry and the requirements are predominantly prescriptive. Chapter 13 discusses fire testing of materials such as conveyor belts, cables, hoses, hydraulic fluids and ventilation control devices in the underground coal mining industry in the USA.

Flammability testing of materials and components used in passenger rail carriages is reviewed in Chapter 14. This chapter focuses on the work undertaken at the National Institute of Standards and Technology, USA, although, some European studies are also summarised. Chapter 15 discusses the codes and standards for materials used in ships and submarines, the emphasis being to minimise the use of combustible materials in this industry. Finally, Chapter 16 describes the tests for assessing the auto-ignition behaviour of solids, with a focus on coal, which is a major source of spontaneous combustion in underground mines.

As new materials and components are being developed prompting new applications, there is an ongoing need to re-examine the existing flammability test methods, and how the data from these tests are interpreted. The current trend in fire research is to have a limited set of bench-scale flammability tests, which can comprehensively measure all aspects of flammability under a variety of fire conditions – fuel controlled to ventilation controlled. The information from these tests would then be used in conjunction with mathematical fire models to predict large-scale fire behaviour, validated against large-scale fire experiments. Such validated fire models will enable fire engineers to better design effective fire safety systems.

Compiling and editing this book has been a privilege and a great learning experience. The editor would like to sincerely thank all the authors, who are international experts in their respective fields, for their valuable time and efforts in writing the chapters. This has been a wonderful opportunity for me to interact with my colleagues in fire science and engineering, and to improve my understanding of this vast, evolving subject.

I am grateful to Woodhead Publishing for initiating the idea of this book, and for their sustained support and patience throughout its publication.

I hope the book will be useful for fire scientists, fire safety engineers, people from the construction, transport and mining industries, regulators, standards organisations, fire brigades, plastic manufacturers, fire retardant manufacturers and so on. The heat release rate data from large-scale experiments will provide useful guidelines for choosing appropriate and realistic design fires.

The editor and the authors would welcome readers' feedback on the errors, omissions and shortcomings in the contents of the book. We will endeavour to take on board constructive suggestions and comments.

1

Understanding materials flammability

C LAUTENBERGER, University of California, USA,
J TORERO, University of Edinburgh, UK and
C FERNANDEZ-PELLO, University of California, USA

1.1   Introduction: definition of flammability

As it is currently used by the fire community, 'flammability' may be loosely associated with a material's combustibility or its inherent fire hazard. However, this definition is ambiguous because both combustibility and fire hazard are complex and depend on many parameters related to the material, its end use configuration, and the environmental conditions. It is more useful to define flammability in terms of characteristics that may be directly measured or inferred from laboratory tests and then used to assign relative rankings to different materials. With this in mind, a better yet still imperfect definition of flammability is: the ease with which a material is ignited, the intensity with which it burns and releases heat once ignited, its propensity to spread fire, and the rate at which it generates smoke and toxic combustion products during gasification and burning. A comprehensive evaluation of a material's overall flammability may require data from several laboratory tests, perhaps combined with some form of analysis or modeling to interpret the results properly.

1.2   Key aspects of flammability

In the following sections, simplified analysis is used to identify the 'fire properties' that affect a material's flammability as defined above. Emphasis is given to those properties that may be directly measured or calculated from bench-scale flammability tests.

1.2.1   Ignitability

A material's ignition resistance is a critical measure of flammability because there is no fire hazard if ignition does not occur, although under some circumstances prolonged non-flaming gasification or smoldering may represent a threat to life safety. In terms of fire safety, the most important questions to ask are: 'Given a certain thermal exposure, will the item ignite? If so, how long will it take?' It is not yet possible to answer these questions based solely on the chemical composition or structure of a material, particularly because ignition occurs in the gas phase and is influenced by geometrical and environmental factors. Several ignition tests have been developed to measure the ignitability of a material under different exposures, usually involving some form of radiative heating.[1–3] Since auto ignition is highly dependent on the geometry and ambient conditions, most ignition tests use a pilot (flame, spark, hot wire) to induce ignition. In addition to being a more conservative ignition mode than auto ignition, piloted ignition may be more relevant to compartment fires where sparks or embers from already burning objects can serve as a pilot. For these reasons, the discussion here is limited to the piloted ignition of radiatively heated solids. Readers looking for a more fundamental or detailed treatment of ignition are referred to the reviews of solid ignition, including Janssens,[4] Kashiwagi,[5] Fernandez-Pello,[6] Atreya,[7] Drysdale,[8] Kanury,[9] and Babrauskas.[10]

A sequence of physical and chemical events takes place in the piloted ignition of a solid. In general, three stages can be identified: (i) inert heating, (ii) thermal decomposition, and (iii) gaseous mixing leading to the onset of gas-phase reactions. In the inert heating stage, the rate of temperature rise depends primarily on the applied heat flux, heat losses to the surroundings, and thermo-physical properties of the fuel. Also important are the latent heat of any phase changes and the spectral characteristics of the radiation source and the fuel sample.[11]

After the initial inert heating period, a decomposition zone begins to develop near the surface and slowly propagates to the interior of the material. The rate of material decomposition (pyrolysis) is affected by the kinetics and the endo-thermic or exothermic nature of the decomposition reaction. The gasified solid fuel (pyrolyzate) flows out of the condensed phase and enters the gas-phase to form a fuel-oxidizer mixture. At the 'flashpoint' (see e.g. Drysdale and Thomson[12] or Drysdale[8]) the fuel concentration in the vicinity of the igniter corresponds approximately to the lower flammable limit, and a premixed flame may propagate from the igniter through the flammable mixture toward the surface of the solid fuel. A diffusion flame becomes anchored at the fuel surface if the generation rate of pyrolysate vapors is sufficient to sustain the flame. This corresponds to the 'firepoint', and is considered the moment of ignition. Although in many cases the firepoint is attained soon after the flashpoint, under some conditions (particularly low O_2 concentration or elevated flow velocity), flashing may occur without sustained ignition ever occurring.

It is evident from the above description that both solid and gas-phase phenomena are relevant to ignition. The basic thermophysical processes involved are complicated because practical materials have complex composition and may contain fire retardants that serve as a heat sink, promote charring/intumescence, or inhibit gas-phase combustion reactions. Additionally, anisotropy or inhomo-geneity (wood, composites) may influence the heat and mass transfer processes within the solid. Explicitly modeling these physical processes is difficult. Although a few detailed numerical calculations treat both the gaseous and condensed phases,[13,14] most simplified theoretical analyses consider only the solid-phase, and relate the ignition event to attainment of some critical condition in the solid-phase, such as a critical surface temperature or critical pyrolysis rate. Among these, the most physically correct ignition criterion is 'a critical mass flow of volatiles (pyrolyzate) sufficient to support a nascent flame capable of losing heat to the surface without the flame temperature being reduced to a value below which the flame is extinguished'.[15] Although the mass flux at ignition is notoriously difficult to determine experimentally, the available measurements[12,15] show that for a particular fuel the mass flux at ignition varies slightly with the applied heat flux. In vitiated atmospheres, it has been found to decrease slightly with increasing oxygen concentration.[16]

Due to the difficulties associated with determining the gas phase ignition or the critical mass flux for ignition, most theoretical analyses relate the moment of ignition to attainment of a critical surface temperature, normally called the 'ignition temperature' (T_{ig}). Although it has often been assumed that materials have a well-defined and constant ignition temperature, there is no physical basis to support this suggestion because ignition is a gas-phase process. For example, experimental data indicate that the T_{ig} decreases with increasing heat flux for Douglas fir,[17] but increases with increasing heat flux for common polymers (PMMA, POM, PE, PP)[18,19] as well as a polypropylene/glass composite.[20] The same trend has been predicted by numerical models.[13,14] The data reported by Petrella[21] show that the ambient oxygen concentration had a minor effect on the measured time to ignition, but T_{ig} was not directly measured. Similarly, Cordova et al.[19] found that the piloted ignition time of PMMA increased weakly as the oxygen concentration was increased from a lower ignition limit (approximately 18%) to 25%, beyond which the ignition time was independent of the oxygen concentration. Actual measurements of T_{ig} have shown that it increases with decreasing oxygen concentration and increasing flow velocity near the sample surface.[16] However, because the rate of pyrolysis is very sensitive to the surface temperature, unless there are extreme differences in environmental conditions (ambient oxygen concentration or flow velocity) ignition will occur within a narrow range of surface temperatures. For these reasons, simplified thermal ignition theories based on inert solid heating are successful at correlating experimental ignition time data using a critical surface temperature as the ignition criterion. Thus, the time to ignition (or ignition delay time, t_{ig}) can be determined by modeling the inert heating of a solid to determine when its surfaces reaches T_{ig}. Several variations of the same basic analysis have been presented in the literature,[22–25] and reviewed by Babrauskas.[10]

The starting point of a simplified ignition analysis is usually the transient one-dimensional heat conduction equation without considering in-depth fuel pyrolysis:

$$\rho c \frac{\partial T(x,t)}{\partial t} = k \frac{\partial^2 T(x,t)}{\partial x^2} \qquad \text{1.1a}$$

where ρ and c are the material's specific heat and density, k is the thermal conductivity, and x is the spatial coordinate. Thus, the problem of determining when ignition occurs is reduced to solving eqn 1.1a subject to the appropriate boundary conditions to determine t_{ig}, the time at which the surface temperature reaches the ignition temperature T_{ig}. The front-face boundary condition says that the rate of heat conduction into the solid balances the net heat flux at the material's surface:

$$-k \frac{\partial T(0,t)}{\partial x} = \dot{q}''_{net}(t) \qquad \text{1.1b}$$

The second boundary condition is the 'back face' boundary condition, which is dictated by the particular configuration under consideration. The net heat flux to the fuel surface may be broken into separate contributions from any external sources ($\dot{q}''_e$) as well as convective and reradiative losses ($\dot{q}''_{conv}$, $\dot{q}''_{rr}$):

$$\dot{q}''_{net} = \alpha \dot{q}''_e - \dot{q}''_{conv} - \dot{q}''_{rr} \qquad \text{1.2}$$

Here, α is the surface radiative absorptivity. For a thermally thin solid, the temperature is considered uniform across its thickness δ, and the lumped thermal capacitance form (see e.g. ref. 26) of eqn 1.1a can be applied. When the external heat flux ($\dot{q}''_e$) is much larger than convective and radiative losses, the net heat flux to the fuel is approximately constant and equal to $\alpha \dot{q}''_e$ and the following expression is obtained for the ignition time (t_{ig}) of a thermally thin fuel:

$$t_{ig} = \rho c \delta \frac{T_{ig} - T_\infty}{\alpha \dot{q}''_e} \qquad \text{1.3}$$

where T_∞ is the ambient temperature (also equal to the solid's initial temperature). However, most practical materials are 'thermally thick',[22] and the analysis is more complicated because the solid temperature is not uniform. In the limiting case of high external heat flux (where the surface heat losses are small in comparison to the external heat flux), the ignition time of a thermally thick solid can be found from the solution of eqn 1.1a:[27,28]

$$t_{ig} = \frac{\pi}{4} k \rho c \left(\frac{T_{ig} - T_\infty}{\alpha \dot{q}''_e} \right)^2 \qquad \text{1.4a}$$

Here, the product $k\rho c$ is usually referred to as the thermal inertia. Equation 1.4a is strictly valid only for high external heat flux levels (more specifically, when $t_{ig} \ll k\rho c / h_T^2$). At low heat flux levels, surface losses become important and a slightly more complicated solution for the ignition time is obtained:[28,29]

$$t_{ig} = \frac{1}{\pi} \frac{k\rho c}{h_T^2} \left(1 - \frac{h_T(T_{ig} - T_\infty)}{\alpha \dot{q}''_e} \right)^{-2} \qquad \text{1.4b}$$

where h_T is the total heat transfer coefficient, which includes convective cooling and a linearized approximation to surface reradiation. Equation 1.4b is valid for low heat flux levels when $t_{ig} \gg k\rho c/h_T^2$. These equations predict very well experimental ignition time data for a variety of fuels[30] when the environmental conditions are such that gas phase chemistry is fast (elevated oxygen concentration and low flow velocity) and does not affect the ignition process. When gas phase chemistry starts to play a role in the process, the above equations have to be modified to incorporate the gas phase induction time.[6,31]

By setting $t_{ig} \to \infty$ in eqn 1.4b and assuming the total heat transfer coefficient is known, the effective ignition temperature T_{ig} can be inferred from the measured critical heat flux for ignition ($\dot{q}''_{cr}$):

$$T_{ig} = T_\infty + \alpha \dot{q}''_{cr}/h_T \qquad\qquad 1.5$$

Here, $\dot{q}''_{cr}$ is value of $\dot{q}''_e$ at which the ignition time becomes infinite. The reader should be cautioned that the critical heat flux for ignition is not defined consistently throughout the literature, and it is prudent to check a particular worker's definition to avoid confusion.

The primary 'material fire properties' that emerge from this simplified ignition analysis are T_{ig}, $k\rho c$, and $\dot{q}''_{cr}$. The product $k\rho c$ is the apparent thermal inertia; it is determined experimentally by plotting $t_{ig}^{-1/2}$ as a function of $\dot{q}''_e$ for $\dot{q}''_e \gg \dot{q}''_{cr}$. The thermal inertia $k\rho c$ is an apparent value that is different from the product of the actual (temperature-dependent) values of k, ρ, and c because it is deduced from an inert heating equation with no losses or a linearized approximation to the total heat transfer coefficient,[32] and without considering the endothermic decomposition reaction of the solid[19] (pyrolysis in most cases). Furthermore, the apparent thermal inertia is not an intrinsic property of a material because it is affected by the environmental conditions, test apparatus, and data reduction technique.[10] In spite of these limitations, the results of a particular flammability test can be used to establish relative rankings but should not be extrapolated to scenarios where environmental conditions are significantly different.

1.2.2   Mass burning rate and heat release rate

The mass burning rate ($\dot{m}''$), defined as the mass of solid fuel pyrolyzed per unit area per unit time, is an important parameter since it determines a fire's heat release rate (HRR) and the mass of combustion products generated. During steady state burning, the mass burning rate can be calculated from the ratio of the net heat transfer to the fuel surface $\dot{q}''_{net}$ and the fuel's effective heat of gasification ΔH_g:

$$\dot{m}'' = \frac{\dot{q}''_{net}}{\Delta H_g} \qquad\qquad 1.6$$

A material's effective heat of gasification is dependent to some degree on the burning conditions.[33] The net heat flux in eqn 1.6 is slightly different than that defined in eqn 1.2 because the convective and radiative flame heat feedback ($\dot{q}''_{fc}$, $\dot{q}''_{fr}$) must be considered:

$$\dot{q}''_{net} = \alpha \dot{q}''_e + \dot{q}''_{fc} + \dot{q}''_{fr} - \dot{q}''_{rr} \qquad\qquad 1.7$$

In writing eqn 1.6 with $\dot{q}''_{net}$ defined as in eqn 1.7, it is tacitly assumed that the heat conducted to the solid interior is small in comparison to that absorbed at the surface, although this may not always be the case.[34] $\dot{q}''_e$ may be attributed to remote flames, a hot gas layer, or compartment surfaces. $\dot{q}''_{rr}$ depends primarily on the surface temperature of the burning material, typically in the range 500 °C to 700 °C.[35] The absolute and relative magnitudes of the flame radiation $\dot{q}''_{fr}$ and convection $\dot{q}''_{fc}$ terms depend on the material's adiabatic flame temperature, sooting propensity, physical size, and orientation relative to gravity. As a fire increases in size, the radiative feedback from the flame to the fuel surface dominates over the convective heat transfer.[36]

Closely related to the mass burning rate, the heat release rate (HRR) per unit area ($\dot{Q}''_{HRR}$) is an important predictor of fire hazard because it determines a fire's growth rate. Furthermore, the total HRR can be directly linked to the mass of air entrained (smoke production), plume temperatures and velocities, and whether flashover will occur. The HRR per unit area is the product of the mass burning rate and the chemical (or actual) heat of combustion (ΔH_c):

$$\dot{Q}''_{HRR} = \dot{m}'' \Delta H_c \qquad\qquad 1.8$$

In eqn 1.8, ΔH_c is the amount of heat released per unit mass of fuel consumed under actual fire conditions. It is the product of the net heat of complete combustion ΔH_T, a thermodynamic quantity determined by the fuel's chemical composition, and the combustion efficiency χ:

$$\Delta H_c = \chi \Delta H_T \qquad\qquad 1.9$$

The combustion efficiency may be close to 1 for clean fuels burning under well-ventilated conditions. As the degree of oxygen vitiation is increased, the combustion efficiency is decreased and may approach 0.4 for ventilation-limited burning.[25]

After combining eqns 1.8 and 1.6, it can be seen that under steady-state burning the HRR per unit area is proportional to the net rate of heat transfer to the fuel surface and the ratio $\Delta H_c / \Delta H_g$:

$$\dot{Q}''_{HRR} = \frac{\Delta H_c}{\Delta H_g} \dot{q}''_{net} \qquad\qquad 1.10$$

The ratio $\Delta H_c / \Delta H_g$ has been called the 'heat release parameter'[25] or 'combustibility ratio'.[8] It is a fire property that controls the steady-state rate of heat release given a particular net rate of heat transfer to the fuel surface. Since HRR is one of the most important indicators of fire hazard,[37] the ratio $\Delta H_c / \Delta H_g$ emerges also as one of the most important material fire properties. It is an important component of the mass transfer or 'B' number which is sometimes also treated as a fire property. Details of the 'B' number's definition and of relevant applications can be found in Drysdale.[8]

1.2.3 Propensity for fire propagation or flame spread

A material's tendency to support flame spread is an important indicator of fire hazard because it affects the total HRR through an increase in the burning area. Reviews of flame spread have been given by Williams,[38] Fernandez-Pello and Hirano,[39] Fernandez Pello,[6] and Quintiere.[40] By treating the flame spread process as a series of piloted ignitions,[31] the flame spread rate V can be thought of as a ratio of the solid heating length (Δ) ahead of the pyrolysis front to its ignition time (t_{ig}):

$$V = \frac{\Delta}{t_{ig}} \qquad 1.11$$

Simplified expressions for the flame spread velocity on thermally thin and thick materials can be obtained by substituting eqn 1.3 or eqn 1.4 into eqn 1.11 and replacing the external heat flux with the net heat flux defined in eqn 1.7:

$$V = \frac{(\dot{q}''_e + \dot{q}''_{fc} + \dot{q}''_{fr} - \dot{q}''_{rr})\Delta}{\rho c \delta (T_{ig} - T_s)} \quad \text{(thermally thin)} \qquad 1.12a$$

$$V = \frac{4}{\pi} \frac{(\dot{q}''_e + \dot{q}''_{fc} + \dot{q}''_{fr} - \dot{q}''_{rr})^2 \Delta}{k\rho c (T_{ig} - T_s)^2} \quad \text{(thermally thick)} \qquad 1.12b$$

Here, T_s is the surface temperature of the material ahead of the heat transfer region and it has been assumed that $\alpha = 1$ because most materials blacken near the flame during flame spread. Though greatly simplified, eqn 1.12 illustrates the key physics and properties involved in flame spread. The same properties that control piloted ignition also control flame spread when gas-phase chemical kinetics are not controlling.[6] The magnitude of the heat flux terms and length scale of the forward heat transfer region Δ depend on the geometry of the flame spread process and environmental conditions.

In downward spread or forced opposed flow flame spread, the flame propagates against the oncoming flow. The heating length scale ahead of the flame front (Δ) is usually no larger than several mm unless there is a large radiation view factor between the flame and the forward heating zone, as may be the case with horizontal spread. For opposed spread on a flat plate, the convection term in eqn 1.7 may be approximated as:[31]

$$\dot{q}''_{fc} = C_1 (k_g \rho_g c_{pg} V_g)^{1/2} (T_f - T_{ig}) \qquad 1.13$$

and the convective forward heating length is approximately proportional to the boundary layer thickness:[31]

$$\Delta \propto \ell \mathrm{Re}_\ell^{-1/2} = C_2 V_g^{-1/2} \qquad\qquad 1.14$$

where ℓ is a characteristic length of the problem, V_g is the velocity of the oncoming gas, and C_2 is a constant. The following expressions for the flame spread velocity are obtained after substituting eqns 1.13 and 1.14 into eqn 1.12:

$$V = \frac{(C_3(k_g\rho_g c_{pg})^{1/2}(T_f - T_{ig}) + (\dot{q}_e'' + \dot{q}_{fr}'' - \dot{q}_{rr}'')\Delta)}{\rho c \delta (T_{ig} - T_s)}$$

$$\text{(thermally thin)} \qquad 1.15a$$

$$V = \frac{4}{\pi} \frac{(C_3(k_g\rho_g c_{pg})^{1/2}(T_f - T_{ig})V_g^{1/4} + (\dot{q}_e'' + \dot{q}_{fr}'' - \dot{q}_{rr}'')\Delta^{1/2})^2}{k\rho c (T_{ig} - T_s)^2}$$

$$\text{(thermally thick)} \qquad 1.15b$$

The constant C_3 can be determined by matching experimental data to the theory.[31] It should be noted that if $\dot{q}_e'' = 0$ and the convective term is larger than both the flame radiation and surface re-radiation terms, the resulting flame spread expressions are basically the same as those derived more formally by de Ris[41] using an Oseen velocity profile (flat velocity profile):

$$V = \frac{\sqrt{2}k_g(T_f - T_s)}{\rho c \delta(T_{ig} - T_s)} \quad \text{(thermally thin)} \qquad\qquad 1.16a$$

$$V = \frac{V_g(k\rho c)_g (T_f - T_{ig})^2}{k\rho c(T_{ig} - T_s)^2} \quad \text{(thermally thick)} \qquad\qquad 1.16b$$

The primary difference is that eqn 1.16b gives that the flame spread velocity is linearly proportional to the velocity of the oncoming flow due to the Oseen approximation, but eqn 1.15b gives a square root dependency. Equation 1.15 is a better predictor of the spread rate velocity dependence on flow velocity than eqn 1.16 because it accounts for the more realistic case of a velocity gradient at the surface.[42]

As was indicated in the ignition section, the above equations are valid for fast gas chemistry (elevated oxygen concentration and low flow velocity). They correlate well with a variety of experimental flame spread data obtained under these conditions.[31,43] At low oxygen concentration and/or high flow velocity, a term corresponding to the induction time must be included, which affects the characteristics of the above equations.[6,31]

Being general, eqn 1.12 also applies for concurrent or wind aided flame spread where the flame spread direction is the same as the ambient oxidizer flow. The rate of concurrent spread is important from a fire safety standpoint because it can be ~1–2 orders of magnitude greater than the rate of opposed spread. Upward flame spread is a special case of concurrent spread where the oxidizer flow velocity is caused by the fire's buoyancy. It is usually the dominant mechanism of fire spread, and will be examined more closely here. Upward spread rates are faster than opposed spread rates because Δ may be of the order of tens of cm to several m (compared to several mm for opposed spread).

For concurrent flame spread, Δ is approximately equal to the flame length ℓ_f. If the flames are laminar, their length is a function of the mass transfer (or 'B' number).[44] The length of turbulent flames is related to the heat release rate per unit area and ℓ_p, the height of the burning region (pyrolysis zone):

$$\Delta = \ell_f \propto (\ell_p \dot{Q}''_{HRR})^n \qquad\qquad 1.17$$

The exponent n is approximately 2/3 for turbulent burning of a vertical wall.[45] After substituting eqn 1.17 into eqn 1.12, the expressions for the upward flame spread rate become:

$$V \propto \frac{(\dot{q}''_e + \dot{q}''_{fc} + \dot{q}''_{fr} - \dot{q}''_{rr})(\ell_p \dot{Q}''_{HRR})^{2/3}}{\rho c \delta (T_{ig} - T_s)} \quad \text{(thermally thin)} \qquad 1.18a$$

$$V \propto \frac{4}{\pi} \frac{(\dot{q}''_e + \dot{q}''_{fc} + \dot{q}''_{fr} - \dot{q}''_{rr})^2 (\ell_p \dot{Q}''_{HRR})^{2/3}}{k \rho c (T_{ig} - T_s)^2} \quad \text{(thermally thick)} \qquad 1.18b$$

It can be seen from these equations that upward flame spread is inherently acceleratory because the height of the pyrolysis zone increases with time. Recall from section 1.2.2 eqn 1.10 that for steady-state burning, $\dot{Q}''_{HRR}$ is proportional to the product of the net heat flux and the heat release parameter $\Delta H_c / \Delta H_g$. Therefore, the rate of upward flame spread is primarily a function of the heat release parameter, the flame and pyrolysis zone length, and the flame heat flux. The convection term $\dot{q}''_{fc}$ is related to the flame temperature and scale of the fire; as the fire size increases, it becomes radiatively dominated and the relative importance of the flame convection term decreases compared to the flame radiation term $\dot{q}''_{fr}$.[36,46,47] In addition to scale, the magnitude of the flame radiation term is related to the adiabatic flame temperature and the fuel's sooting propensity.

To avoid the complexities associated with calculating the flame temperature, gas-phase transport properties, flame height, and flame heat fluxes, the terms in the numerator of eqn 1.12 (or equivalently eqn 1.15 or 1.18) are sometimes lumped together into a fire property known as the flame spread parameter (ϕ for thin fuels or Φ for thick fuels). This parameter was originally developed for lateral flame spread on thermally thick fuels,[22] but a similar analysis can be applied to opposed or concurrent flame spread. This flame spread parameter can be evaluated from experimental data and used to determine relative rankings.[22] It is therefore common to correlate flame spread data as a function of the surface temperature (T_s) and use eqn 1.19 to estimate empirical values of ϕ or Φ:

$$V = \frac{\phi}{\rho c \delta (T_{ig} - T_s)} \quad \text{(thermally thin)} \qquad 1.19a$$

$$V = \frac{\Phi}{k\rho c (T_{ig} - T_s)^2} \quad \text{(thermally thick)} \qquad 1.19b$$

The thermal inertia and ignition temperature obtained from ignition studies appear again as properties controlling flame spread together with the new fire property Φ. It should be noted that for upward flame spread, the flame spread parameter is a function of time, but is constant for lateral flame spread.

1.2.4   Smoke production and toxicity

Nonthermal hazards, rather than heat and burns, are the primary cause of death in fires. The two main nonthermal effects of fire smoke on humans are: (i) reduced visibility leading to a reduction in wayfinding ability, and (ii) asphyxia caused by uptake of toxic gases that can lead to confusion, incapacitation, or death. Irritation or damage to the respiratory tract caused by acrid smoke is also important, but to a lesser degree. Visibility in fire smoke is reviewed by Jin,[48] and Purser[49] gives perhaps the most comprehensive review of fire toxicity.

The reduction in visibility during a fire is caused not only by smoke's opacity, attributed largely to the presence of particulate soot, but also by its irritant nature. The data presented by Jin[48] indicate that visibility levels and walking speed in fire smoke decrease with both the soot density and irritant levels. The most common irritants in fire smoke are hydrogen fluoride (HF), hydrogen chloride (HCl), hydrogen bromide (HBr), nitrogen oxides, phosphoric acid, sulfur dioxide, acrolein, and formaldehyde. Although there are a large number of potentially toxic species in fire smoke, the overall toxicity is usually dominated by carbon monoxide.[49] HCN is the second most important toxicant in fire effluents. Whereas CO intoxication tends to occur quite slowly and is related to the dose accumulated over time, HCN intoxication can be rather sudden and is quite sensitive to the instantaneous HCN concentration in the fire smoke.[49]

It is difficult to identify the material fire properties that contribute to the generation of CO and HCN in fires because species production depends on the solid properties as well as the gas-phase combustion process. HCN only appears in significant quantities during the burning of fuels containing nitrogen, but CO is always present in fires to some degree due to the vitiated nature of most fires. The generation rate of HCN and particularly CO are affected by the ventilation conditions, increasing with the global fuel to air ratio.[50] The generation rate of species i (where i may represent CO, HCN, CO_2, soot, etc.) is related to the mass burning rate and that species' 'yield' y_i. A simplified expression for the mass generation rate of species i is:

$$\dot{m}_i'' = y_i \dot{m}'' \qquad 1.20$$

In eqn 1.20, y_i is the mass of species i released by the fire per unit mass of fuel liberated to the gas phase. Species yields are fuel-specific and vary with the degree of incomplete combustion, usually a function of the fire size relative to the available ventilation. Tewarson gives CO, CO_2, and soot yields for well-ventilated turbulent fires.[25]

A fuel's CO yield, soot yield, radiative fraction, and incompleteness of combustion are all well correlated by the laminar smoke point height,[25,51] defined as the height of a laminar candle-like diffusion flame burning in air at which smoke is first released from the flame tip. Fuels with high CO or soot generation potentials and radiant fractions have a short laminar smoke point height, whereas clean-burning fuels have a much taller smoke point height. de Ris[52,53] has long advocated use of the laminar smoke point height as a flammability property for characterizing a fuel's radiative emission, but it may also be thought of as a fire property that characterizes relative CO and smoke generation potentials.

1.3   Standardized bench-scale flammability tests

Several of the fire properties identified earlier can be determined from bench-scale flammability tests. This is useful for establishing relative rankings (e.g. T_{ig}, $k\rho c$) or for developing input data for predictions of large-scale fire behavior. The cost of small-scale fire testing is considerably less than that of large-scale fire tests because relatively small quantities of sample material are required, and the setup and breakdown time is much shorter. These factors make bench-scale flammability testing a cost-effective screening tool and can reduce a new material's time-to-market. This section gives a brief overview of common or representative flammability tests. Where possible, these tests are connected to the material properties identified earlier.

1.3.1 Ignition

Some of the most widely used small-scale ignition tests are the Cone Calorimeter,[1] LIFT apparatus,[2] and FM Flammability Apparatus or Fire Propagation Apparatus (FPA).[3] In each test, a small sample (e.g. 100 mm by 100 mm in the Cone Calorimeter, 155 mm by 155 mm in the LIFT) is irradiated at a prescribed radiant flux and the time to piloted ignition in the presence of a small pilot is recorded. A particular material's ignition time at a certain heat flux will not necessarily be the same in all three tests. This is attributed to differences in the sample preparation protocol (the materials are blackened in the FPA), the source of thermal radiation (coiled resistance heater in the Cone Calorimeter, methane-fired radiant panel in the LIFT, tungsten filament heaters in the FPA), the pilot (small flame in the LIFT and FPA, electrical spark in the Cone Calorimeter), and the convective conditions. The LIFT and Cone Calorimeter are normal-air tests, whereas the FPA is usually run with a 40% oxygen concentration because it is believed data obtained in an enhanced oxygen atmosphere are a better indicator of full-scale fire behavior.[54] Directly measured quantities include the time to ignition as a function of the applied heat flux, and the critical heat flux for ignition $\dot{q}_{cr}''$.

A material's apparent thermal inertia $k\rho c$ and effective ignition temperature T_{ig} may be inferred from the measured ignition delay time curve using one of several data reduction techniques[22–25] based on an analysis similar to that in section 1.2.1. However, the derived parameters are apparatus-specific and are affected by the environmental conditions such as airflow velocity and oxygen concentration;[19] they are not true material properties and must be interpreted as apparent values. Derived values of thermal inertia and ignition temperature depend on the particular data reduction technique used. It is for this reason that care must be taken when using literature values of thermal inertia or ignition temperature due to differences in the experimental apparatus or data analysis methods.

1.3.2   Heat release rate

Modern heat release calorimeters use the oxygen consumption principle[55] to calculate the rate of heat release from a limited number of gas concentration measurements (O_2, CO, CO_2). Calculation of HRR on the basis of oxygen consumption is possible because the quantity of heat released per unit mass of oxygen consumed is approximately constant and independent of fuel at 13.1 MJ/kg.[56] Most small-scale measurements of HRR are now made using the Cone Calorimeter,[1] although a large number of measurements have been made in the FM Fire Propagation Apparatus,[3] and will continue to be made with its closely related successor, the Advanced Flammability Measurements Apparatus (AFM).[57] These tests provide transient measurements of the sample's mass loss rate, HRR, and species generation rates/yields (CO, CO_2, soot). Additionally, the fuel's effective heat of gasification ΔH_g can be determined by plotting the mass loss rate during steady-state burning as a function of irradiance level and finding the slope of a linear fit to the data. Despite the potential usefulness of bench scale HRR measurements, there have been only a few successful attempts at using these measurements to predict fire growth.

1.3.3   Flame spread

The LIFT apparatus[2] is widely used for evaluating a material's propensity for lateral flame spread. A 155 mm by 800 mm sample is exposed to a decaying heat flux from a methane fired radiant panel. The sample is allowed to reach thermal equilibrium and then ignited in the high heat flux region. The rate of lateral flame spread is then determined as a function of the incident radiant heat flux.

By igniting the material after it has achieved thermal equilibrium, the denominator of eqn 1.19 can be expressed as a function of the external heat flux and the critical heat flux for ignition. This information is used to determine the lateral flame spread parameter presented in eqn 1.19 using the procedure developed by Quintiere and Harkelroad.[22]

The ASTM E2058 Fire Propagation Apparatus[3] includes an upward fire propagation test with a sample having a width of 0.10 m and a height of 0.60 m. The bottom 120 to 200 mm of the sample is ignited at a heat flux of 50 kW/m^2 with a pilot flame. The coflowing oxidizer stream is enriched to 40% oxygen because a flame radiation scaling technique[58] is used to simulate the enhanced flame radiation that occurs in large-scale fires. During the test, the pyrolysis front is tracked as a function of time while the chemical heat release rate measured. The fire propagation index (FPI) is calculated from the measured heat release rate, apparent thermal inertia, and effective ignition temperature. A material may be grouped into one of several propagation regimes (non-propagating, decelerating propagation, non-accelerating propagation, or accelerating propagation) by the numerical value of the FPI.[25] Consistent with the simplified theory in section 1.2.3, this suggests that a material's tendency for upward flame spread may be explained by its heat release rate (related to the heat release parameter), apparent thermal inertia, and effective ignition temperature.

The Limiting Oxygen Index (LOI) test, ASTM D 2863,[59] is frequently applied to polymers. The LOI is the lowest oxygen mole fraction that will just support burning on a small sample immersed in a flowing atmosphere. A material's LOI can be a misleading indicator of full-scale fire behavior because the flame retardants that are so effective at limiting creeping flame spread may not be effective at limiting the rate of upward flame spread at hazardous scales due to increased flame residence times.[52] Since the LOI test is conducted at room temperature, care must be taken when extrapolating its results to elevated temperatures (e.g. during a real fire) where flame propagation is possible at lower oxygen concentrations due to preheating.

1.3.4   Smoke production and toxicity

The bench-scale heat release rate tests discussed earlier can be used to evaluate a material's CO and soot yields for well-ventilated fires, as can the laminar smoke point height.[25,53] However, these tests give no information about the other potentially toxic products generated during gasification or combustion, primarily because of their limited instrumentation. Therefore, specialized tests for determining the total toxic potency of a material's decomposition products have been developed by several organizations, including the National Bureau of Standards[60] and the University of Pittsburgh.[61] In these tests, rats are exposed to the atmosphere created by gasification or combustion of the test material. In the NBS test method, the material is gasified under non-flaming conditions at 25 °C less than its ignition temperature. The University of Pittsburgh Test Method[61] is similar, with the principal difference being that the sample is heated at a rate of 20 °C per minute and usually transitions to flaming at some point during test. In principle, the University of Pittsburgh test[61] provides an indication of a material's toxicity over a broader range of decomposition temperatures (including flaming) than the NBS test.[60] The chemical composition of the atmosphere is measured, and the LC_{50} concentration (the mass of material per unit volume that causes death in 50% of the exposed animals) is determined. Whereas this type of test can be used to establish relative rankings of material toxicity under different decomposition regimes, it provides no material fire properties analogous to species yields.

1.4   Standardized real-scale flammability tests

Due to potential time and cost savings, combined with an increased recognition of the importance of material fire properties, there is considerable interest in using data obtained from small-scale flammability tests in conjunction with correlations or models to predict large-scale fire behavior. However, large-scale tests will likely remain an important part of flammability testing until the primary difficulties associated with extrapolating bench-scale data to the prediction of large-scale fire behavior are addressed or better models of fire behavior are developed. These difficulties are related to differences in the flame residence times and the relative importance of flame radiation in small-scale and large-scale fires. Halogenated fire retardants that are effective at inhibiting combustion in bench-scale tests may not be as effective in large scale fires where the flames are larger, giving longer flame residence times that can overwhelm halogenated fire retardants.[52] Direct application of bench-scale convectively dominated heat release data to prediction of large-scale radiatively dominated fires may provide misleading results.

In the United States, the most widely used surface flammability test is the 'Steiner tunnel' test (ASTM E-84,[62] NFPA 255,[63] UL 723[64]). A test specimen 20 inches (0.5 m) wide and 25 feet (7.5 m) long forming the ceiling of the test tunnel is ignited at one end by a gaseous diffusion flame. The flame propagation distance is recorded as a function of time in the presence of a forced air flow. The test data are used to assign a material a flame spread index relative to a special type of red oak flooring. The Steiner tunnel test provides only a relative flame spread index, and there is not necessarily a correlation between a material's performance in the test and its behavior in real fires.

Another flammability test widely used to classify wall lining materials is the 'room/corner' test (ISO 9705,[65] NFPA 265,[66] NT FIRE 025[67]). A standard-sized room (nominally 2.4 m wide by 3.6 m deep by 2.4 m high with a single 0.8 m by 2.0 m door in one of the shorter walls) is lined with the lining material to be tested. An ignition burner is placed in a corner on the wall opposite the door and the fire products are collected in a hood. Oxygen consumption calorimetry is used to measure the rate of heat release. The burner size and HRR history vary from standard to standard. The ISO 9705[65] burner is 0.17 m on edge with a HRR of 100 kW for the first ten minutes followed by 300 kW for the second ten minutes. The corner configuration is used because flames tend to be taller and hotter in a corner compared to a flat wall due to reduced air entrainment and radiation interaction between surfaces. Some materials that perform well in the Steiner tunnel may perform poorly in the room/corner test, particularly textile wall coverings[68] and foams.

1.5   Customized flammability tests

Several groups have developed customized flammability tests. As an example, the bench-scale Forced Ignition and Spread Test (FIST)[19,69] has been developed to evaluate the ignitability and flame spread characteristics of solid combustibles in a spacecraft environment. Although several large-scale room fire tests have been standardized, so-called 'free burn' fire tests remain largely unstandardized due to the variety of objects encountered in practice. Many commodities and products have been burned at large-scale under HRR calorimeters; much of the available data have been summarized by Babrauskas.[70] Most fire tests of large-scale fuel packages are also unstandardized, e.g., rail[71] and highway vehicles.

1.6   Fire modeling as a complement to flammability testing

Zone fire models[72–74] are based on the experimental finding that during the initial stages of a compartment fire, thermal stratification due to buoyancy leads to the formation of two distinct and homogenous zones: a hot upper layer, and a relatively cool lower layer. Exploiting this observation, the compartment is divided into two spatially uniform zones, or control volumes, corresponding to the upper and lower layers. By applying conservation equations to each control volume, temperature and species concentrations in each zone are determined as a function of time. Comparisons of zone model predictions with experimental compartment fire data have shown that the agreement between prediction and experiment can be good.[75] On modern desktop computers, a typical zone model simulation requires a few seconds of CPU time.

CFD-based fire models[76–78] are often called 'field models' by the fire community to differentiate them from zone models. The space under consideration is divided into a large number (50,000–2,000,000) of control volumes, and the conservation equations are solved for each control volume. The result is a detailed spatial description of the flowfield and fire effects. When used properly, they provide the most accurate prediction of smoke and heat transport from a fire currently available. The required CPU time ranges from hours to weeks. Both zone and field models usually do not calculate flame spread and fire growth. Rather, they model the consequences of a user-specified fire on a space.

1.6.1 Fire consequence modeling with full-scale fire test data as input

Fire consequence modeling is particularly useful for performance-based design of the built environment. Customized fire protection strategies are developed to mitigate hazards expected in an occupancy, allowing for the same or better levels of fire safety than traditional prescriptive design but at a lower cost. Zone or field models may be used, and the fire is idealized as a heat release rate history with associated species yields. Sometimes this information can be obtained from full-scale fire testing of the fuel package or assembly. In other cases, this may not be practical and it is necessary to use bench-scale test data to estimate the large-scale fire behavior. The material fire properties identified in earlier sections can be used in conjunction with correlations or simplified models of fire growth to estimate the heat release rate curves.

1.6.2 Predicting real-scale burn behavior from bench-scale test data

Fire growth modeling is potentially useful for forensic fire reconstruction or estimating the heat release rate history of fuel packages that are not practical to burn at full scale due to cost or size constraints (e.g. rail vehicles, aircraft, boats, etc.). Computer fire models can be used to extrapolate normal gravity flammability data to reduced gravity environments.[79] Particularly in the early phases of product development, fire modeling provides a cost-effective alternative to expensive full-scale fire testing[62,65] required by building codes and other regulations. Simplified models of fire growth have been used in conjunction with zone fire models to simulate real-scale fire tests.[80–82] The test material is characterized by properties measured from bench-scale fire tests (e.g. apparent thermal inertia, effective ignition temperature, and heat of gasification). This approach is limited because it can be applied only to a particular geometry with a particular ignition scenario.

Field models have matured to the point where it is possible to calculate flame spread and fire growth in practical geometries. The basic physics are included in NIST's freely available Fire Dynamics Simulator,[76] putting detailed fire growth modeling within the reach of most practitioners. By building a three-dimensional representation of the space or object being modeled, the fire growth and heat release rates can be modeled for different geometries and ignition scenarios. At a minimum, each surface must be characterized by material fire properties obtained from bench-scale flammability tests, including thermal properties, ignition temperature, heat of gasification, and total heat content (to model burnout). The gaseous combustion reaction is characterized primarily by the heat of combustion and species yields (particularly CO and soot) as well as the stoichiometric coefficients. To date, field modeling of fire growth has been used mostly as a research tool for calculating flame spread in simple geometries;[83–85] however, in the coming years it will likely be used with increasingly greater frequency for practical fire safety assessments.

1.7   Future trends

In the coming years, the industry may start to move away from *ad hoc* flammability tests that may not be a good indicator of large-scale end-use fire performance. The most appealing alternative is to rely on small-scale and scientifically sound laboratory fire tests. By deriving material fire properties from these tests, large-scale fire behavior can be characterized with correlations or models; large-scale fire testing would still be required to confirm the capabilities of predictive models. This could lead to an increased reliance on a smaller number of flammability tests.

1.8   References

1. ASTM E1354-03, *Standard Test Method for Heat and Visible Smoke Release Rates for Materials and Products Using an Oxygen Consumption Calorimeter.*
2. ASTM E1321-97a, *Standard Test Method for Determining Material Ignition and Flame Spread Properties.*
3. ASTM E2058-03, *Standard Test Methods for Measurement of Synthetic Polymer Material Flammability Using a Fire Propagation Apparatus (FPA).*
4. Janssens M, 'Piloted Ignition of Wood: a Review', *Fire and Materials,* 1991 15 151–167.
5. Kashiwagi T, 'Polymer Combustion and Flammability-Role of the Condensed Phase', *Proceedings of the Combustion Institute,* 1994 25 1423–1437.
6. Fernandez-Pello A C, 'The Solid-phase', in *Combustion Fundamentals of Fire,* ed. Cox G, 31–100, London, Academic Press Limited, 1995.
7. Atreya A, 'Ignition of fires', *Philosophical Transactions of the Royal Society of London,* 1998 356 2787–2813.
8. Drysdale D D, *An Introduction to Fire Dynamics,* 2nd edn, John Wiley & Sons, 1999.
9. Kanury A M, 'Flaming Ignition of Solid Fuels', in *SFPE Handbook of Fire Protection Engineering,* 3rd edn, Ed. DiNenno P, 2-229–2-245, National Fire Protection Association, Quincy, MA, 2002.
10. Babrauskas V, *Ignition Handbook,* Issaquah, WA, Fire Science Publishers, 2003.
11. Thomson H E and Drysdale D D, 'Flammability of Plastics I: Ignition Temperatures', *Fire and Materials,* 1987 11 163–172.
12. Drysdale D D and Thomson H E, 'Flammability of Plastics. II. Critical Mass Flux at the Firepoint', *Fire Safety Journal,* 1989 14 179–188.
13. Zhou Y Y, Walther D C, and Fernandez-Pello A C, 'Numerical Analysis of Piloted Ignition of Polymeric Materials', *Combustion and Flame,* 2002 131 147–158.

14. Lautenberger C W, Zhou Y Y, and Fernandez-Pello A C, 'Numerical Modeling of Convective Effects on Piloted Ignition of Composite Materials', *Combustion Science and Technology*, 2005 177 1231–1252.
15. Rasbash D J, Drysdale D D, and Deepak D, 'Critical Heat and Mass Transfer at Pilot Ignition and Extinction of a Material', *Fire Safety Journal*, 1986 10 1–10.
16. Delichatsios M A, 'Piloted Ignition Times, Critical Heat Fluxes and Mass Loss Rates at Reduced Oxygen Atmospheres', *Fire Safety Journal*, 2005 40 197–212.
17. Atreya A and Abu-Zaid M, 'Effect of Environmental Variables on Piloted Ignition', *Proceedings of the Third International Symposium on Fire Safety Science*, 1991 177–186.
18. Thomson H E, Drysdale D D, and Beyler C L, 'An Experimental Evaluation of Critical Surface Temperature as a Criterion for Piloted Ignition of Solid Fuels', *Fire Safety Journal*, 1988 13 185–196.
19. Cordova J L, Walther D C, Torero J L and Ferndandez-Pello A C, 'Oxidizer Flow Effects on the Flammability of Solid Combustibles', *Combustion Science and Technology*, 2001 164 253–278.
20. Stevanovic A, Mehta S, Walther D C, and Fernandez-Pello A C, 'The Effect of Fiberglass Concentration on the Piloted Ignition of Polypropylene/fiberglass Composites', *Combustion Science and Technology*, 2002 174 171–186.
21. Petrella R V, 'The Assessment of Full-Scale Fire Hazards from Cone Calorimeter Data', *Journal of Fire Sciences*, 1994 12 14–43.
22. Quintiere J, and Harkelroad M, 'New Concepts for Measuring Flame Spread Properties', NBS Report NBSIR 84-2943, 1984.
23. Delichatsios M A, Panagiotou Th, and Kiley F, 'The Use of Time to Ignition Data for Characterizing the Thermal Inertial and the Minimum (Critical) Heat Flux for Ignition or Pyrolysis', *Combustion and Flame*, 1991 84 323–332.
24. Janssens M L, 'Improved Method of Analysis for the LIFT Apparatus. Part I: Ignition', *Proceedings of the Second Fire and Materials Conference, Interscience Communications*, 1993.
25. Tewarson A, 'Generation of Heat and Chemical Compounds in Fires', in *SFPE Handbook of Fire Protection Engineering*, 3rd edn, ed. DiNenno P, 3-82–3-161, National Fire Protection Association, Quincy, MA, 2002.
26. Incropera F P and DeWitt D P, *Fundamentals of Heat and Mass Transfer*, 5th edn, John Wiley & Sons, 2002.
27. Simms D L, 'On the Pilot Ignition of Wood By Radiation', *Combustion and Flame*, 1963 7 253–261.
28. Long R T, Torero J L, and Quintiere J G, 'Scale and Transport Considerations on Piloted Ignition of PMMA', *Fire Safety Science – Proceedings of the Sixth International Symposium* 1999 567–578.
29. Lawson D I and Simms D L, 'The Ignition of Wood by Radiation', *British Journal of Applied Physics*, 1952 3 288–292.
30. Hopkins D and Quintiere J G, 'Material Fire Properties and Predictions for Thermoplastics', *Fire Safety Journal* 1996 26 241–268.
31. Fernandez-Pello C, 'Modeling Flame Spread as a Flame Induced Solid Ignition Process', *Proceedings of the Fourth International Seminar on Fire and Explosion Hazards,* Londonderry, UK, Sept 8–12 2003.
32. Mowrer F W, 'An Analysis of Effective Thermal Properties of Thermally Thick Materials', *Fire Safety Journal*, 2005 40 395–410.

33. Staggs J E J, 'The heat of gasification of polymers', *Fire Safety Journal*, 2004 39 711–720.

34. Fernandez-Pello, A C, 'Pool and Wall Fires: Some Fundamental Aspects', *Proceedings of the ASME/JSME Thermal Engineering Joint Conference*, Book No. I0309E, 1991, 261–268.

35. Urbas J, Parker W J, and Luebbers, G E 'Surface Temperature Measurements on Burning Materials Using an Infrared Pyrometer: Accounting for Emissivity and Reflection of External Radiation', *Fire and Materials*, 2004 28 33–53.

36. Apte V B, Bilger R W, Green A R, and Quintiere J G, 'Wind-aided Turbulent Flame Spread and Burning Over Large-Scale Horizontal PMMA Surfaces', *Combustion and Flame*, 1991 85 169–184.

37. Babrauskas V and Peacock R D, 'Heat Release Rate: the Single Most Important Variable in Fire Hazard', *Fire Safety Journal*, 1992 18 255–272.

38. Williams F A, 'Mechanisms of Fire Spread', *Proceedings of the Combustion Institute*, 1976 16 1281–1294.

39. Fernandez-Pello A C and Hirano T, 'Controlling Mechanisms of Flame Spread', *Combustion Science and Technology*, 1983 32 1–31.

40. Quintiere J G, 'Surface Flame Spread', in *SFPE Handbook of Fire Protection Engineering*, 3rd edn, ed. DiNenno P, 2-246–2-257, National Fire Protection Association, Quincy, MA, 2002.

41. de Ris J, 'Spread of a Laminar Diffusion Flame', *Proceedings of the Combustion Institute*, 1969 12 241–249.

42. Wichman I S, 'Flame Spread in an Opposed Flow with a Linear Velocity Gradient', *Combustion and Flame*, 1983 50 287–304.

43. Fernandez-Pello A C, Ray S R, and Glassman I, 'Flame Spread in an Opposed Forced Flow: The Effect of Ambient Oxygen Concentration', *Proceedings of the Combustion Institute*, 1981 18 579–589.

44. Pagni P J and Shih T M, 'Excess Pyrolysate', *Proceedings of the Combustion Institute*, 1976 16 1329–1343.

45. Tu K M and Quintiere J G, 'Wall Flame Heights with External Radiation', *Fire Technology*, 1991 27 195–203.

46. de Ris J, 'Fire Radiation – a Review', *Proceedings of the Combustion Institute*, 1979 17 1003–1015.

47. Orloff L, Modak A T, and Alpert R L, 'Burning of Large-scale Vertical Surfaces', *Proceedings of the Combustion Institute*, 1977 16 1345–1352.

48. Jin T, 'Visibility and Human Behavior in Fire Smoke', in *SFPE Handbook of Fire Protection Engineering*, 3rd edn, ed. DiNenno P, 2-42–2-53, National Fire Protection Association, Quincy, MA, 2002.

49. Purser D A, 'Toxicity Assessment of Combustion Products', in *SFPE Handbook of Fire Protection Engineering*, 3rd edn, ed. DiNenno P, 2-83–2-171, National Fire Protection Association, Quincy, MA, 2002.

50. Gottuk D T and Lattimer B Y, 'Effect of Combustion Conditions on Species Production', in *SFPE Handbook of Fire Protection Engineering*, 3rd edn, ed. DiNenno P, 2-54–2-82, National Fire Protection Association, Quincy, MA, 2002.

51. Markstein G H, 'Correlations for Smoke Points and Radiant Emission of Laminar Hydrocarbon Diffusion Flames', *Proceedings of the Combustion Institute*, 1986 22 363–370.

52. de Ris J, 'Flammability Testing State-of-the Art', *Fire and Materials*, 1985 9 75–80.

53. de Ris J, and Cheng F, 'The Role of Smoke-Point in Material Flammability Testing', *Fire Safety Science – Proceedings of the Fourth International Symposium,* 1994 301–312.

54. Tewarson A, Lee J L, and Pion R F, 'The Influence of Oxygen Concentration on Fuel Parameters for Fire Modeling', *Proceedings of the Combustion Institute,* 1981 18 563–570.

55. Parker W, 'Calculations of the Heat Release Rate by Oxygen Consumption for Various Applications', NBS Report NBSIR 81-2427, 1982.

56. Huggett C, 'Estimation of the Rate of Heat Release by Means of Oxygen Consumption', *Fire and Materials,* 1980 12 61–65.

57. Beaulieu P, Dembsey N, and Alpert A, 'A New Material Flammability Apparatus and Measurement Techniques', *Proceedings of SAMPE2003*, Society for the Advancement of Materials and Process Engineering, Covina, CA, 2003.

58. Tewarson A, 'Flammability Parameters of Materials: Ignition, Combustion, and Fire Propagation', *Journal of Fire Sciences,* 1994 10 188–241.

59. ASTM D 2863-00, *Standard Test Method for Measuring the Minimum Oxygen Concentration to Support Candle-Like Combustion of Plastics (Oxygen Index).*

60. Levin B C, Fowell A J, Birky M M, Paabo, M, Stolte S, and Malek D, NBSIR 82-2532, National Bureau of Standards, Washington, DC (1982).

61. Alarie Y and Anderson R C, *Toxicology and Applied Pharmacology*, 1979 51 341–345.

62. ASTM E84-03, *Standard Test Method for Surface Burning Characteristics of Building Materials.*

63. NFPA 255, *Surface Burning Characteristics of Building Materials*, 2000 edn, National Fire Protection Association, Quincy, MA.

64. UL 723, *Standard for Test for Surface Burning Characteristics of Building Materials*, 1996.

65. ISO 9705, *International Standard – Fire Tests – Full Scale Room Test for Surface Products*, 1993 International Organization for Standardization, Geneva (1993).

66. NFPA 265, *Standard Methods of Fire Tests for Evaluating Room Fire Growth Contribution of Textile Coverings on Full Height Panels and Walls*, 2000 edn, National Fire Protection Association, Quincy, MA.

67. NT FIRE 025, 'Surface Products: Room Fire Tests in Full Scale', NORDTEST, Espoo, Finland (1986).

68. Fisher F L, MacCracken W, and Williamson R B, 'Room Fire Experiments of Textile Wall Coverings, A Final Report of All Materials Tested Between March 1985 and January 1986', ES-7853, 1986, Service to Industry Report No. 86-2, Fire Research Laboratory.

69. Cordova J L and Fernandez-Pello, A C, 'Convection Effects on the Endothermic Gasification and Piloted Ignition of a Radiatively Heated Combustible Solid', *Combustion Science and Technology*, 2000 156 271–289.

70. Babrauskas V, 'Heat Release Rates', in *SFPE Handbook of Fire Protection Engineering*, 3rd edn, ed. DiNenno P, 3-1–3-37, National Fire Protection Association, Quincy, MA, 2002.

71. Peacock R D, Averill J D, Madrzykowski D, Stroup D, Reneke P A, and Bukowski R W, 'Fire Safety of Passenger Trains; Phase III: Evaluation of Fire Hazard Analysis Using Full-Scale Passenger Rail Car Tests', NISTIR 6563, 2004.

72. Jones W W, Forney G P, Peacock R D, Reneke P A, 'A Technical Reference for

CFAST: An Engineering Tool for Estimating Fire and Smoke Transport', NIST TN 1431 2003.

73. Mitler H E and Rockett J A, 'Users Guide to FIRST, a Comprehensive Single-room Fire Model', NBSIR 87-3595, Center for Fire Research, National Bureau of Standards, Gaithersburg, MD, USA, 1987.

74. Wade C A, 'BRANZFIRE Technical Reference Guide', *BRANZ Study Report 92* (revised). Building Research Association of New Zealand. Judgeford, Porirua City, New Zealand, 2003.

75. Dembsey N A, Pagni P J, and Williamson R B, 'Compartment Fire Experiments: Comparison with Models', *Fire Safety Journal*, 1995 25 187–227.

76. McGrattan K, 'Fire Dynamics Simulator (Version 4) Technical Reference Guide', NIST Special Publication 1018, 2004.

77. Ewer J, Galea E R, Patel M K, Taylor S, Knight B, Petridis M, 'SMARTFIRE: An Intelligent CFD Based Fire Model', *Journal of Fire Protection Engineering*, 1999 10 13–27.

78. Rubini P A, 'SOFIE – Simulation of Fires in Enclosures', *Proceedings of the Fifth International Symposium on Fire Safety Science*, 1997 1326.

79. Zhou Y Y, Walther D C, Fernandez-Pello A C, Torero J L, and Ross H D, 'Theoretical Prediction of Piloted Ignition of Polymeric Fuels in Microgravity at Low Velocity Flows', *Microgravity Science Technology*, 2003 XIV 44–50.

80. Janssens M L, 'Modeling the E84 Tunnel Test for Wood Products', *Proceedings of the First International Fire and Materials Conference*, pp 33–42, 1992.

81. Quintiere J G, 'A Simulation Model for Fire Growth on Materials Subject to a Room-Corner Test', *Fire Safety Journal,* 1993 20 313–339.

82. Sorathia U, Long G, Gracik T, Blum M, and Ness J, 'Screening Tests for Fire Safety of Composites for Marine Applications', *Fire and Materials*, 2001 25 215–222.

83. Liang K M, Ma T, Quintiere J G, and Rouson D, 'Application of CFD Modeling to Room Fire Growth on Walls', NIST GCR 03-849 2003.

84. Hietaniemi J, Hostikka S, and Vaari J, 'FDS Simulation of Fire Spread – Comparison of Model Results with Experimental Data', *Technical Report VTT Working Paper 4*, VTT Building and Transport, Espoo, Finland 2004.

85. Brehob E G, Kim C I, and Kulkarni A K, 'Numerical Model of Upward Flame Spread on Practical Wall Materials', *Fire Safety Journal* 2001 36 225–240.

2

Fundamental measurement techniques

M J A N S S E N S , Southwest Research Institute, USA

2.1   Introduction

Fire testing relies on fundamental measurements of physical quantities such as temperature, pressure, flow rate, mass, etc. This chapter provides an overview of the fundamental measurement techniques that are commonly used in support of flammability testing of materials. Each section begins with a description of the sensors that are used to measure a particular physical quantity. This is followed by a discussion of some of the challenges that need to be overcome to perform accurate measurements. The chapter concludes with a brief introduction to the concepts of measurement error and uncertainty.

2.2   Temperature measurements

2.2.1   Instrumentation to measure temperature

Thermocouples

Thermocouples are by far the most commonly used sensors to measure temperature in fire tests. A basic thermocouple circuit consists of two wires of dissimilar metals or metal alloys that are joined at one end and left open at the other end (see Fig. 2.1). When the junction is heated, an electromotive force is generated that results in a voltage differential between the open ends. This is referred to as the Seebeck effect. The Seebeck voltage is not affected by the temperature gradient in the wires. Thermocouple wires can be extended with other types of metal wires (e.g., lower grade extension wire) provided that both ends of the thermocouple wires are at the same temperature. Figure 2.2 shows a variation of the basic thermocouple circuit. In this case the voltage differential between the open wire terminals is a function of the temperature difference between the two junctions. The measuring junction is referred to as the hot junction while the reference junction is also called the cold junction. A thermopile consists of several alternating hot and cold junctions and is typically used to measure the average temperature difference between two regions instead of two points.

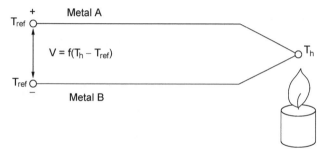

2.1 Basic thermocouple circuit.

Although the Seebeck effect is observed for any pair of dissimilar metals or metal alloys, only a limited number of standard combinations are used in practice. The most common standard thermocouple types are described in Table 2.1. The application determines which type is most suitable, based primarily on the temperature range. Temperatures in fire tests are typically measured with type K thermocouples. The Seebeck voltage is a non-linear function of temperature. The temperature that corresponds to a certain voltage can be determined from tables or polynomial functions that have been developed for the standard type thermocouples. The tables and polynomials are based on a reference temperature of $0\,°C$. Modern data acquisition systems that accept standard thermocouple inputs include electronic circuits to linearize the thermocouple signal and to compensate for cold junction temperatures different from $0\,°C$.

Standard thermocouple wire can be obtained with or without electrical insulation. A wide range of insulation materials are commercially available for specific uses, for example, high-temperature or corrosive environments. Wire diameters vary between 0.013 and 1.63 mm. Thermocouples can be ordered with

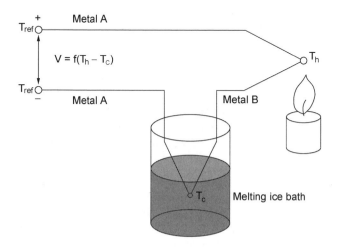

2.2 Basic thermocouple circuit with cold junction.

Table 2.1 Standard thermocouple types

Type	Positive metal	Negative metal	Temperature range[1]
B	Platinum-6% rhodium	Platinum-30% rhodium	0 °C to 1820 °C
C	Tungsten-5% rhenium	Tungsten-26% rhenium	0 °C to 2320 °C
E	Nickel-10% chromium[2]	Copper-45% nickel[3]	−270 °C to 1000 °C
J	Iron	Copper-45% nickel[3]	−210 °C to 760 °C
K	Nickel-10% chromium[2]	Nickel-2% aluminum[4]	−270 °C to 1372 °C
N	Nickel-14.2% chromium- 1.4% silicon[5]	Nickel-4.4% silicon- 0.1% magnesium[6]	−270 °C to 1300 °C
R	Platinum-13% rhodium	Platinum	−50 °C to 1768 °C
S	Platinum-10% rhodium	Platinum	−50 °C to 1768 °C
T	Copper	Copper-nickel[3]	−270 °C to 400 °C

[1] Lower maximum for wire diameter < 3 mm
[2] Also referred to as chromel
[3] Also referred to as constantan
[4] Also referred to as alumel
[5] Also referred to as nicrosil
[6] Also referred to as nisil

a factory-welded hot junction. This is particularly useful for fine gages that require specialized welding equipment. Sometimes it is desirable to butt-weld the thermocouple wires to minimize the size of the junction and its intrusion on the environment in which it is used. Butt-welded thermocouples down to a wire diameter of 0.25 mm are commercially available.

Metal-sheathed thermocouple probes consist of thermocouple wires inside a metal sheath that withstands mechanical impact, vibration, high temperature and high pressure. The sheath is filled with ceramic powder (usually MgO) to electrically insulate the thermocouple wires. Inconel sheath material may be specified instead of stainless steel for corrosive applications. The sheath may be open at the end to expose the junction, or it may be capped to protect the junction (see Fig. 2.3). If the sheath is capped at the end, the junction may be isolated from the sheath (ungrounded junction) or connected to the sheath (grounded junction). There is a trade-off between ruggedness and response time characteristics. Exposed-junction metal-sheathed thermocouple probes are the most fragile, but have the fastest response time. Ungrounded-junction metal-sheathed thermocouple probes are the most rugged and have the slowest response time. Grounded-junction thermocouple probes offer a compromise, but may be unsuitable if the sheath is connected to a high-voltage source.

Other types of temperature sensors

Other types of temperature sensors that are used in fire testing are liquid-in-glass thermometers, resistance temperature detectors (RTDs), thermistors and optical pyrometers. Liquid-in-glass thermometers consist of a sealed bulb connected to a glass capillary inside a vertical stem. Due to thermal expansion, the liquid rises

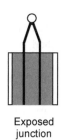 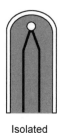 

| Exposed junction | Isolated junction | Grounded junction |

2.3 Types of sheathed thermocouples.

in the capillary with increasing temperature. A scale is marked on the stem to indicate how temperature relates to the height of the liquid column. Mercury is the most commonly used liquid, but organic liquids are used when there are environmental concerns or if cost is an issue.

All metals produce a positive change in electrical resistance when temperature increases. RTDs are based on this effect. The most common metal used for this application is platinum because it can withstand high temperatures, is chemically stable and has a high resistivity. Nickel or nickel alloys are sometimes used because they are less expensive than platinum. RTDs are more accurate and stable than thermocouples but are also more expensive and cumbersome to work with. Thermistors are closely related to RTDs and consist of a semiconductor material that has an electrical resistance which usually decreases with increasing temperature. Thermistors are more sensitive than RTDs but have a limited temperature range (maximum 100 °C for thermistors vs. 850 °C for RTDs).

Optical pyrometers make it possible to measure the temperature of a solid surface without contact. The sensor inside an optical pyrometer actually measures the thermal radiation emitted by an area of the target surface that is in the pyrometer's field of view. The area is a function of the distance between the pyrometer and the target and by the pyrometer's viewing angle. The latter is determined by the optics of the pyrometer. If the emissivity of the target surface is known, the average temperature of the area within the pyrometer's field of view can be calculated from the measured radiation based on Boltzmann's law. Optical pyrometers typically measure radiation at wavelengths between 0.7 and 20 micrometers and are therefore also referred to as infra-red (IR) thermometers.

2.2.2 Challenges of measuring temperatures in fire tests

Gas temperatures

Gas temperature measurements are made in nearly every room fire experiment. Thermocouples are used for this purpose due to their simplicity and low cost. In standard tests it is not so critical that gas temperature measurements be accurate, but they must be repeatable and reproducible. For example, US model building codes require that textile wall coverings for use in spaces that are not protected by automatic sprinklers pass a room/corner test. The test method is described in *NFPA 265 – Standard methods of fire tests for evaluating room fire growth contribution of textile wall coverings*. A wall covering fails the test if flashover occurs during the 15-minute test duration. One of the flashover criteria is based on the upper layer temperature, which is defined as the average reading of five thermocouples that are located 100 mm below the ceiling at the four quadrants and the center of the room. The thermocouples actually do not measure the temperature at these locations with great accuracy. However, the fact that the standard prescribes which thermocouples to use and where they have to be located ensures good precision.

Gas temperature measurements in room fire experiments are also often obtained and used for the validation of computer models of compartment fires. This involves a comparison of experimental temperature data and calculations to assess the predictive capability of the model. A standard methodology for assessing the predictive capability of fire models is described in *ASTM E 1355 – Standard guide for evaluating the predictive capability of deterministic fire models*. It is essential in this case that the temperature measurements be accurate. Unfortunately, a thermocouple does not measure the temperature of the surrounding gas but that of its hot junction. The difference between the two temperatures is often significant so that a correction is necessary.

Figure 2.4 shows a schematic of a thermocouple that is used to measure the upper layer gas temperature at a particular location below the ceiling in a room fire experiment. Heat is transferred to or from the junction by conduction

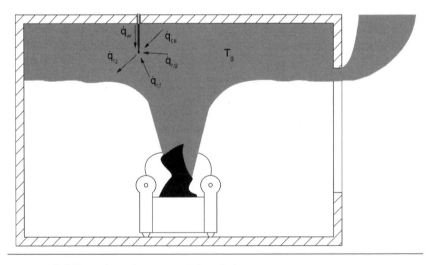

2.4 Upper layer thermocouple heat balance.

through the thermocouple wires. The junction also exchanges heat with the surrounding gas by convection and is heated by radiation from flames, hot surfaces and hot combustion gases. The response of the thermocouple junction to the transient environment of a room fire may also need to be accounted for. Since thermocouple junctions are small in size (~ 1 mm) and have a high thermal conductivity, temperature gradients inside the junction can be ignored. The heat balance for a thermocouple junction as shown in Fig. 2.4 can therefore be written as

$$m_j c_j \frac{dT_j}{dt} = \dot{q}_w + \dot{q}_c + \dot{q}_{r,f} + \dot{q}_{r,s} + \dot{q}_{r,g} - \dot{q}_{r,j} \qquad 2.1$$

where m_j = mass of the thermocouple junction (kg);
c_j = specific heat of the junction material (J/kgK);
T_j = temperature of the junction (K);
t = time (s);
$\dot{q}_w$ = heat flow to the junction by conduction via the thermocouple wires (W);
$\dot{q}_c$ = heat flow to the junction by convection from the surrounding gas (W);
$\dot{q}_{r,f}$ = heat flow to the junction by radiation from flames (W);
$\dot{q}_{r,s}$ = heat flow to the junction by radiation from surfaces (W);
$\dot{q}_{r,g}$ = heat flow to the junction by radiation from combustion gases (W);
$\dot{q}_{r,j}$ = radiative heat losses from the junction (W).

The term on the left-hand side of eqn 2.1 is usually ignored. Alternatively, it is not too difficult to calculate the term on the left-hand side of eqn 2.1 since the time derivative can be estimated from T_j, which is measured as a function of time, and the properties of thermocouple materials are known. For example, Omega Engineering, Inc. lists a specific heat of 448 J/kg·K and 523 J/kg·K for chromel and alumel respectively (Anon., 2002). The density of these type K thermocouple materials is reported in the same reference as 8730 kg/m^3 and 8600 kg/m^3 respectively. The mass of a thermocouple junction can be determined on the basis of the density of the materials and the volume of the junction, which is assumed to be spherical or cylindrical in shape.

The first term on the right-hand side of eqn 2.1 can be eliminated by running the thermocouple wires along an expected isotherm over a certain distance from the junction. For example, the wires of thermocouples that are used to measure gas temperatures in a room test are typically in a horizontal plane for the first 50–100 mm from the junction. The conduction term in eqn 2.1 can then be ignored. If the transient term on the left-hand side is neglected as well, the heat balance of the junction takes the following simplified form after dividing the remaining terms by the exposed area of the junction:

$$\dot{q}_c'' + \dot{q}_r'' = h(T_g - T_j) + \dot{q}_r'' = 0 \qquad\qquad 2.2$$

where $\dot{q}_r''$ = net radiation heat flux to the junction (W/m^2);
         h = convective heat transfer coefficient (W/m^2·K);
        T_g = gas temperature at the thermocouple junction (K).

For a given thermocouple size, the convection coefficient is primarily a function of the gas velocity at the junction. Velocities in fire-induced flows through vertical openings and buoyant flows in flames and fire plumes can be as high as a few meters per second. In other regions of the room velocities are well below one meter per second. The convection coefficient can be calculated from the following correlation for forced flow over a sphere (Whitaker, 1972):

$$\mathrm{Nu} \equiv \frac{hD}{k} = 2 + 0.4\,\mathrm{Re}^{1/2} + 0.06\,\mathrm{Re}^{2/3})\mathrm{Pr}^{0.4} \qquad\qquad 2.3$$

where Nu = Nusselt number;
     D = diameter of the sphere (m);
     k = thermal conductivity of the gas (W/mK);
    Re = Reynolds number ($D \cdot v/\nu$);
     v = free stream velocity of the gas at the thermocouple junction (m/s);
     ν = kinemantic viscosity of the gas (m^2/s);
    Pr = Prandtl number.

The properties of the gas (k, ν, and Pr) have to be evaluated at the 'film temperature', which is the average between the junction and free stream gas temperatures (Atreya, 2002).

Other investigators who studied this problem have used this or a similar equation to calculate the convection coefficient (Blevins, 1999; Blevins and Pitts, 1999; Brohez et al., 2004; Francis and Yau, 2004; Janssens and Tran, 1992; Jones, 2002; Pitts et al., 2002; Steckler et al., 1982). Some also considered a different correlation for natural convection at low velocities (Janssens and Tran, 1992; Jones, 2002). The convection coefficient calculated according to eqn 2.3 for a sphere with a diameter of 1 mm is shown in Fig. 2.5 as a function of gas velocity and junction temperature. As expected, the convection coefficient increases with increasing velocity and temperature. Note that tabulated properties for air were used in these calculations. Moreover, the property values were determined at the junction temperature instead of the film temperature because the gas temperature is not known. Figure 2.6 shows the convection coefficient as a function of gas velocity and diameter for a fixed junction temperature of 300 K. This figure shows that convective heat transfer also increases with decreasing size of the junction.

Several investigators have used a radiation network approach to express the irradiance in eqn 2.2 as a function of flame, surface and smoke layer temperatures, emissivities and view factors (Jones, 2002; Luo, 1997). In practice it is not possible to correct a gas thermocouple reading for radiation errors using

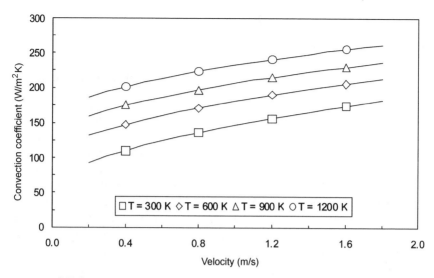

2.5 Convection coefficient as a function of gas velocity and junction temperature.

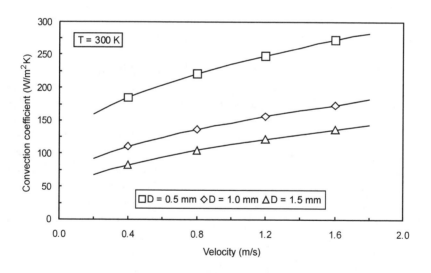

2.6 Convection coefficient as a function of gas velocity and junction diameter.

this approach because the temperatures and view factors vary with time and are not known. However, this approach could be used to adjust the calculation so that a model predicts the junction temperature instead of the surrounding gas temperature. A more useful approach involves the introduction of an average temperature of the surroundings that is defined by rewriting eqn 2.2 as follows:

$$h(T_j - T_g) = \dot{q}''_r \equiv \epsilon\sigma(T_e^4 - T_j^4) \qquad\qquad 2.4$$

where ϵ = emissivity of the thermocouple junction;
σ = Boltzmann constant (5.6710^{-8} W/m^{2}K^4);
T_e = effective temperature of the environment surrounding the junction (K).

Equation 2.4 indicates that if a thermocouple junction is immersed in a black gas, the radiation error is zero and the junction temperature is equal to the gas temperature. In practice, thermocouples are usually immersed in a (semi-) transparent gas and radiation errors can be significant. Jones has often criticized experimental room fire gas temperature data and has indicated that the errors in some cases were in excess of 100 °C (Jones, 1995, 1998, 2004).

Three strategies have been used to reduce or eliminate radiation errors. The first strategy relies on a thermocouple measurement in a location where the gas temperature is known (Janssens and Tran, 1992; Steckler et al., 1982). For example, the air temperature close to the floor in a vertical opening should be equal to ambient. However, if the thermocouple junction picks up radiation from the flame, hot surfaces and combustion gases in the compartment, it will read a higher temperature. The known radiation error for this thermocouple can then be used to estimate the error for other thermocouples in the doorway. Since the incident radiation on a bare-bead thermocouple is a function of its location vis-à-vis the sources of radiant heat, the correction may not be very accurate unless the effect of location can be accounted for.

Equation 2.4 indicates that the error can be reduced by increasing h and by decreasing the difference between T_e and T_j. The former can be accomplished by increasing the velocity of the gases flowing over the junction as shown in Figs 2.5 and 2.6. The latter can be obtained by placing a shield around the thermocouple junction. Aspirated thermocouples combine these two effects. These types of thermocouples are also referred to as suction pyrometers and have been used for a long time in combustion research, but were not very common in fire testing until the mid-1970s. This changed with the development of a simple aspirated thermocouple for use in the Harvard University Home Fire Project bedroom fire tests (Croce, 1975). A detailed description of the probe can be found in the literature (Newman and Croce, 1979). Figure 2.7 shows a schematic. Newman and Croce recommended an aspiration velocity of 7.65 m/s based on measurements that are reported in their seminal paper. Recent investigations at the National Institute of Standards and Technology (NIST) indicate that the radiation error may still be significant (20% or higher) at this velocity, in particular when the gas temperature is near ambient (thermocouple in the lower layer) and the surrounding temperature is high (Blevins, 1999; Blevins and Pitts, 1999). NIST recommended using much higher aspiration velocities and a double shield to further reduce radiation error. This obviously adds to the complexity of using aspirated thermocouples.

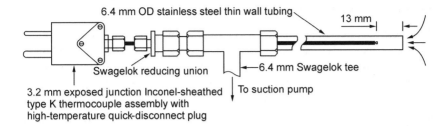

6.4 mm OD stainless steel thin wall tubing

13 mm

Swagelok reducing union

6.4 mm Swagelok tee

3.2 mm exposed junction Inconel-sheathed
type K thermocouple assembly with
high-temperature quick-disconnect plug

To suction pump

2.7 Aspirated thermocouple assembly.

The third strategy involves measuring the temperature at a particular location with two thermocouples of different size. Figure 2.6 shows that the convection coefficient increases with decreasing junction diameter. This is consistent with the observation that the radiation error of a bare-bead thermocouple exposed to a constant radiant heat flux decreases with decreasing junction size. In fact, it has been demonstrated that the relationship between radiation error and bead diameter is approximately linear and that the error can be eliminated by extrapolating the size to zero (Quintiere and McCaffrey, 1980). Thus, the radiation error can be eliminated by simply measuring the temperature with two thermocouples of different diameter and by extrapolating the measured temperatures to zero diameter. Quintiere and McCaffrey found that the slope of the linear relationship between radiation error and bead diameter increases as a function of incident irradiance at the junction.

Recent work performed in France indicates that the linear relationship between radiation error and bead size does not change very much for T_s ranging from 300 K to 1200 K, but that the line does not go through the origin (Brohez *et al.*, 2004). The French study recommends measuring the gas temperature with two thermocouples with bead diameters of 0.25 mm and 1 mm respectively. The actual gas temperature can then be calculated from the following simple equation:

$$T_g \approx T_{1\,mm} - 1.8(T_{1\,mm} - T_{0.25\,mm}) \qquad\qquad 2.5$$

where $T_{1\,mm}$ = temperature measured with the 1 mm diameter thermocouple (K); and

$T_{0.25\,mm}$ = temperature measured with the 0.25 mm diameter thermocouple (K).

More work is needed to modify the equation for different thermocouple sizes and to determine the limits of this simple approach. Moreover, NIST found that the two-thermocouple approach may not work if the gas temperature fluctuates rapidly with time (Pitts *et al.*, 2002). This needs further investigation as well.

Surface temperature of solids

A critical surface temperature, T_{ig}, is a commonly used criterion for piloted ignition of materials (Thomson *et al.*, 1988). Typically T_{ig} is determined on the basis of an analysis of experimental data of the time to ignition at different heat fluxes (Janssens *et al.*, 2003). The analysis is based on a simple model that describes the heat transfer to the specimen prior to ignition. T_{ig} can also be determined experimentally by measuring the surface temperature of the specimen during an ignition test. After the specimen ignites, the surface temperature is needed to calculate the heat of gasification, which is a function of the net heat flux into the specimen (Urbas, 1993). The heat of gasification is equal to the energy that is required to turn one mass unit of the material into volatiles. It is an important material property for input into flame spread and fire growth models. Unfortunately it is very difficult to measure accurately the surface temperature of a specimen, in particular while it is burning. This section provides a brief overview of three techniques that have been used to measure surface temperature in fire tests.

The most common approach involves the use of fine thermocouples. The wire diameter has to be as small as possible to avoid the thermocouple altering the material's response in the test. Although pre-welded type K unsheathed thermocouples are available with wire diameters down to 0.013 mm, it is extremely tedious to handle wires that are less than 0.13 mm in diameter. Butt-welded thermocouples are preferred because they have no bead. Since the smallest diameter of commercially available butt-welded thermocouples is 0.25 mm, it is recommended that these be used instead of 0.13 mm standard beaded wire thermocouples. Thermocouples are installed on the surface by drilling two small holes through the specimen at 5–10 mm from opposite sides of its center. The wires are pulled through the holes and taped to the back side of the specimen so that the thermocouple junction is in the middle between the holes and in contact with the specimen surface. It is beneficial to make a small incision between the holes so that the exposed part of the thermocouple wire is partially below the surface (see Fig. 2.8(a)). It is critical to apply the right tension so that the wire is not pulled into the material (see Fig. 2.8(b)) or loses contact with the surface (see Fig. 2.8(c)). This arrangement is adequate for surface temperature measurements prior to ignition. However, as soon as the material ignites, the surface starts to recede and the thermocouple will lose contact with the surface. To avoid this problem, the wires can be attached to a mechanism that keeps the tension constant instead of taping the wires to the back surface (Urbas and Parker, 1993).

From the description in the previous paragraph the reader can appreciate that it is very difficult and time-consuming to measure accurately the surface temperature of a specimen in a fire test with a thermocouple. The problems of this technique can be avoided by using a non-contact method that relies on an

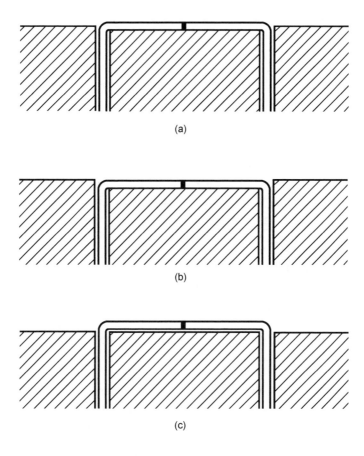

2.8 Measuring surface temperature with a thermocouple.

optical pyrometer. However, this approach is not without problems either. First of all, it may not be possible to position the pyrometer so that the instrument has a clear unobstructed view of the target surface. Often the radiant panel of the test apparatus is in the way and the pyrometer has to be positioned at an angle. Secondly, if the absorptivity of the target surface is less than unity, part of the incident heat flux from the radiant panel is reflected. The pyrometer signal has to be corrected to account for this reflection. Finally, to measure accurately the surface temperature of a burning specimen the radiation from the flame and interference of the flame with the radiation from the surface have to be accounted for. The last challenge has been successfully addressed by using a narrow-band pyrometer that operates in the 8–10 μm range of the IR spectrum, i.e., outside the absorption/emission bands of carbon dioxide and water vapor (Urbas *et al.*, 2004).

Investigators in Sweden have recently experimented with the use of thermographic phosphors to measure the surface temperature in fire tests

(Göransson and Omrane, 2004; Omrane *et al.*, 2003). This technique relies on the fact that the phosphorescence lifetime and spectral properties of UV laser-induced emissions from a thermographic phosphor applied to the surface of a test specimen are a function of the temperature of the phosphor. This method is still in its infancy and more work is needed to demonstrate that it can be used for a wide range of materials and fire test conditions.

2.3 Heat flux

Objects that are exposed in a room fire environment are heated by the radiation from flames, combustion gases and hot surfaces. If an object is in contact with a flame or a hot gas, it is also heated by convection. These heating conditions are simulated in fire tests by exposing specimens to a radiant panel or a gas burner or pool fire flame. The incident heat flux, i.e., the incident heat flow per unit exposed surface area of the receiving object, is a measure of the severity of the fire exposure. It is essential to have the capability of accurately measuring heat fluxes in fire tests so that experiments can be conducted in a repeatable and reproducible manner and the results can be related to real fire performance.

2.3.1 Instrumentation to measure heat flux

Slug calorimeters

A slug calorimeter is the most basic device to measure heat flux in fire tests. It consists of a slug of metal, usually copper, that is insulated on all sides except the front side (see Fig. 2.9). A thermocouple is connected to the slug to measure its temperature. When a slug calorimeter is exposed to a constant irradiance, its temperature rises at a linear rate after an initial transient. When the heat losses

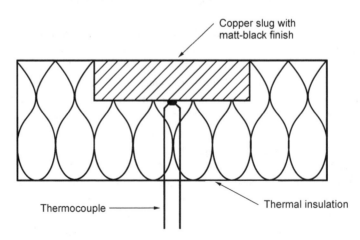

2.9 Slug calorimeter.

through the insulation and from the front face become significant, the temperature levels off and eventually reaches an asymptotic value. The incident heat flux can be determined from the rate of temperature rise during the linear period or from the asymptotic temperature. Slug calorimeters cannot be used when the incident heat flux varies with time due to their slow response. They are suitable for the calibration of radiant panels in fire tests, which emit radiation at a constant rate. Slug calorimeters can be modified to measure heat fluxes under dynamic conditions (Morris *et al.*, 1996). In this case thermocouples are inserted at known distances from the exposed surface. Inverse heat transfer calculations are used to determine the net heat flux at the exposed surface on the basis of the measured temperature profile.

Gardon gauges (Gardon, 1953)

A Gardon gauge consists of a thin circular foil of constantan (see Fig. 2.10). A copper wire is connected to the center of the foil, which in turn is connected to a water-cooled copper jacket at its perimeter. This forms a copper-constantan thermocouple. When a Gardon gauge is exposed to a constant heat flux, a temperature gradient develops between the center and the edge of the foil. The resulting thermocouple signal is proportional to the incident heat flux. The sensitivity of a Gardon gauge is determined by the thickness of the foil. The

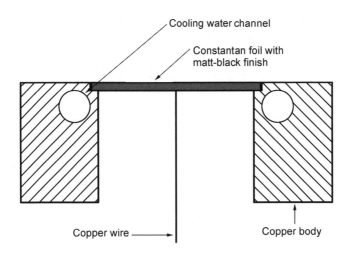

2.10 Gardon gauge.

thickness usually results in an output signal of 10 mV at the maximum heat flux for which the gauge is designed. This implies, based on the reference tables for type T thermocouples, that the maximum center temperature is approximately 230 °C. Because the foil is relatively thin and has low thermal inertia, Gardon gauges are suitable for use under dynamic conditions.

Schmidt-Boelter gauges

A Schmidt-Boelter gauge consists of a small piece of anodized aluminum potted in a circular water-cooled housing (see Fig. 2.11). A thermopile is used to measure the temperature difference between the front and back face of the aluminum piece. The measured temperature difference is a linear function of the incident radiant heat flux. The surface temperature distribution on the exposed face of a Schmidt-Boelter gauge is relatively uniform. Like Gardon gauges, Schmidt-Boelter gauges are also typically designed to generate an output signal of 10 mV at the maximum heat flux for which the gauge is designed. Assuming that a thermopile of five alternating hot and cold junctions is used, this implies, based on the reference tables for type K thermocouples, that the maximum exposed surface temperature is approximately 70 °C. Because the piece of aluminum is small, a Schmidt-Boelter gauge responds quite quickly to incident flux fluctuations.

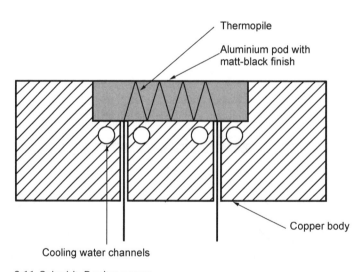

2.11 Schmidt-Boelter gauge.

Hemispherical radiometers (Gunners, 1967)

Hemispherical radiometers consist of a gold-plated ellipsoidal cavity with a small aperture at one focal point and a thermopile at the other focal point (see

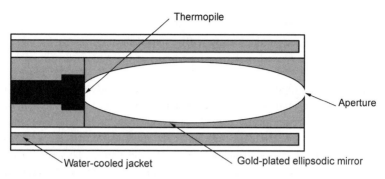

2.12 Hemispherical radiometer.

Fig. 2.12). Rays that enter the cavity through the opening are reflected by the walls and hit the thermopile. This gauge only measures radiation and is not sensitive to convection. It is therefore referred to as a 'radiometer'. This type of heat flux sensor is rather expensive and is used as a primary calibration standard for Schmidt-Boelter and Gardon gauges.

2.3.2   Challenges of measuring heat flux in fire tests

Gardon and Schmidt-Boelter gauges are by far the most commonly used instruments to measure heat fluxes in fire tests. Figure 2.13 shows a schematic of a typical arrangement for measuring the incident heat flux to the surface of an object using a Schmidt-Boelter gauge. The conclusions from the following discussion apply to Gardon gauges as well, but it is easier to explain the concepts with the assumption that a Schmidt-Boelter gauge is used. The surface of the gauge is coated with a matt-black finish so that close to 100% of the incident radiation is absorbed by the gauge. The incident heat flux to the gauge is not measured directly, but is determined on the basis of measured conduction heat transfer into the gauge. Since the surface temperature of the gauge is different from the gas temperature, there is some convective heat transfer at the surface. The heat balance at the surface of the gauge can therefore be written as follows:

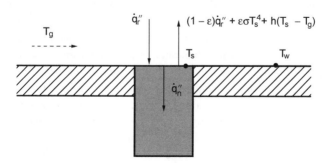

2.13 Heat flux gauge surface heat balance.

$$\dot{q}_n'' = \epsilon(\dot{q}_r'' - \sigma T_s^4) - h(T_s - T_g) \qquad\qquad 2.6$$

where $\dot{q}_n''$ = net conduction heat flux into the gauge (W/m^2); and
$\quad\quad\; T_s$ = exposed surface temperature of the gauge (K).

Suppose Fig. 2.13 shows the setup for measuring incident irradiance to a dummy specimen in a small-scale fire test such as the cone calorimeter or the lateral ignition and flame spread test. T_g is lower than T_s in this case and eqn 2.6 indicates that the incident irradiance is underestimated as a result of the convective heat losses from the surface of the gauge. This error can be very significant at low irradiance levels. The convection error can be reduced (but not completely eliminated) by lowering the temperature of the cooling water and by extending the gauge beyond the surface of the dummy board (Robertson and Ohlemiller, 1995). Convection errors can also be avoided by using a quartz window in front of the gauge. The window transmits radiation and eliminates convective heat transfer.

Suppose Fig. 2.13 shows the setup for measuring the incident heat flux to a wall that is heated in a room fire test by a flame. In this case the total incident heat flux is of interest. Since T_g is higher than T_s, eqn 2.6 indicates that in addition to the irradiance, heat is also transferred by convection to the gauge. However, the measured total heat flux to the gauge is different from the total heat flux to the wall because T_s and T_w are different. This problem has been discussed in detail in the literature (Wickström and Wetterlund, 2004). In the assumption that T_s is known, the gauge has to be supplemented with a surface thermocouple to measure T_w in order to determine the irradiance and the total heat flux to the wall. The incident irradiance to the wall can also be determined by using two heat flux gauges with different emissivities (Lennon and Silcock, 2001).

The calibration of Gardon and Schmidt-Boelter gauges presents another challenge. A number of standard methods have been developed, e.g., 'BS 6809 – British standard method for calibration of radiometers for use in fire testing', 'Nordtest Method NT FIRE 050 – Heat flux meters: calibration', and 'EGOLF Standard Method SM/4 – Method for the use and calibration of radiometers to be used for the measurement of irradiation and/or total heat flux in fire resistance and reaction to fire testing'. In addition, a number of fire testing and calibration laboratories have developed their own method. The FORUM, an international group of directors of fire research laboratories, recently conducted two round robins in which five fire research laboratories performed independent calibrations of two sets of Gardon and Schmidt-Boelter gauges. One set of gauges was also calibrated by two additional general measurement and calibration laboratories. Although systematic variations between calibrations from different laboratories were identified, the degree of agreement was viewed as being satisfactory for most fire testing and research purposes. Convection errors were suspected as being the main reason for the discrepancies. ISO/TC92/SC1/WG10 recommends an elaborate system that involves a primary hemispherical radiometer and primary, secondary and working heat flux gauges. International standards are under development to describe the calibration methods for the three different types of heat flux sensors.

2.4   Mass

2.4.1   Instrumentation to measure mass in fire tests

Load cells

The most commonly used type of sensors to measure the mass of a burning object is the load cell. Load cells come in many forms and shapes to facilitate their use for a wide range of applications. A load cell consists of a metal beam, which deflects under load. The force that acts on the beam due to the weight of a burning object is proportional to the displacement measured by a strain gauge bonded to the beam.

Linear variable displacement transducers (LVDTs)

Linear variable displacement transducers are also used to measure mass in fire tests. An LVDT consists of a metal rod that moves inside an electrical coil. A force applied to the rod results in a displacement that affects the inductance of the coil. An LVDT balance typically has a feature to adjust the tare by changing the initial location of the rod. A more accurate mass loss measurement can be performed as a result because the total mass loss during a test is often much smaller than the initial mass of the specimen, substrate and specimen holder.

2.4.2   Challenges of measuring mass loss rate of burning objects

The mass of a burning specimen is often measured in a fire test to determine the mass loss rate, i.e., the time derivative of the mass loss curve. The mass loss rate can be estimated by sliding a window of n data points along the mass vs. time curve and by calculating the time derivative of a polynomial of degree $n - 1$ that goes through the n data points inside the window. It is very common to use a window that consists of five data points ($n = 5$). This leads to the following five-point formula:

$$-\left[\frac{dm}{dt}\right]_i = \frac{-m_{i-2} + 8m_{i-1} - 8m_{i+1} + m_{i+2}}{12\Delta t}$$

2.7

where m = specimen mass (g);
     t = time (s);
     i = data scan number;

m_{i-2} = specimen mass at two scans prior to the ith data scan (g);
m_{i-1} = specimen mass at the scan prior to the ith data scan (g);
m_{i+1} = specimen mass at the scan following the ith data scan (g);
m_{i+2} = specimen mass at two scans following the ith data scan (g);
Δt = time period between two data scans (s).

It is assumed that the data acquisition system scans and records measurements at constant intervals. Slightly different equations are used near the start and the end of the test. Unfortunately, this type of numerical differentiation often results in noisy mass loss rates. A numerical smoothing method can be used to alleviate this problem (Savitzky and Golay, 1964).

2.5   Pressure

2.5.1   Instrumentation to measure pressure

Capacitive pressure sensors

Hydrostatic pressure differences across a vertical opening in a room fire are of the order of 1–10 pascals (McCaffrey and Rockett, 1977). These differential pressures are actually driving the fire-induced flows through the opening. One pascal (Pa) is the pressure exerted on an area of one square meter (m^2) by a force of one newton (N). This is a very small pressure, i.e., approximately one 100,000th of atmospheric pressure! It is remarkable that such a low pressure can be measured accurately with a capacitive pressure transducer. This type of pressure sensor is a capacitor that consists of an electrically conductive membrane inside a housing that is partly coated with a metal (see Fig. 2.14). The capacitance is a function of the location of the membrane, which in turn depends

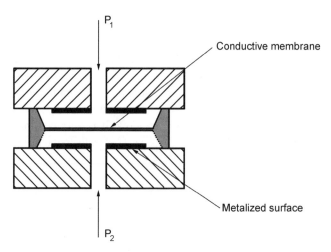

2.14 Capacitive pressure sensor.

on the difference in pressure between the chambers on both sides of the membrane. Although capacitive transducers are available for measuring pressures up to 100 bar, piezoresistive sensors are more common for high-pressure use because they are more robust.

Other types of manometers

Other types of manometers are used in fire tests to provide a visual indication of a pressure. A U-tube manometer consists of a U-shaped glass or plastic tube that is partially filled with a liquid. If there is a difference in pressure between the two ends of the tube, a height differential develops between the liquid surfaces on the low and high pressure sides. The height differential is proportional to the difference in pressures, which can be read from the scale that is provided alongside the tube. To increase sensitivity of the manometer one of the legs may be inclined. A U-tube manometer can be used to measure absolute pressure by sealing one end of the tube and evacuating the head space above the liquid column in that leg. Bourdon gauges consist of a flexible metallic element that deflects when subjected to pressure. The deflection of the element actuates a pointer on a scale through a system of mechanical connections and/or gears.

2.5.2 Challenges of measuring pressure in fire tests

Pressures in room fires are affected by turbulence and are typically very noisy. The signal from a pressure transducer can be smoothed electronically with an RC circuit. The optimum time constant of the RC circuit is typically of the order of five seconds. That seems to eliminate most of the unwanted noise while largely preserving the slower dynamics of the pressure signal. Pressure measurements can also be smoothed numerically with a Fast Fourier Transform (FTT) low pass filter (Janssens, 1991).

2.6   Gas velocity

2.6.1   Instrumentation to measure gas velocity in fire tests

Pitot tubes

The velocity in a gas stream can be determined by measuring the differential pressure of a Pitot tube. A Pitot tube actually consists of two concentric tubes as shown in Fig. 2.15. Theoretically, the difference in pressure in the two tubes is equal to the dynamic pressure, so that the velocity can be calculated as follows:

$$v = \sqrt{\frac{2\Delta p}{\rho}}$$

2.8

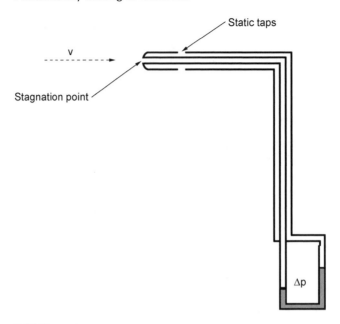

2.15 Pitot tube.

where v = velocity of the gas (m/s);
    Δp = differential pressure between the inner and outer tubes (Pa);
    ρ = density of the gas (kg/m^3).

In practice, the velocity is slightly different and a small correction is necessary. Equation 2.8 implies that the density of the gas must be known. In fire tests it can be assumed that the gas has the properties of air, so that $\rho \approx 352/T$ where T is the temperature of the gas in K. In other words, to determine the velocity of a gas in a particular location it is also necessary to know or measure the gas temperature at that location.

Bi-directional probes

The main problem with using a Pitot tube to measure the velocity in a flame or in fire effluents is that the holes quickly get clogged with soot. A robust bi-directional probe as shown in Fig. 2.16 was developed to address this problem (McCaffrey and Heskestad, 1976.) The probe is based on the same concept as a Pitot tube, i.e., the velocity is determined on the basis of the difference between total pressure (measured at the upstream side) and static pressure (measured at the downstream side.) The probe is 'bi-directional' because it can be used to measure velocity regardless of flow direction, provided the pressure transducer that is connected to the pressure taps is capable of measuring negative pressure differentials. Only part of the dynamic pressure is recovered. McCaffrey and

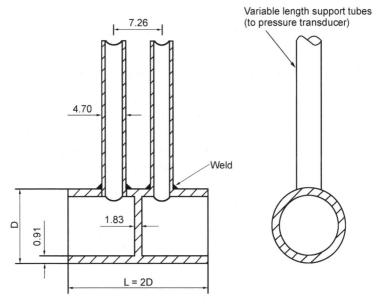

2.16 Bi-directional probe.

Heskestad developed the following correction as a function of the Reynolds number:

$$v = \frac{1}{f(\text{Re})} \sqrt{\frac{2\Delta p}{\rho}}$$ 2.9

where $f(\text{Re}) = 1.533 - 1.366 \cdot 10^{-3}\text{Re} + 1.688 \cdot 10^{-6}\text{Re}^2 - 9.706 \cdot 10^{-10}\text{Re}^3 + 2.555 \cdot 10^{-13}\text{Re}^4 - 2.484 \cdot 10^{-17}\text{Re}^5$;

Re   = Reynolds number;

Δp = differential pressure between the pressure taps (Pa).

The characteristic length for the Reynolds number is the diameter of the probe, D. Equation 2.9 is valid for $40 < \text{Re} < 3800$. For $\text{Re} \geq 3800$ the correction is equal to 1.08. Another advantage of bi-directional probes over Pitot tubes is that the former are much more insensitive to the angle between the flow direction and the probe axis. The mean velocity is measured within $\pm 10\%$ for an angle of up to 50 degrees. An assembly consisting of two bi-directional probes has been used to obtain multi-directional velocity data (Newman, 1987).

Other types of velocity sensors

The velocity of cool combustion air can also be measured with a hot wire anemometer. This type of velocity sensor is based on the fact that convective cooling of a hot wire is a direct function of the velocity of the gas that flows over the wire. Hot wire anemometer measurements are corrected for the temperature of the gas. Some types can be used in environments up to 150 °C.

2.6.2   Challenges of measuring gas velocity in fire tests

The bi-directional probe is the most commonly used device to measure velocity in fire tests. However, this involves measuring the differential pressure between the pressure taps and the gas temperature at the probe. Hence, measuring velocity with a bi-directional probe faces the same challenges as measuring pressure (see section 2.5.2) and gas temperature (see section 2.2.2).

2.7   Gas flow rate

2.7.1   Instrumentation to measure gas flow rate in fire tests

Flow rate measurements based on velocity distribution

The flow rate in a duct can be determined on the basis of a velocity measurement along the centerline of the duct and a coefficient that relates the mean velocity over the cross-section of the duct to the centerline velocity. The mean velocity can be determined by measuring the velocity profile along a diameter of the duct. The log-linear method is the most accurate method to determine the velocity profile (Ower and Pankhurst, 1977). For fully developed pipe flow, the four-point method ($N = 4$) gives an error of less than 0.5%. A six-point method ($N = 6$) is needed to obtain the same accuracy for pipe flow that is not fully developed and for irregular velocity distributions. In this case it may also be necessary to measure the velocity profile in two directions that are perpendicular. Figure 2.17 shows the location of the measuring points for the four-point and six-point log-linear methods. The coefficient then follows from:

$$k_c = \frac{\sum_{i=1}^{N} v_i}{N v_c} \qquad\qquad 2.10$$

where k_c = ratio of mean velocity over the duct cross section to centerline velocity;

v_i = velocity measured at location i (m/s);

N = number of measuring points;

v_c = centerline velocity (m/s).

The flow in the exhaust duct of most fire tests is turbulent, so that k_c is typically between 0.9 and 0.95. The calibration to determine k_c can be conducted at ambient temperature with a Pitot tube, bi-directional probe or hot-wire anemometer. Detailed procedures and recommendations for performing this type of calibration can be found in the literature. The mass flow rate in the exhaust

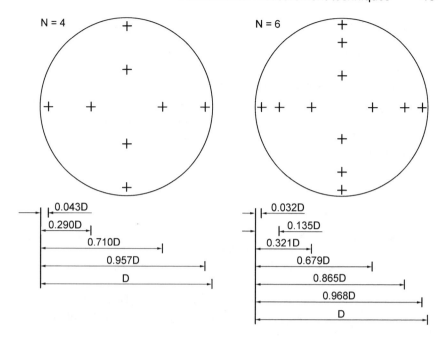

2.17 Location of measuring points for the log-linear method.

duct during a test can then be determined based on the centerline velocity as follows:

$$\dot{m}_e = \rho_e k_c v_c A \qquad 2.11$$

where $\dot{m}_e$ = mass flow rate in the exhaust duct (kg/s);
ρ_e = density of the exhaust gases (kg/m^3);
A = cross-sectional area of the duct (m^2).

If the centerline velocity is measured during a test with a bi-directional probe and a thermocouple in the exhaust duct, substitution of eqn 2.9 in eqn 2.11 to eliminate v_c leads to:

$$\dot{m}_e = \frac{\rho_e k_c A}{f(\mathrm{Re})} \sqrt{\frac{2\Delta p}{\rho_e}} = \frac{k_c A}{f(\mathrm{Re})} \sqrt{2\rho_e \Delta p} \approx 26.5 \frac{k_c A}{f(\mathrm{Re})} \sqrt{\frac{\Delta p}{T_e}} \equiv C \sqrt{\frac{\Delta p}{T_e}}$$

$$2.12$$

where T_e = temperature of the exhaust gases at the bidirectional probe (K);
C = constant (K$^{1/2}$kg$^{1/2}$m$^{1/2}$).

With $k_c = 0.9$ and $f(\mathrm{Re}) = 1.08$ the constant is approximately equal to $22.1 \times A$.

Sharp-edge orifice flowmeters

Another commonly used method to measure flow rate in a duct is based on the pressure drop across a sharp-edge orifice. Figure 2.18 shows a schematic of a sharp-edge orifice flowmeter. The stream emerges from the orifice as a jet. The jet converges to a 'vena contracta' just downstream of the orifice plate (section 2 in Fig. 2.18), and then breaks up into a turbulent flow region. Applying the continuity equation and Bernouilli's theorem between sections 1 and 2 leads to the following equation:

$$\dot{m}_e = C\sqrt{\frac{\Delta p}{T_e}}$$
2.13

where C = constant ($K^{1/2}kg^{1/2}m^{1/2}$);
Δp = differential pressure between the pressure taps (Pa);
T_e = temperature of the exhaust gases at the orifice flowmeter (K).

This equation is identical to the final form of eqn 2.12. However, the orifice constant is a function of the ratio of the diameter of the orifice to that of the duct, the location of the pressure taps and the physical characteristics of the gas. Several standards have been developed that describe in detail how to perform accurate flow rate measurements with sharp-edge orifice plates (e.g., 'ISO 5167-1 – Measurement of fluid flow by means of pressure differential devices – Part 1: Orifice plates, nozzles and Venturi tubes inserted in circular cross-section conduits').

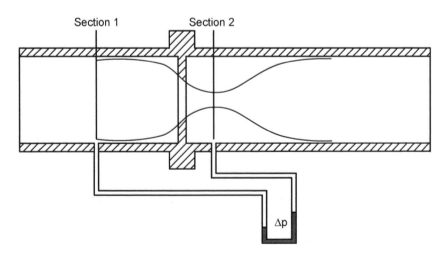

2.18 Sharp-edge orifice flowmeter.

Other types of flowmeters

The flow rate of gaseous fuels and combustion air is often measured in fire test apparatuses with a rotameter. This type of flowmeter consists of a glass or clear plastic tapered vertical tube. The metered gas is supplied at the bottom of the tube and is removed at the top. A needle valve is usually located upstream of the flowmeter to control the flow rate. A 'float' assumes a vertical position inside the tube that is a direct, usually linear, function of the flow rate. The latter can be determined from the scale on the tube, sometimes in conjunction with a calibration curve. Rotameters are available with floats of different types of materials so that a variety of gases can be accommodated or a wider measuring range can be provided.

Thermal mass flowmeters are used to measure the flow rate of gaseous fuel supplied to radiant panels, flaming ignition sources and calibration gas burners. These flowmeters typically consist of a laminar flow element. The flow rate is measured on the basis of the temperature rise of a small fraction of the flow that is diverted through a capillary bypass. The capillary tube contains a heating element and upstream and downstream temperature sensors. For a known heat input, the temperature rise is only a function of the mass flow rate and the specific heat of the gas, i.e., the type of gas that flows through the meter. In other words, if the gas is known, the mass flow rate can be determined from the measured temperature rise. Thermal mass flowmeters are also available with a built-in valve so that they can be used not only to measure but also to control the flow rate.

2.7.2   Challenges of measuring gas flow rate in fire tests

Accurately measuring the flow rate of hot gases in the exhaust duct of a fire test is difficult. Since either a bidirectional probe or a sharp-edge orifice plate is used for this purpose, the primary challenges are again due to the problems associated with measuring differential pressure (see section 2.5.2) and gas temperature (see section 2.2.2).

2.8   Light extinction

One of the hazards associated with smoke generated in fires is the attenuation of light, which affects visibility (Jin, 1978). Standard fire test methods are available to characterize the light extinction characteristics of smoke under a wide range of fire conditions. Full-scale room fire experiments are often supplemented with light extinction measurements in the room or in the exhaust duct through which all fire effluents are removed. This section describes some of the basic features of light extinction measurement systems and addresses the main challenges in using these systems. For a detailed discussion of design parameters for light-extinction measurement devices, the reader is referred to the literature (Putorti, 1999).

2.8.1   Instrumentation to measure light extinction in fire tests

White light source systems

Traditionally a white light source has been used in systems to measure light extinction in fire tests. A schematic of such a system for measuring light extinction of smoke flowing through a duct is shown in Fig. 2.19. The light source, typically a gas-filled tungsten filament bulb with a color temperature of approximately 2900 K, and a collimating lens are mounted on one side of the duct. A second collimating lens, aperture and detector are located at the opposite side of the duct. The detector typically includes a CEI 1924 photopic filter so that its sensitivity matches that of the human eye (Clark, 1985). This setup can also be used to measure the light extinction over a specified path length when smoke accumulates in a fixed volume, but the discussion in this and the following sections assumes a dynamic flow-through system.

The intensity of light passing through smoke is attenuated due to scattering and absorption by the suspended particles. Light attenuation by smoke is described by the following equation, which is known as the Bouguer or Lambert-Beer law:

$$I = I_0 \exp(-k\ell) \qquad\qquad 2.14$$

where I = light intensity measured at the detector (mW);
I_0 = light intensity at the source (mW);
k = extinction coefficient (1/m);
ℓ = path length (m).

Assuming that the light source is stable, I_0 can be measured at the start of a test. The path length is equal to the diameter of the duct. The extinction coefficient can be calculated from the measured light intensity during a test, and is a direct

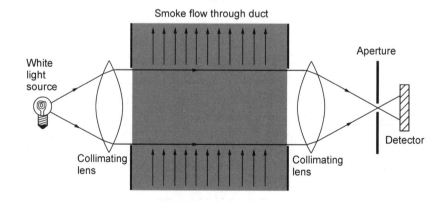

2.19 White light source smoke measuring system.

measure of the concentration of particles in the exhaust duct. This implies that k is a function of the generation rate of particles in the fire and the volumetric flow rate in the duct. The smoke production rate is independent of the duct flow rate and is defined as follows:

$$\dot{s} \equiv k\dot{V}_s \qquad\qquad 2.15$$

where $\dot{s}$ = smoke production rate (m²/s);
$\dot{V}_s$ = actual volumetric flow rate at the smoke meter (m³/s).

Laser light source systems

Strictly speaking eqn 2.14 applies only to monochromatic light, although some studies have indicated that it is approximately valid and can be used in practice for white light systems also (Östman *et al.*, 1985). More importantly, there are some practical problems with white light systems, as will be discussed in the next section. As a result, monochromatic light extinction measurement devices have been developed. Figure 2.20 shows a schematic of such a system. Monochromatic systems typically use a He-Ne laser light source, which operates at a wavelength of 632.8 nm. The laser, beam splitter and reference detector are mounted on one side of the duct. A second beam splitter and measuring detector are located on the opposite side of the duct. The beam splitters transmit approximately 90% of the incident light and the remaining 10% is diverted to a detector. Two beam splitters are used so that the maximum light intensity at each detector is comparable and the same type detector can be used to measure I_0 and I. Lenses are not necessary because the laser beam is highly collimated.

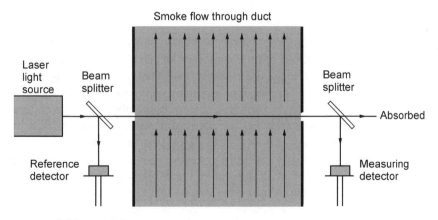

2.20 Laser light source smoke measuring system.

2.8.2   Challenges of measuring light extinction in fire tests

A major practical problem with white light systems is soot deposit on the lenses. Corrections can be made on the basis of the reduction of I_0 after a test, but these are only approximate because it is not known how the soot is deposited on the lenses during the test. The problem can be alleviated by purging the lenses with air, but this may affect the path length if the purging air enters the duct. This is not an issue for laser systems because there are no lenses and the holes necessary for transmission of the laser beam are very small so that air entering the duct has negligible impact on the smoke flow. Note that the extraction fan is located downstream of the smoke meter, so that the pressure at the smoke meter is below ambient.

The calibration of light extinction measurement devices also presents some major challenges. This is typically done with a set of neutral density filters with known light absorption characteristics. Several types of neutral density filters are commercially available. One type consists of a glass disc with a surface coating. The coating is known to deteriorate over time and the use of this type of filter is therefore not recommended (Babrauskas and Wetterlund, 1995). The best type of neutral density filter has uniformly distributed light-absorbent material contained within the filter. Standard procedures (e.g., EGOLF Standard Method SM/3 – Calibration of Smoke Opacity Measuring Systems) indicate that smoke meters have to be calibrated with great care to minimize systematic measurement errors. For example, it has been demonstrated that alignment is critical and that a slight angular deviation in filter position can result in significant calibration errors (Elliot and Whiteley, 1994).

2.9   Gas analysis

2.9.1   Instrumentation to analyze gas composition in fire tests

Oxygen analyzers

Oxygen consumption calorimetry is the most common technique to measure the heat release rate in fire tests. This technique relies on an accurate measurement of oxygen concentration in the fire effluents. Two types of oxygen analyzers are commercially available. The first type uses a high-temperature zirconia sensor. Zirconia (or zirconium-oxide) heated to a temperature above 600 °C conducts oxygen ions. Therefore, when the two faces of a heated zirconia disc are in contact with gas mixtures that have different oxygen concentrations, a current will flow toward the mixture with the lowest oxygen content. The current is proportional to the logarithm of the ratio of the oxygen concentrations on each side of the disc. Thus, the oxygen concentration in a sample gas applied to one side of the disc can be determined if a reference gas with known oxygen concentration (usually dry air) is applied to the other side of the disc. The

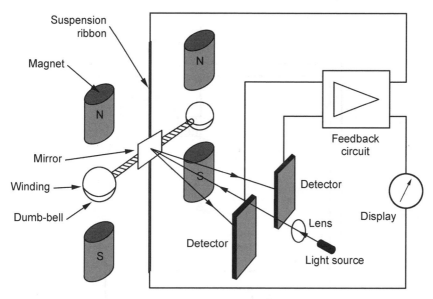

2.21 Paramagnetic oxygen analyzer cell (first design).

advantage of zirconia analyzers is that the sample does not need to be cleaned and dried before it is introduced into the cell. Unfortunately, zirconia analyzers have a slow response time and a number of other disadvantages which make them unsuitable for use in fire testing (Nussbaum, 1987).

The second type of analyzer is based on the fact that oxygen is paramagnetic, i.e., oxygen molecules are attracted into a magnetic field. Two distinct designs of paramagnetic oxygen analyzer cells are available. The first design consists of a dumb-bell-shaped quartz body that is suspended in a non-uniform magnetic field (see Fig. 2.21). When a gas mixture containing oxygen is introduced into the cell, the oxygen molecules are attracted to the strongest part of the magnetic field. This results in a rotation of the dumb-bell, which is sensed by a photocell assembly that receives light from a mirror attached to the dumb-bell. The output from the photocell assembly is used to generate a restoring current that flows through a conductor wound around the dumb-bell. This results in a torque that rotates the dumb-bell back to its original position. The restoring current is proportional to the oxygen concentration of the gas mixture in the cell.

The second design consists of two flow channels, each equipped with a hot filament (see Fig. 2.22). The filaments are connected in a Wheatstone bridge circuit that is fed by a constant voltage source. A reference gas (typically dry air) is supplied to the flow channels at a constant rate. A restriction in one of the flow channels is adjusted to obtain a symmetrical reference gas flow in the measuring section when reference gas is supplied through the sample inlet port. The oxygen molecules in the sample gas are attracted to the magnetic field and cause a back-pressure at the outlet port of the reference gas in the magnetic field.

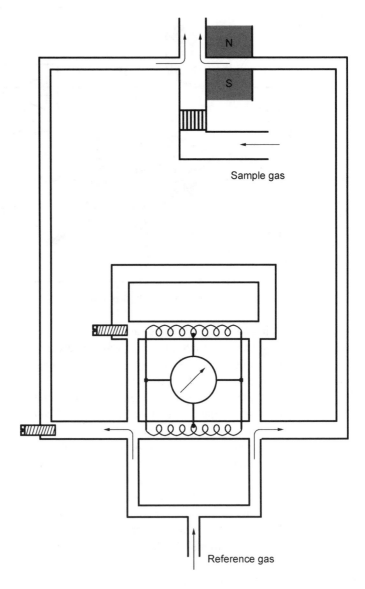

2.22 Paramagnetic oxygen analyzer cell (second design).

If the oxygen concentration in the sample gas is different from that in the reference gas, the reference gas flow through the measuring system is not symmetrical. This results in different cooling rates of the filaments and an imbalance of the Wheatstone bridge due to differences in the resistance of the filaments. The oxygen concentration of the sample gas is a linear function of the bridge signal.

Both designs of paramagnetic oxygen analyzer cells have been used successfully in fire testing. Each design has advantages and disadvantages. Cells of the first design have a faster response and do not require a reference gas. Advantages of the second design are the absence of moving parts and the fact that the measuring system is exposed to a clean reference gas and not in contact with the sample gas.

Non-dispersive infrared (NDIR) analyzers

The molecules of many chemical species that are of interest in fire gas analysis absorb IR radiation. Binary molecules such as diatomic nitrogen and oxygen are an exception to this rule. Each IR-active gas has a unique absorption spectrum. An extensive collection of IR spectra is available from the following National Institute of Standards and Technology (NIST) web site: http://webbook.nist.gov (Stein, 2003). Figures 2.23 and 2.24 depict the absorption spectra from the NIST database for CO_2 and CO respectively. These figures show the absorbance in arbitrary units as a function of wavenumber. The wavenumber of an electromagnetic wave is equal to the inverse of its wavelength and is proportional to its frequency. It is more convenient to use wavenumbers because the corresponding frequencies have very high values.

Figures 2.23 and 2.24 show that both CO_2 and CO absorb IR radiation in one or more distinct wavenumber regions or 'frequency bands'. This is typical for any IR-active gas. NDIR analyzers are based on this observation. Figure 2.25 shows a schematic of a NDIR CO analyzer. The source produces a broad

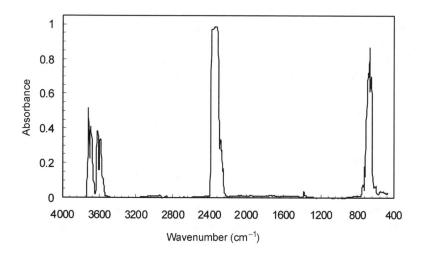

2.23 Infrared absorption spectrum for carbon dioxide.

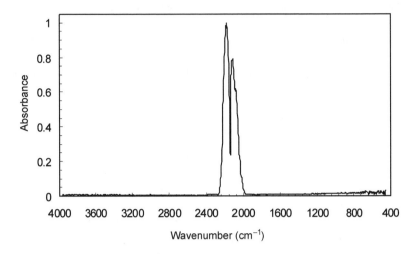

2.24 Infrared absorption spectrum for carbon monoxide.

spectrum of IR energy. A lens is used to collimate the beam, which then passes through an optical filter that only transmits IR radiation over a limited range of frequencies that are within the absorption band of CO. According to Fig. 2.24, such a filter would transmit radiation over a narrow range of wavenumbers between 2300 and 2000 cm^{-1}. The beam then passes through a gas-filled sample cell, where some of the IR energy is absorbed, and is funneled through a condenser before it hits a detector. The reduced intensity measured by the detector is due to the absorption in the sample cell and is a direct function of the concentration of the absorbing gases in the sample. The path length of the cell determines the lowest concentration that can be detected and a wide range of cells are available to suit most fire testing and research needs.

The problem with this setup is that besides the gas that needs to be measured, in this case CO, the sample may contain other gases that absorb IR radiation at the same frequencies. The resulting interference can be eliminated with the gas filter correlation technique. Two glass cells, one containing nitrogen and another containing CO are mounted on a wheel that rotates at constant speed between the collimating lens and the optical filter. When the nitrogen filter is in position, no absorption takes place and all IR radiation passes through to the sample cell. When the other filter is in position, absorption takes place and the intensity of the wavelengths characteristic to CO is reduced. The difference between the two signals accounts only for the absorption due to CO and eliminates the effect of interfering gases.

NDIR analyzers are widely used to measure concentrations of CO and CO_2. NDIR analyzers are also commercially available to measure other IR-active

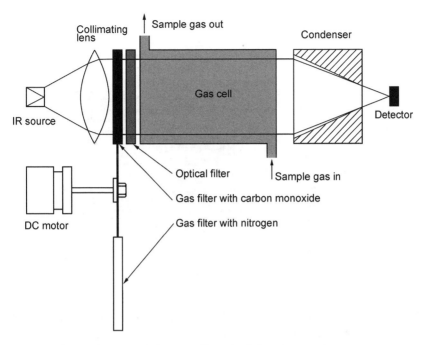

2.25 Gas filter correlation non-dispersive infrared gas analyzer.

gases and vapors such as NO, NO_2, HCN, HCl and H_2O. The latter two are very difficult to measure because the gas sampling lines and analyzer must be heated to avoid condensation.

Fourier transform infrared (FTIR) spectroscopy

FTIR spectrometers rely on the same basic principle as NDIR analyzers, i.e., the fact that many gases absorb IR radiation at species-specific frequencies. However, FTIR spectroscopy is a disperse method, which means that measurements are performed over a broad spectrum instead of a narrow band of frequencies. Figure 2.26 shows a schematic of a FTIR spectrometer. A blackbody source emits IR radiation over a range of wavenumbers, typically between 4000 and 400 cm^{-1} for (fire) gas analysis. The IR beams passes through an interferometer, which consists of a beam splitter, a fixed mirror and a moving mirror. The interferometer is a cleverly designed optical device that separates the spectral components of the beam in time. The beam then passes through the sample gas cell before it hits the detector. To increase the sensitivity, a system of mirrors in the cell significantly increases the path length. An interferogram is recorded consisting of the detector signal as a function of time. The corresponding absorbance spectrum is obtained from a Fourier transform (a mathematical transformation from the time to the frequency domain) of the

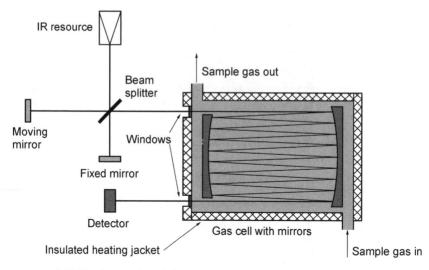

IR resource

Beam splitter

Sample gas out

Moving mirror

Windows

Fixed mirror

Detector

Insulated heating jacket

Gas cell with mirrors

Sample gas in

2.26 Fourier transform infrared spectrometer.

interferogram. The concentration of IR-active gases in the sample can be determined from the absorbance spectrum. In order to measure the concentration of a specific gas it is necessary first to obtain reference spectra for certified mixtures with different concentrations of the gas in N_2.

FTIR spectroscopy has the advantage that a large number of gases can be measured with one analyzer. The technique is commonly used to measure simultaneously the concentrations of CO, HBr, HCN, HCl, HF, NO and SO_2 because these gases form the basis for smoke toxicity regulations of the International Maritime Organization (Orvis and Janssens, 1999). However, many other gases and vapors such as CH_4, C_2H_2, C_2H_4, C_2H_6, C_3H_4O (acrolein), CO_2, $COCl_2$ (phosgene), COF_2 (carbonyl fluoride), H_2O and NO_2 have been routinely measured as well (Speitel, 2002). Another advantage of FTIR spectroscopy is that it is semi-continuous since an interferogram can typically be obtained in less than five seconds.

The analysis of FTIR spectra is very difficult because the interference between overlapping species, the effect of unknown components in the gas mixture and errors due to noise all have to be accounted for. At least a dozen mathematical techniques of varying complexity have been proposed to accomplish the task (Pottel, 1996). Classical multivariate chemometrical techniques such as classical least squares (CLS), partial least squares (PLS) and implicit non-linear regression (INLR) appear to be most widely used but alternative techniques such as quantitative target factor analysis (QTFA) have also been found suitable (Hakkarainen *et al.*, 2000). The accuracy of the measurements is strongly affected by system characteristics such as cell volume, path length and the type of detector and by operating parameters such as sample flow rate;

sample line temperature; cell temperature and pressure; spectrum resolution, etc. A first attempt at specifying optimum operating conditions ('Nordtest Method NT FIRE 047 – Combustible products smoke gas concentrations, continuous FTIR analysis') lacked detail in several areas (Bulien, 1996, 1997). Subsequent research in Europe (Hakkarainen, 1999) recently resulted in the development of an international standard guide for analyzing fire gases with FTIR spectroscopy ('ISO DIS 19702 – Toxicity testing of fire effluents – Guide for analysis of gases and vapours in fire effluents using FTIR gas analysis').

Other approaches for measuring gas concentrations

Gas detector tubes are very popular for measuring toxic gas concentrations in fire tests because they are simple and inexpensive. A gas detector tube consists of a glass vial that is filled with a chemical substance that changes color when it reacts with a particular gas. A constant volume of sample gas is drawn through the vial and the length of color change is a direct indication of the concentration of the measured gas. Gas detector tubes are available for the major toxic gases that are generated in fires, except HBr. The advantages of the method are offset by a number of significant limitations. The method has limited accuracy (10–15% at best) and is not suitable for continuous online analysis because it can take several minutes to pull the required sample volume through the tube. Some gas detector tubes might be sensitive to other gas species in the sample. Finally, the color change is also affected by the temperature of the sample.

Ion-selective electrodes can be used to determine indirectly the concentration of hydrogen halides such as HBr, HF, and HCl in a gas sample. The sample is bubbled through an impinger solution. An electrode that is sensitive to a specific ion (e.g., F^-, Br^- or Cl^-) is used to measure the concentration of the ion in the solution. The concentration of the corresponding hydrogen halide in the sample can then be determined on the basis of the ion concentration and the known volumes of the sample and the solution. With the appropriate electrode, the method has also been used to measure the concentration of HCN. This type of analysis has been used in batch as well as continuous mode, although the latter is much more difficult.

One of the most accurate and complete methods to analyze fire gases uses a combination of a gas chromatograph (GC) and a mass spectrometer (MS.) A gas chromatograph consists of a packed or lined column to separate the different components in a gas sample. A detector at the end of the column is used to quantify the amount of each component. In a GC/MS combination, the mass spectrometer is used to identify the components that are present in the sample. A mass spectrometer converts the molecules in the sample to ions and then separates and identifies the ions according to their mass-to-charge ratio. The GC/MS method is suitable only for batch analysis due to the long retention time in the gas chromatograph. Mass spectrometers can also be used in quantitative mode. This has been proposed as an improvement over other methods for analyzing fire gases, but the idea still needs to be tested (Enright and Vandevelde, 2001).

Sometimes it might be of interest to know the concentration of unburnt hydrocarbons in fire effluents. A continuous hydrocarbon analyzer with a flame ionization detector can be used for this purpose. The detector cannot make a distinction between different types of hydrocarbons and the results are therefore expressed in the form of parts per million of carbon.

Measuring smoke particle concentration

The concentration of particles in the exhaust duct of a fire test can be measured by diverting a fixed fraction of the duct flow through a filter. Based on this technique it is relatively easy to determine the average soot concentration over the duration of a test by measuring the mass of the filter before and after the test. The average size distribution can be determined by using multiple filters with different porosities in series. With this technique it is very difficult to measure the soot concentration as a function of time during a test, because that requires recording the mass of the filter(s) as the test progresses. A more practical method is based on the observation that the ratio of the light extinction coefficient of flame-generated smoke to the mass concentration of soot in the smoke is approximately constant (Mulholland and Croarkin, 2000). The estimated value of the constant is $\sigma_s = 8.7 \pm 1.1 \, \text{m}^2/\text{g}$. The implication of this nearly universal value is that mass concentration of smoke can be determined from light extinction measurements (Mulholland *et al.*, 2000).

2.9.2 Challenges of measuring gas composition in fire tests

'ASTM E 800, Standard guide for measurement of gases present or generated during fires' states the following: 'More errors in analysis result from poor or incorrect sampling than from any other part of the measurement process. It is therefore essential to devote special attention to sampling, sample transfer, and presentation aspects of the analysis procedures.' The guide then continues with nearly three pages of detailed discussion on sampling. 'ISO 19701 – Methods for analysis and sampling of fire effluents' also provides detailed recommendations for proper sampling techniques.

2.10 Error and uncertainty of measurements

The objective of a measurement is to determine the value of the measurand, i.e., the physical quantity that needs to be measured. Every measurement is subject to error, no matter how carefully it is conducted. The (absolute) error of a measurement is defined as follows:

$$\epsilon \equiv x_m - x_t \qquad\qquad 2.16$$

where ϵ = measurement error;

x_m = measured value of the measurand;

x_t = true value of the measurand.

All terms in eqn 2.16 have the units of the physical quantity that is measured. This equation cannot be used to determine the error of a measurement because the true value is unknown, otherwise a measurement would not be needed. In fact, the true value of a measurand is unknowable because it cannot be measured without error. However, it is possible to estimate, with some confidence, the expected limits of error. This estimate is referred to as the uncertainty of the measurement and provides a quantitative indication of its quality.

Errors of measurement may have two components, a random component and a systematic component. The former is due to a number of sources that affect a measurement in a random and uncontrolled manner. Random errors cannot be eliminated, but their effect on uncertainty may be reduced by increasing the number of repeat measurements and by applying a statistical analysis to the results. Systematic errors remain unchanged when a measurement is repeated under the same conditions. Their effect on uncertainty cannot be completely eliminated either, but it can be reduced by applying corrections to account for the error contribution due to recognized systematic effects.

An internationally accepted guide, referred to as the GUM, has been developed for estimating the uncertainty of measurements (Anon., 1993). The uncertainty of a measurement is first expressed as a weighted sum of uncertainties for all sources of error that contribute to the overall uncertainty. The GUM groups uncertainty components into two categories based on their method of evaluation. 'Type A' uncertainty components are determined on the basis of a statistical analysis of repeated observations. 'Type B' uncertainty components are determined by other means, e.g., based on manufacturer specifications, calibration data, experience, etc. Each uncertainty component is expressed as a standard deviation and the resulting standard uncertainty components are combined to produce an estimate of the overall standard uncertainty. The standard uncertainty is then multiplied with a coverage factor k to adjust the value for the desired level of confidence (typically $k = 2$ for a confidence level of 95%).

Space limitations prohibit a more detailed discussion of the important topic of measurement uncertainty in this book. The reader is referred to the GUM for an in-depth explanation of the subject. NIST applied the general concepts of the GUM to determine the uncertainty of radiative heat flux measurements to the floor of a room/corner fire test (Bryant et al., 2003). Those who are interested in the subject of uncertainty of fundamental measurements in fire tests are encouraged to review the NIST paper.

2.11   References

Anon. (1993). *Guide to the Expression of Uncertainty in Measurement*. International Organization for Standardization, Geneva, Switzerland.

Anon. (2002). *The Temperature Handbook*. Omega Engineering, Inc., Stamford, CT, Z-16.

Atreya, A. (2002). 'Convection Heat Transfer.' In the *SFPE Handbook of Fire Protection Engineering*, Quincy, MA, National Fire Protection Association, Section 1, 44–72.

Babrauskas, V., and Wetterlund, I. (1995). 'Choice of Optical Calibration Filters for Laser Photometers.' *Fire Safety Journal*, 24, 197–199.

Blevins, L. (1999). 'Behavior of Bare and Aspirated Thermocouples in Compartment Fires.' *Paper HTD99-280, 33rd National Heat Transfer Conference*, August 15–17, 1999, Albuquerque, NM.

Blevins, L., and Pitts, W. (1999). 'Modeling of Bare and Aspirated Thermocouples in Compartment Fires.' *Fire Safety Journal*, 33, 239–259.

Brohez, S., Delvosalle, C., and Marlair, G. (2004). 'A Two-Thermocouples Probe for Radiation Corrections of Measured Temperatures in Compartment Fires.' *Fire Safety Journal*, 39, 399–411.

Bryant, R., Womeldorf, C., Johnsson, E., and Ohlemiller, T. (2003). 'Radiative Heat Flux Measurement Uncertainty.' *Fire and Materials*, 27, 209–222.

Bulien, O. (1996). 'FTIR Spectrometer for Measuring Toxic Smoke Components in Fire Testing – Review of Equipment and Calibration Routines in NT FIRE 047.' *Fire and Materials*, 20, 225–233.

Bulien, O. (1997). 'Erratum to "FTIR Spectrometer for Measuring Toxic Smoke Components in Fire Testing – Review of Equipment and Calibration Routines in NT FIRE 047".' *Fire and Materials*, 21, 195 (original paper published 1996).

Clark, F. (1985). 'Assessment of Smoke Density with a Helium-Neon Laser.' *Fire and Materials*, 9, 30–35.

Croce, P. (1975). 'A Study of Room Fire Development: The Second Full-Scale Bedroom Fire Test of the Home Fire Project (July 24, 1974.) Volume I and II.' *FMRC Technical Report RC75-T-31, Serial No. 21011.4*, Factory Mutual Research Corporation, Norwood, MA.

Elliot, P., and Whiteley, R. (1994). 'Calibration of the Smoke Measurement System on the Cone Calorimeter.' *Fire Safety Journal*, 23, 103–107.

Enright, P., and Vandevelde, P. (2001). 'Use of the Mass Spectrometer for Heat Release Rate Analysis in Fire Calorimeters.' *Fire and Materials*, 25, 203–207.

Francis, J., and Yau, T. (2004). 'On Radiant Network Models of Thermocouple Error in Pre and Post Flashover Compartment Fires.' *Fire Technology*, 40, 277–294.

Gardon, R. (1953). 'An Instrument for the Direct Measurement of Thermal Radiation.' *The Review of Scientific Instruments*, 24, 366–370.

Göransson, U., and Omrane, A. (2004). 'Surface Temperature Measurements in the Cone Calorimeter Using Phosphorescence.' *Interflam 2004*, July 5–7, 2004, Edinburgh, Scotland, 1431–1442.

Gunners, N. (1967). 'Methods of Measurement and Measuring Equipment for Fire Tests.' *Acta Polytechnica Scandinavica*, Civil Engineering and Building Construction Series Nr. 43, 1–45.

Hakkarainen, T. (1999). 'Smoke Gas Analysis by Fourier Transform Infrared Spectroscopy: The SAFIR Project.' *VTT Research Notes 1981*, VTT-Technical Research Center of Finland, Espoo, Finland.

Hakkarainen, T., Mikkola, E., Laperre, J., Gensous, F., Fardell, P., LeTallec, Y., Baiocchi, C., Paul, K., Simonson, M., Deleu, C., and Metcalfe, E. (2000). 'Smoke Gas Analysis by Fourier Transform Infrared Spectroscopy: Summary of the SAFIR Project Results.' *Fire and Materials*, 24, 101–112.

Janssens, M. (1991). *Thermophysical Properties of Wood and Their Role in Enclosure Fire Growth*. Ph.D. Thesis, University of Ghent, Ghent, Belgium.

Janssens, M., and Tran, H. (1992). 'Data Reduction of Room Tests for Zone Model Validation.' *Journal of Fire Sciences*, 10, 528–555.

Janssens, M., Kimble, J., and Murphy, D. (2003). 'Computer Tools to Determine Material Properties for Fire Growth Modeling from Cone Calorimeter Data.' *Fire and Materials 2003, 8th International Conference and Exhibition*, January 27–28, 2003, San Francisco, CA.

Jin, T. (1978). 'Visibility Through Smoke.' *Journal of Fire and Flammability*, 9, 135–155.

Jones, J. (1995). 'On the Use of Metal Sheathed Thermocouples in a Hot Gas Layer Originating From a Room Fire.' *Journal of Fire Sciences*, 13, 257–260.

Jones, J. (1998). 'On the Measurement of Temperatures in Simulated Room Fires.' *Journal of Fire Sciences*, 16, 3–6.

Jones, J. (2002). 'Suggestions Towards Improved Reliability of Thermocouple Temperature Measurement in Combustion Tests.' *ASTM STP 1427, Thermal Measurements: The Foundation of Fire Standards*, December 3, 2001, Dallas, TX, 16–31.

Jones, J. (2004). 'Further Comments on Temperature Measurements in Simulated Room Fires.' *Fire Safety Journal*, 39, 157–158.

Lennon, P., and Silcock, G. (2001). 'An Investigation of the Ability of a Thin Plate Heat Flux Device to Determine the Incident Heat Fluxes During Enclosure Fires.' *International Journal on Engineering Performance-Based Fire Codes*, 3, 1–15.

Luo, M. (1997). 'Effects of Radiation on Temperature Measurement in a Fire Environment.' *Journal of Fire Sciences*, 15, 443–461.

McCaffrey, B., and Heskestad, G. (1976). 'A Robust Bidirectional Low-Velocity Probe for Flame and Fire Application.' *Combustion and Flame*, 26, 125-127.

McCaffrey, B., and Rockett, J. (1977). 'Static Pressure Measurements of Enclosure Fires.' *Journal of Research of the National Bureau of Standards*, 82, 107–117.

Morris, B., O'Conner, D., and Silcock, G. (1996). 'Incident Heat Flux Evaluation in Fire Tests and the Use of an Instrumented Steel Billet for Measurement.' *Journal of Applied Fire Science*, 6, 27–41.

Mulholland, G., and Croarkin, C. (2000). 'Specific Extinction Coefficient of Flame Generated Smoke.' *Fire and Materials*, 24, 227–230.

Mulholland, G., Johnsson, E., Fernandez, M., and Shear, D. (2000). 'Design and Testing of a New Smoke Concentration Meter.' *Fire and Materials*, 24, 231–243.

Newman, J. (1987). 'Multi-Directional Flow Probe Assembly for Fire Application.' *Journal of Fire Sciences*, 5, 50–56.

Newman, J., and Croce, P. (1979). 'Simple Aspirated Thermocouple for Use in Fires.' *Journal of Fire and Flammability*, 10, 326–336.

Nussbaum, R. (1987). 'Oxygen Consumption Measurements in the Cone Calorimeter. A Direct Comparison Between a Paramagnetic Cell and a High-Temperature Cell.' *Fire and Materials*, 11, 201–203.

Omrane, A., Ossler, F., Alden, M., Göransson, U., and Holmstedt, G. (2003). 'Surface Temperature Measurement of Flame Spread Using Thermographic Phosphors.'

Seventh (7th) International Symposium, June 16-21, 2003, Worcester, MA, 141–152.

Orvis, A., and Janssens, M. (1999). 'Trends in Evaluating Toxicity of Fire Effluents.' *Fire and Materials 1999, 6th International Conference and Exhibition*, February 22–23, 1999, San Antonio, TX, 95–106.

Östman, B., Svensson, G., and Blomqvist, J. (1985). 'Comparison of Three Test Methods for Measuring Rate of Heat Release.' *Fire and Materials*, 9, 176–184.

Ower, E., and Pankhurst, R. (1977). *The Measurement of Air Flow*, Pergamon Press.

Pitts, W., Braun, E., Peacock, R., Mitler, H., Johnsson, E., Reneke, P., and Blevins, L. (2002) 'Temperature Uncertainties for Bare-Bead and Aspirated Thermocouple Measurements in Fire Environments.' *ASTM STP 1427, Thermal Measurements: The Foundation of Fire Standards*, December 3, 2001, Dallas, TX, 3–15.

Pottel, H. (1996). 'Quantitative Models for Prediction of Toxic Component Concentrations in Smoke Gases from FTIR Spectra.' *Fire and Materials*, 20, 273–291.

Putorti, A. (1999). 'Design Parameters for Stack-Mounted Light Extinction Measurement Devices.' *NISTIIR 6215*, National Institute of Standards and Technology, Gaithersburg, MD.

Quintiere, J., and McCaffrey, B. (1980). 'The Burning of Wood and Plastic Cribs in an Enclosure: Volume I.' *NBSIR 80-2054*, National Bureau of Standards, Gaithersburg, MD.

Robertson, A., and Ohlemiller, T. (1995). 'Low Heat-Flux Measurements: Some Precautions.' *Fire Safety Journal*, 25, 109–124.

Savitzky, A., and Golay, M. (1964). 'Smoothing and Differentiation of Data by Simplified Least Squares Procedures.' *Analytical Chemistry*, 36, 1627–1639.

Speitel, L. (2002). 'Fourier Transform Infrared Analysis of Combustion Gases.' *Journal of Fire Sciences*, 20, 349–371.

Steckler, K., Quintiere, J., and Rinkinen, W. (1982). 'Flow Induced by Fire in a Compartment.' *NBSIR 82-2520*, National Bureau of Standards, Gaithersburg, MD.

Stein, S. (2003). 'Infrared Spectra.' in P. Linstrom and W. Mallard, NIST Chemistry WebBook, NIST Standard Reference Database Number 69, Gaithersburg, MD, National Institute of Standards and Technology.

Thomson, H., Drysdale, D., and Beyler, C. (1988). 'An Experimental Evaluation of Critical Surface Temperature as a Criterion for Piloted Ignition of Solid Fuels.' *Fire Safety Journal*, 13, 185–196.

Urbas, J. (1993). 'Non-Dimensional Heat of Gasification Measurements in the Intermediate Scale Rate of Heat Release Apparatus.' *Fire and Materials*, 17, 119–123.

Urbas, J., and Parker, W. (1993). 'Surface Temperature Measurements on Burning Wood Specimens in the Cone Calorimeter and the Effect of Grain Orientation.' *Fire and Materials*, 17, 205–257.

Urbas, J., Parker, W., and Luebbers, G. (2004). 'Surface Temperature Measurements on Burning Materials Using an Infrared Pyrometer: Accounting for Emissivity and Reflection of External Radiation.' *Fire and Materials*, 28, 33–53.

Whitaker, S. (1972). 'Forced Convection Heat Transfer Calculations for Flow in Pipes, Past Flat Plates, Single Cylinders and for Flow in Packed Beds and Tube Bundles.' *AIChE Journal*, 18, 261–371.

Wickström, U., and Wetterlund, I. (2004). 'Total Heat Flux Cannot Be Measured.' *Interflam 2004*, July 5-7, 2004, Edinburgh, Scotland, 269–276.

Part I

Flammability tests for buildings
and their contents

3

Flammability of wood products

B A-L ÖSTMAN, S P Trätek/Wood Technology, Sweden

3.1 Introduction

The new classification system for the reaction to fire performance of building products in Europe, the so-called Euroclass system, has been applied to five different types of wood products: wood-based panels, structural timber, glued laminated timber, solid wood panelling and cladding and wood flooring as being 'products with known and stable fire performance'. The European classification system includes two sub-systems, one main system for all construction products except floorings and the other for flooring products. Wood properties such as density, thickness, joints and types of end use application including different substrates have been studied thoroughly and are included in the classification. Most wood products fall in classes D-s2,d0 or D_{fl}-s1 (for floorings). Some products may also fall in the main classes C or E. Testing has been performed according to EN 13823 SBI Single Burning Item test, EN ISO 9239-1 Radiant panel test, and EN ISO 11925-2 Small flame test. In all, more than one hundred wood products in different end use applications have been studied. With the experimental evidence, clear relationships between the main Euroclass fire performance parameters and product parameters (such as density and thickness) have been demonstrated. Tables with reaction to fire classification of different wood products and end use applications have been developed, approved by the European Commission and published in their official journal. This procedure is ongoing with further official decisions to be published.

3.2 New European classes for the reaction to fire performance of building products

A new classification system for the reaction to fire properties of building construction products has recently been introduced in Europe.[1] It is often called the Euroclass system and consists of two sub-systems, one for construction products excluding floorings, i.e. mainly wall and ceiling surface linings and the other for floorings. Both sub-systems have classes A to F of which classes A1 and A2 are non-combustible products. The new system will replace the present individual European national classification systems, which have formed obstacles to trade.[2]

The new European classification system for reaction to fire performance was published in the *Official Journal* in 2000[1] and is based on a set of EN standards for different test methods[3–6] and for classification systems.[7] The new European system has to be used for all construction products in order to get the CE- mark, which is the official mandatory mark to be used for all construction products on the European market. Different product properties have to be declared and may vary for different products, but the reaction to fire properties are mandatory for all construction products. The normal route is that each manufacturer tests and declares their own products individually. Products with known and stable performance may be classified as groups according to an initiative from the EC.[8] This is a possibility for wood products that have a fairly predictable fire performance. Properties such as density, thickness, joints and type of end use application may influence the classification. If no common rules are available, each producer has to test their products in order to fulfil requirements in product standards and to get the CE-mark.

3.2.1   CWFT – classification without further testing

The procedure for CWFT, Classification without further testing, is described in a document from DG Enterprise.[8] CWFT corresponds to the definition 'Products which have been proven to be stable in a given European class (on the basis of testing to the appropriate EN test method(s)) within the scope of their variability in manufacturing allowed by the product specification (harmonised standard or ETA)'. CWFT is a list of generic products, not a list of proprietary products. CWFT lists will be established by Commission Decision(s) in consultation with the Standing Committee on Construction (SCC). The Fire Regulators Group (FRG) European Commission Expert Group on Fire Issues Regarding Construction Products (EGF), advised by its CWFT Working Group made up of representatives of regulators and experts on fire performance of building products, CEPMC (Council of European Producers of Materials for Construction), Notified Bodies Group, CEN TC127 WG4 and CEN/EOTA TCs (invited for specific cases, as applicants), will consider all requests made and forward recommendations to the SCC for final opinion. Products claiming CWFT must be clearly above the lower class limits, to provide a safety margin. This should be determined on a statistical basis in relation to the scattering of results. In general terms, each classification parameter (as defined in the relevant classification standard in the EN 13501 series) should be either 20% above the class limit (although some relaxation of this may be possible) or shown by statistical means to have a satisfactory safety level for a request to be accepted. Due account will be taken of the likely variability in the production process of products.

The SCC will make the final decision based upon the recommendations from the FRG/EGF. All requests and related data will be submitted to the FRG/EGF for discussion. The advice of the FRG/EGF will largely determine whether the request is forwarded to the SCC for opinion.

The CWFT approach has been applied to several wood products. Full reports are available, one for wood products excluding floorings[9] and another for floorings.[10]

3.2.2 Classification system

The new European system for the reaction to fire performance classes consists of two sub-systems, one for construction products, i.e., mainly wall and ceiling surface linings, see Table 3.1, and another similar system for floorings. Three test methods are used for determining the classes of combustible building products, see Table 3.2. For non-combustible products additional test methods are also used.[1]

3.3   Applications for different wood products

Five different types of wood products have been included:

- wood-based panels – e.g. particle board
- structural timber
- glued laminated timber
- solid wood panelling and claddings
- wood floorings.

The application of the new European system differs among the various wood products studied as described below. End-use applications of the products are essential for the reaction to fire classification, e.g., substrates or air gaps behind the wood product, joints and surface profiles, see Fig. 3.4. The classification is always related to harmonised product standards in which a CWFT reaction to fire classification table is included. Mounting and fixing conditions have to be specified.

3.3.1 Harmonised European product standards

The harmonised product specifications which will include the fire classifications are:

- EN 13 986 Wood-based panels for use in construction – characteristics, evaluation of conformity and marking
- EN 14 080 Timber structures – glued laminated timber – requirements
- prEN 14 081 Timber structures – strength graded structural timber with rectangular cross section

Table 3.1 Overview of the European reaction to fire classes for building products excl. floorings[1]

Euroclass	Smoke class	Burning droplets class	Requirements according to Non comb	SBI	Small flame	FIGRA W/s	Typical products
A1	—	—	X	—	—	—	Stone, concrete
A2	s1, s2 or s3	d0, d1 or d2	X	X	—	≤ 120	Gypsum boards, mineral wool
B	s1, s2 or s3	d0, d1 or d2	—	X	X	≤ 120	Gypsum boards, fire-retardant wood
C	s1, s2 or s3	d0, d1 or d2	—	X	X	≤ 250	Coverings on gypsum boards
D	s1, s2 or s3	d0, d1 or d2	—	X	X	≤ 750	Wood, wood-based panels
E	—	— or d2	—	—	X	—	Some synthetic polymers
F	—	—	—	—	—	—	No performance determined

SBI = Single Burning Item: main test for the reaction to fire classes for building products.
FIGRA = Fire Growth Rate, main parameter for the main fire class according to the SBI test.

Table 3.2 European test methods used for determining the reaction to fire classes of combustible building products[1]

Test method	Construction products excl. floorings	Floorings	Main fire properties measured and used for the classification
Small flame test EN ISO 11925-2	X	X	Flame spread within 60 or 20 s.
Single Burning Item test, SBI EN 13823	X	—	• FIGRA, FIre Growth RAte; • SMOGRA, SMOke Growth RAte; • Flaming droplets or particles
Radiant panel test EN ISO 9239-1	—	X	• CHF, Critical Heat Flux; • Smoke production

- EN 14 250 Timber structures – product requirements for prefabricated structural members assembled with punched metal plate fasteners
- EN 14 342 Wood flooring – characteristics, evaluation of conformity and marking
- prEN 14 544 Timber structures – strength graded structural timber with round cross-section – requirements
- prEN 14 915 Solid wood panelling and cladding – characteristics, evaluation of conformity and marking.

All these harmonised product specifications form the basis for the CE-marking of construction products and prescribe which product properties have to be documented.

3.2.2   End-use applications

Wood-based panels are mainly used as wall and ceiling linings. End-use conditions without air gaps behind the panel are included. An extension to further end uses took place during 2005.

Structural timber and glued laminated timber products are used as parts of wall, roof or floor systems, that may or may not be load bearing. Studs and beams in timber frame systems are usually covered by wood-based or gypsum boards and members of similar systems in solid wood may not be covered. End-use applications also include free-standing structural elements, see Figs 3.1 and 3.2. A minimum thickness of 22 mm is specified in the product standards for structural timber and 40 mm for glued laminated timber. Structural timber and glued laminated timber products are illustrated in Fig. 3.3.

Solid wood panellings and claddings are used in interior and exterior applications. It includes vapour barriers with or without an air gap behind the wood product and vertical parts of stairs. End use applications such as free-

3.1 Examples of end uses as studs, beams and free-standing elements for structural timber.

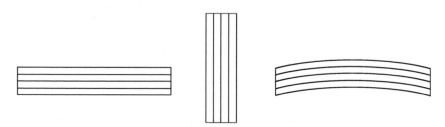

3.2 Examples of end uses as beams and columns for glulam.

3.3 Structural timber and glulam.

standing ribbon elements are also included. Figure 3.4 illustrates the surface profiles and joint types included for solid wood panelling and cladding products. Examples of end uses such as interior panelling, exterior claddings and free-standing ribbon elements are given in Figs 3.5 and 3.6. Wood floorings include

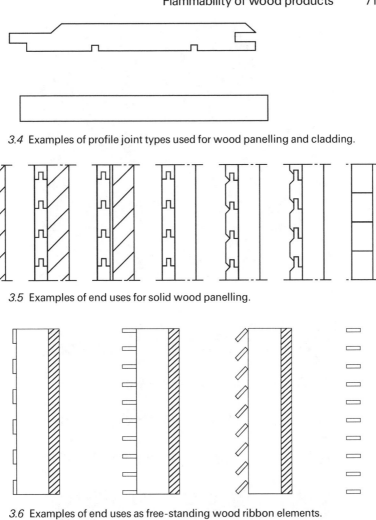

3.4 Examples of profile joint types used for wood panelling and cladding.

3.5 Examples of end uses for solid wood panelling.

3.6 Examples of end uses as free-standing wood ribbon elements.

3.7 Examples of end uses for wood flooring. From the left: flooring directly on a substrate; with an interlayer between the flooring and the substrate; with a closed air gap in between and, to the right, with an open air gap without anything underneath (e.g. as in an open staircase).

both homogeneous and multilayer products. The use as steps in stairs with or without an air gap behind are also included, see Fig. 3.7. Solid wood panellings and claddings are illustrated in Fig. 3.8 and free-standing ribbon elements in Fig. 3.9.

3.8 Solid wood panelling.

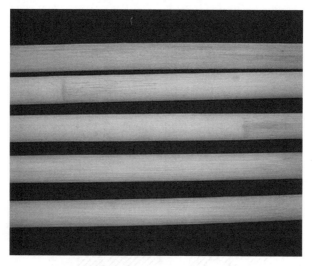

3.9 Free-standing ribbon elements.

3.3.3   Mounting and fixing conditions

Wood-based panels may form major parts of the total surfaces of building elements. For the purpose of testing and classification of wood-based panels, the whole area for exposure in the SBI apparatus, $1.5 \times 1.5\,\mathrm{m}$, was covered with the panel pieces mounted with different specified joints and oriented horizontally or vertically as in the end-use application. Different thicknesses and types of panels

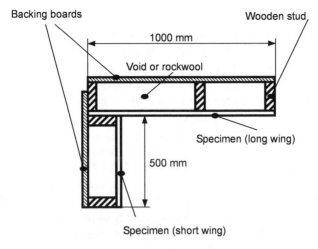

3.10 Schematic view of specimen mounting in the SBI test (as seen from above).

with insulation or other substrates behind the wood panel were included. A schematic sketch of how the sample was mounted in the SBI test is given in Fig. 3.10. Figure 3.11 shows a typical setup of testing in the SBI apparatus.

Structural timber and glued laminated timber products are not generally used to form the major part of the total surfaces of a room and the number of possible applications is very large. Thus a material-testing interpretation has been used. The whole area for exposure of the test specimen in the SBI apparatus, 1.5 ×

3.11 SBI test set up with a wooden product: before fire (left) and during fire exposure (right).

1.5 m, was covered with timber pieces mounted edge to edge (butt jointed), without jointing or bonding and orientated horizontally or vertically. This includes the use of timber battens, minimum 40 × 40 mm, fixed to the test backing boards at 400–600 mm centres horizontally or vertically (perpendicular to the orientation of the timber pieces). Different thicknesses of timber with and without air gaps or with thermal insulation behind were investigated to ensure that the fire behaviour is fully independent of the underlying layers.

Solid wood panelling and cladding usually forms major parts of the total surfaces of a room, similar to the wood-based panels described above. The whole area for exposure in the SBI apparatus, 1.5 × 1.5 m, was covered with the wood pieces mounted with different specified joints and oriented horizontally or vertically. Different thickness of wood, air gaps, vapour barriers and substrates behind the wood products are included. Examples of joint and surface profiles used are illustrated in Fig. 3.4 and examples of interior use of solid wood panelling in Fig. 3.5. Free-standing wood ribbon elements were mounted on a wood batten frame and surrounded by air on all sides as illustrated in Fig. 3.6.

Wood floorings have been tested in the radiant panel test with and without an air gap underneath. Different thicknesses of flooring, surface coatings, air gaps, interlayers and substrates behind the wood flooring are included, see Fig. 3.9. The testing was performed mainly at the fire laboratories of VTT Building and Transport in Finland and SP Trätek/Wood Technology in Sweden.

3.4   Test results for different wood products

For most wood-based panels, the safety margins for FIGRA to the class D limit are in the order of 20–60%. Exceptions are only a few low-density products and a rather thin free-standing product tested with an air gap behind. Thin 9 mm panels on studs with an open air gap behind did not pass class D. Horizontal or vertical joints and different types of substrates did not influence the fire performance significantly. For smoke, all products pass the SMOGRA limit with very high margins to class s1, > 60%. However, for the TSP (total smoke production) limit to class s1, some products have safety margins in the order of 30–60%, while other products do not pass the s1 limit. The FIGRA values have also been analysed in terms of density of the wood-based panels without any surface or other treatments. Data are presented in Fig. 3.12.

A trend of higher FIGRA values at lower density is obvious. The FIGRA limit to the lower class E, 750 W/s, is approximately at a density of 350 kg/m^3. With a safety margin of 20%, the density limit is about 400 kg/m^3. No similar general trend has been observed for smoke parameters as TSP. Figure 3.12 also shows that beyond a density of 500 kg/m^3, FIGRA for the wood-based panels was found to be relatively independent of the density.

For structural timber products, 21–22 mm thick, of different wood species and with different densities tested with an open air gap behind, FIGRA values

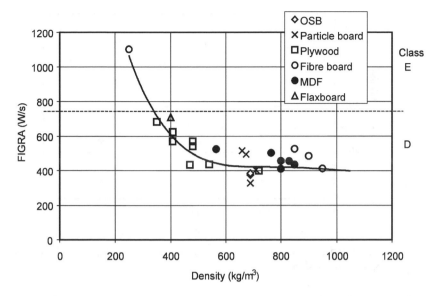

3.12 FIGRA value as a function of density for wood-based panels attached to a calcium silicate substrate. Class D is obtained for all products except for a low-density fiberboard.

are illustrated in Fig. 3.13. The size of the air gap is not important for these quite thick wood products. The FIGRA values decrease with increasing timber density and all values are well below the upper limit, 750 W/s, for class D.

For glued laminated timber products, 40 mm thick, of different wood species and with different densities and different glues tested with an open air gap behind, FIGRA values are illustrated in Fig. 3.14. The FIGRA values decrease

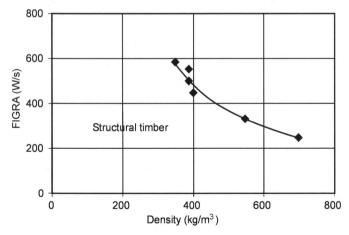

3.13 FIGRA value as a function of density for 22 mm structural timber tested with an open air gap behind.

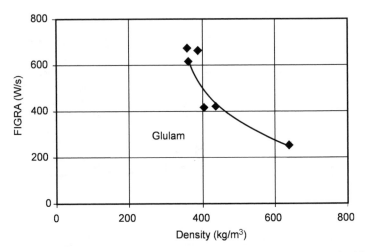

3.14 FIGRA values as a function of density for 40 mm glulam tested with an open air gap behind.

with increasing timber density and all values are well below the upper limit, 750 W/s, for class D.

For solid wood panelling and claddings, the main product parameters influencing the reaction to fire characteristics are thickness, substrate and density. The influence of density is illustrated in Fig. 3.15. For wood densities of at least 390 kg/m^3, all FIGRA values are below 600 W/s, i.e., well below the limit to a lower main class, 750 W/s. The influences of thickness and substrates

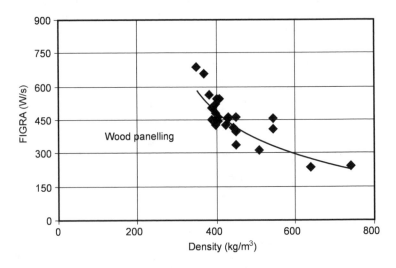

3.15 FIGRA values as a function of density for all solid wood panellings and claddings with thickness 9–21 mm and different profile types, tested with and without an air gap behind.

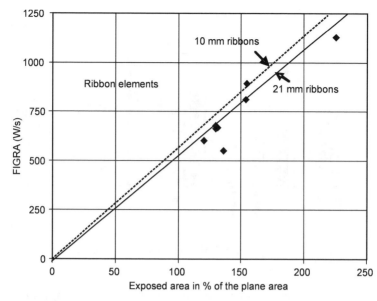

3.16 FIGRA as a function of relative exposed area for wood ribbon exposed to fire from all four sides.

are more indirect and evident mainly from the experience that thinner products burn through and release excessive heat when tested with a void behind. Class D may not be reached.[9] With a substrate directly behind the wood product, burn through does not occur and class D may thus also be reached for thin products. Panel joints and surface profiled areas at the exposed side of the panel, not more than 20% of the plane area, or 25% if measured at both exposed and unexposed sides of the panel, have been shown not to change the fire performance.[9] These surface characteristics are illustrated in Fig. 3.4.

For wood ribbon elements with all sides of the panels exposed to fire, the fire performance is a function of the area exposed in relation to the nominal area in the SBI test method ($2.25\,\mathrm{m}^2$). This relationship is illustrated in Fig. 3.16.

For wood floorings, the flame spread in the orientation along the wood grain is more rapid than transverse grain.[9] The orientation along the grain has therefore been used as the worst-case scenario. The influence of surface coatings have been determined in a systematic study with well defined uncoated products and coating systems including all major systems used by industry, i.e., UV cured acrylic, PU (polyurethane foam) and oil coating systems used by the parquet industry and in addition ordinary wood oil and soap mainly used for solid wood floorings. The results show that all coating systems improve or at least maintain their fire-resistant performance in the radiant panel test, i.e., a higher critical heat flux is reached, see Fig. 3.17.

No clear trend with density has been found as for SBI testing of wood products. The lack of a trend with density for the wood floorings may be

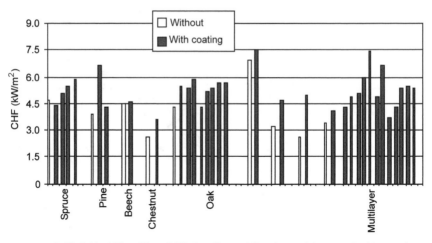

3.17 Critical Heat Flux, CHF, for all wood floorings without and with a surface coat.

explained by the much lower heat flux in the testing of floorings, allowing gases to be released and influencing the flame spread differently for different wood species, depending, e.g., on the wood permeability. The absence of a trend is true both for uncoated and surface coated homogeneous wood floorings, see Fig. 3.18. However, if the test data are analysed by wood species, a certain pattern is obvious. Data for spruce flooring without a surface coating are illustrated in Fig. 3.19. A trend with density is found mainly for surface coated multilayer wood floorings, see Fig. 3.20. Examples of wood floorings after fire tests in the radiant panel are illustrated in Fig. 3.21.

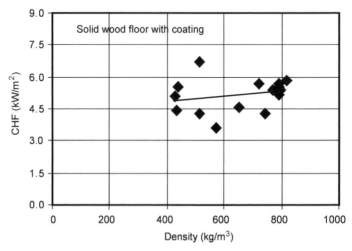

3.18 Critical Heat Flux, CHF, as a function of density for solid wood floorings with a surface coat.

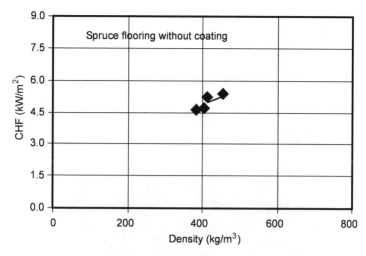

3.19 Critical Heat Flux, CHF, as a function of density for solid spruce floorings without a surface coat.

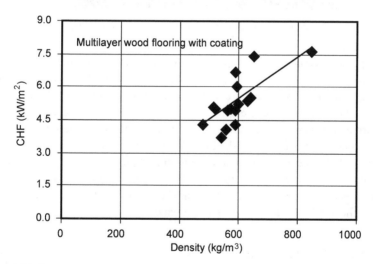

3.20 Critical Heat Flux, CHF, as a function of density for multilayer wood floorings with a surface coat.

3.5   Classification without further testing for wood products

3.5.1   Wood-based panels

Main class D has been verified for most wood-based panels. Panel density and thickness are decisive for the FIGRA values while panel joints and choice of

3.21 Wood floorings after fire tests in the Radiant panel.

standard substrates do not influence the class. Only low density products are class E. Smoke class s2 should be used for all wood-based panels as surface linings in order to achieve a robust classification. Flaming droplets/particles class d0 have been verified for all wood-based panels in class D. The safety margins for these sub-classes are all very high, 20–90%. The final classification of wood-based panels published in the *Official Journal*[11] is given in Table 3.3. The table has been included in the harmonised product standard for wood-based panels, EN 13 986. An extension of the classification for wood-based panels is under way in order to include other panel types and further end-use applications, e.g., those with an air gap behind the panel.

3.5.2 Structural timber

Structural timber with a minimum mean density of $350 \, \text{kg/m}^3$ and minimum thickness and width of 22 mm can, based on the evidence presented (Fig. 3.12), be classified without further testing as class D-s2, d0. The final classification of structural timber published in the *Official Journal*[12] is given in Table 3.4. It will be included in the harmonised product standards for structural timber.

3.5.3 Glued laminated timber

Glued laminated timber with a minimum mean density of $380 \, \text{kg/m}^3$ and minimum thickness and width of 40 mm can, based on the evidence presented,

Table 3.3 Final table in Commission Decision.[11] Classes of reaction to fire performance for wood-based panels[a]

Wood-based panel products[b]	EN product grade reference	Minimum density (kg/m³)	Minimum thickness (mm)	Class[c] (excluding floorings)	Class[d] Floorings
Particleboards	EN 312	600	9	D-s2, d0	D_{FL}-s1
Fibreboards, hard	EN 622-2	900	6	D-s2, d0	D_{FL}-s1
Fibreboards, medium	EN 622-3	600	9	D-s2, d0	D_{FL}-s1
		400	9	E, pass	E_{FL}
Fibreboards, soft	EN 622-4	250	9	E, pass	E_{FL}
Fibreboards, MDF[e]	EN 622-5	600	9	D-s2, d0	D_{FL}-s1
Cement-bonded particleboard[f]	EN 634-2	1000	10	B-s1, d0	B_{FL}-s1
OSB board[g]	EN 300	600	9	D-s2, d2	D_{FL}-s1
Plywood	EN 636	400	9	D-s2, d0	D_{FL}-s1
Solid wood panels	EN 13353	400	12	D-s2, d0	D_{FL}-s1

[a] EN 13986.
[b] Wood-based panels mounted without an air gap directly against class A1 or A2-s1,d0 products with minimum density 10 kg/m³ or at least class D-s2,d0 products with minimum density 400 kg/m³.
[c] Class as provided for in Table 1 of the Annex to Commission Decision 2000/147/EC.
[d] Class as provided for in Table 2 of the Annex to Commission Decision 2000/147/EC.
[e] Dry process fibreboard.
[f] Cement content at least 75% by mass.
[g] Oriented strand board.

be classified without further testing as class D-s2, d0. The final classification of glued laminated timber approved by the Standing Committee on Construction[13] is given in Table 3.5. The table will be included in the harmonised product standard for glued laminated timber.

Table 3.4 Final table in Commission Decision.[12] Classes of reaction to fire performance for structural timber[a]

Material	Product detail	Minimum mean density[c] (kg/m³)	Minimum overall thickness (mm)	Class[b]
Structural timber	Visual and machine graded structural timber with rectangular cross-sections shaped by sawing, planing or other methods or with round cross-sections.	350	22	D-s2, d0

[a] Applies to all species covered by the product standards.
[b] Class as provided for in Table 1 of the Annex to Commission Decision 2000/147/EC.
[c] Conditioned according to EN 13238.

Table 3.5 Final table in Commission Decision.[13] Classes of reaction to fire performance for glulam[a]

Material	Product detail	Minimum mean density[c] (kg/m^3)	Minimum overall thickness (mm)	Class[b]
Glulam	Glued laminated timber products in accordance with EN 14080.	380	40	D-s2, d0

[a] Applies to all species and glues covered by the product standards.
[b] Class as provided for in Table 1 of the Annex to Commission Decision 2000/147/EC
[c] Conditioned according to EN 13238.

3.5.4   Solid wood panelling and cladding

Solid wood panelling and cladding with a total thickness of at least 9 mm and density of at least 390 kg/m^3 can, based on the evidence presented, be classified as class D-s2,d2, and for products with total thickness of at least 12 mm as class D-s2,d0, if mounted with a closed air gap behind or on a non-combustible substrate, such as fibre cement boards, with minimum density 10 kg/m^3 with or without a vapour barrier, e.g., plastic foils, behind the wood panel. A substrate of cellulose insulation of at least class E or an open air gap of maximum 20 mm behind the panel may also be used. Solid wood panelling and cladding with a thickness of at least 18 mm and density of at least 390 kg/m^3 can, based on the evidence presented, be classified without further testing as class D-s2, d0 without any limitations in end-use conditions.

Wood ribbon elements are defined as rectangular wood pieces, with or without rounded corners, mounted horizontally or vertically on a support frame and surrounded by air on all sides, mainly used close to other building elements, both in interior and exterior applications. They can be classified without further testing as class D-s2, d0, if the maximum exposed area (all sides of rectangular wood pieces and wood support frame) is not more than 110% of the total plane area, see Fig. (a) in Table 3.6. A proposed draft Commission Decision[14] is given in Table 3.6.

3.5.5   Wood flooring

The following general conclusions have been reached for wood floorings:

- along the wood grain is the worst-case orientation
- surface coating generally improves the fire-retardant performance
- most wood floorings fulfil at least class D_{fl}-s1; some also class C_{fl}-s1.

More specific conclusions have been demonstrated about the main influencing parameters; thickness (for the whole product and for the top layer of multilayer products), density, wood species, substrate (including air gap and glueing to the substrate) and interlayer. Limits for these parameters have been determined. Based on the evidence and the arguments presented,[14] a proposed draft Commission Decision for Classification without further testing is given in Table 3.7.

3.6   Conclusions

The results presented clearly demonstrate the stable reaction to fire performance of wood-based products. Classes D-s2,d0, D_{fl}-s1 and C_{fl}-s1 have been verified with the required safety limit of 20%. The main parameters influencing the reaction to the fire-retardant characteristics of all wood products are product thickness, density and end-use conditions, such as substrates or air gaps behind the product.

Work has already resulted in the Commission decisions published in the *Official Journal* of the Commission for wood-based panels and structural timber products. Further results are in the process of being finally approved and published. The classes will also be included in the harmonised product specifications as soon as they become available from the product standard committees and used for CE-marking.

In addition to the CWFT decisions, this new knowledge can be utilised in predicting classifications for wood products not covered by the EC decisions or for new products. Wood products and end-use applications not included in the CWFT classification tables have to be tested and classified in the ordinary way. Better classification may then be reached, since no safety margins have to be fulfilled. Fire-retardant-treated wood products have always to be tested and classified separately, since the treatments may influence their reaction to fire performance.

3.7   Acknowledgements

The work has been initiated by standard bodies within CEN TC 112, TC 124 and TC 175. It has been led and sponsored by European industrial federations via CEI-Bois, mainly EPF European Panel Federation, FEIC European Federation of the Plywood Industry, FEROPA European Federation of the Fibreboard Industry, EOS European Organisation of the Sawmill Industry, the European Glued Laminated Industries and FEP European Federation of the Parquet Industry. Cooperation with research colleagues mainly from VTT Building and Transport in Finland and from CTBA in France is also gratefully acknowledged.

3.8   Nomenclature

CEN   European Committee for Standardisation (fr. Comité Européen de
           Normalisation)

Table 3.6 Proposed table for inclusion in Commission Decision.[14] Classes of reaction to fire performance for solid wood panelling and cladding

Product[k]	Product detail[e]	Minimum mean density[f] (kg/m³)	Minimum thicknesses, total/minimum[g] (mm)	End-use condition[d]	Class[c]
Panelling and cladding[a]	Wood pieces with or without tongue and groove and with or without profiled surface	390	9/6	Without air gap or with closed air gap behind	D-s2, d2
		390	12/8	Without air gap or with closed air gap behind	D-s2, d0
Panelling and cladding[b]	Wood pieces with or without tongue and groove and with or without profiled surface	390	9/6	With open air gap ≤ 20 mm behind	D-s2, d0
		390	18/12	Without air gap or with open air gap behind	D-s2, d0
Wood ribbon elements[h]	Wood pieces mounted on a support frame[i]	390	18	Surrounded by open air on all sides[j]	D-s2, d0

Notes to Table 3.6

[a] Mounted mechanically on a wood batten support frame, with the gap closed or filled with a substrate of at least class A2-s1,d0 with minimum density of 10 kg/m^3 or filled with a substrate of cellulose insulation material of at least class E and with or without a vapour barrier behind. The wood product shall be designed to be mounted without open joints.

[b] Mounted mechanically on a wood batten support frame, with or without an open air gap behind. The wood product shall be designed to be mounted without open joints.

[c] Class as provided for in Commission Decision 2000/147/EC Annex Table 1.

[d] An open air gap may include possibility for ventilation behind the product, while a closed air gap will exclude such ventilation. The substrate behind the air gap must be of at least class A2-s1,d0 with a minimum density of 10 kg/m^3. Behind a closed air gap of maximum 20 mm and with vertical wood pieces, the substrate may be of at least class D-s2,d0.

[e] Joints include all types of joints, e.g., butt joints and tongue-and-groove joints.

[f] Conditioned according to EN 13238.

[g] As illustrated below. Profiled area of the exposed side of the panel not more than 20% of the plane area, or 25% if measured at both exposed and unexposed side of the panel. For butt joints, the larger thickness applies at the joint interface.

[h] Rectangular wood pieces, with or without rounded corners, mounted horizontally or vertically on a support frame and surrounded by air on all sides, mainly used close to other building elements, both in interior and exterior applications.

[i] Maximum exposed area (all sides of rectangular wood pieces and wood support frame) not more than 110% of the total plane area, see figure below.

[j] Other building elements closer than 100 mm from the wood ribbon element (excluding its support frame) must be of at least class A2-s1,d0, at distances 100–300 mm of at least class B-s1,d0 and at distances more than 300 mm of at least class D-s2,d0.

[k] Applies also to stairs.

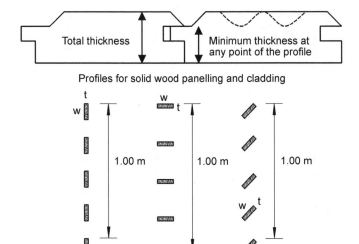

Profiles for solid wood panelling and cladding

Maximum exposed area of wood ribbon element: 2n (t + w) + a = 1.10
n is number of wood pieces per metre.
t is thickness of each wood piece, in metres.
w is width of each wood piece, in metres.
a is exposed area of wood support frame (if any), in m^2, per m^2 of wood ribbon element.

Table 3.7 Proposed table for inclusion in Commission Decision.[14] Classes of reaction to fire performance for wood flooring

Material[a,g]	Product detail[d]	Minimum mean density[e] (kg/m³)	Minimum overall thickness (mm)	End-use condition	Class[c] for floorings
Wood flooring and parquet	Solid flooring of oak or beech with surface coating	Beech: 680 Oak: 650	8	Glued to substrate[f]	C_{fl}-s1
	Solid flooring of oak, beech or spruce and with surface coating	Beech: 680 Oak: 650 Spruce: 450	20	With or without air gap underneath	C_{fl}-s1
	Solid wood flooring with surface coating and not specified above	390	8	Without air gap underneath	D_{fl}-s1
		390	20	With or without air gap underneath	D_{fl}-s1
Wood parquet	Multilayer parquet with a top layer of oak of at least 5 mm thickness and with surface coating	650 (top layer)	10	Glued to substrate[f]	C_{fl}-s1
			14[b]	With or without air gap underneath	C_{fl}-s1

Multilayer parquet with surface coating and not specified above	500	8	Glued to substrate	D_{fl}-s1
		10	Without air gap underneath	D_{fl}-s1
		14[b]	With or without air gap underneath	D_{fl}-s1
Veneered floor covering	Veneered floor covering with surface coating	800	6[b]	Without air gap underneath	D_{fl}-s1

[a] Mounted in accordance with EN ISO 9239-1, on a substrate of at least Class D-s2,d0 and with minimum density of 400 kg/m^3 or with an air gap underneath.

[b] An interlayer of at least Class E and with maximum thickness 3 mm may be included in applications without an air gap, for parquet products with 14 mm thickness or more and for veneered floor coverings.

[c] Class as provided for in Commission Decision 2000/147/EC Annex Table 2.

[d] Type and quantity of surface coatings included are acrylic, polyurethane or soap, 50–100 g/m^2, and oil, 20–60 g/m^2.

[e] Conditioned according to EN 13238.

[f] Substrate at least Class A2-s1,d0.

[g] Applies also to steps of stairs.

CHF   Critical Heat Flux (parameter determined in the RP test and used for classification)

CWFT   Classifcation without further testing (procedure within the EC and also a WG under FRG/EGF)

DG   Directorate General (within the EC)

EC   European Commission

EN   European Norm (Standard); prEN is a preliminary standard

ETA   European Technical Approval

FIGRA   FIre Growth RAte (parameter determined in the SBI test and used for classification)

FRG/EGF   Fire Regulators Group/European Commission Expert Group on Fire Issues Regarding Construction Products (under SCC); Old (–2004)/ New group (2005–)

RP   Radiant Panel test for floorings (EN ISO 9239-1)

SBI   Single Burning Item test (EN 13823)

SCC   Standing Committee on Construction (within the EC)

SMOGRA   SMOke Growth RAte (parameter determined in the SBI test and used for classification)

TC   Technical Committee (within e.g., CEN)

TSP   Total Smoke Production (parameter determined in the SBI test and used for classification)

WG   Working Group (under a TC within e.g., CEN)

3.9   References

1. COMMISSION DECISION of 8 February 2000 implementing Council Directive 89/106/EEC as regards the classification of the reaction to fire performance of construction products. *Official Journal of the European Communities 23.2.2000.*

2. Östman B, Nussbaum R: National standard fire tests in small-scale compared with the full-scale ISO room test. *Trätek Report I* 8702017, 1987.

3. SBI test, EN 13823, Reaction to fire tests for building products – Building products excluding floorings – exposed to the thermal attack by a single burning item, 2002.

4. Small flame test, EN ISO 11925-2, Reaction to fire tests for building products – Ignitability of building products subjected to direct impingement of flame – Part 2: Single-flame source test, 2002.

5. Radiant panel test, EN ISO 9239-1, Reaction to fire tests for floor coverings – Part 1: Determination of the burning behaviour using a radiant heat source, 2002.

6. EN 13238 Reaction to fire tests for building products – Conditioning procedures and general rules for selection of substrates, 2001.

7. Classification system, EN 13501-1, Fire classification of construction products and building elements – Part 1: Classification using test data from reaction to fire tests, 2002.

8. Classification of products of known and stable performance – Procedural aspects. CONSTRUCT 01/491 rev 3, 2004.

9. Östman B, Mikkola E: European classes for the reaction to fire of wood products

(excluding floorings). *Trätek Report I* 0411025, 2004.

10. Tsantaridis L, Östman B: European classes for the reaction to fire of wood floorings. *Trätek Report I* 0411026, 2004.

11. COMMISSION DECISION of 17 January 2003 establishing the classes of reaction-to-fire performance of certain construction products. *Official Journal of the European Communities 18.1.2003.*

12. COMMISSION DECISION of 7 August 2003 establishing the classes of reaction-to-fire performance of certain construction products. *Official Journal of the European Communities 8.8.2003.*

13. COMMISSION DECISION of 9 August 2005 establishing the classes of reaction-to-fire performance of certain construction products. *Official Journal of the European Communities 11.8.2005.*

14. Draft COMMISSION DECISION establishing the classes of reaction-to-fire performance of certain construction products. *Official Journal of the European Communities* CONSTRUCT 05/708, 2005.

4

Flammability tests for external cladding on buildings

E SOJA and C WADE, BRANZ Ltd, New Zealand

4.1 Introduction

Fire spread on external building façades, though being one of the more visible fire phenomena, does not necessarily pose the greatest risk to life safety in a building. This may be due to the historical use of non-combustible materials on façades as is required by many building codes around the world, and to the fire protection offered to buildings where external fire spread may be an issue as in high-rise buildings. In addition, an exterior wall fire usually results from an intense fire within the building and the building fire usually masks the exterior wall fire. Thus, it sometimes can be very difficult to distinguish between the contribution made by the building fire and that made by an exterior wall fire involving combustible cladding material.

A small number of such incidents do not necessarily mean that fire spread on exterior claddings does not present significant risk to both property protection and life safety. This is because:

1. An increasing number of high-rise buildings are being built using potentially combustible exterior claddings. Exterior fire spread problems become increasingly more important as rapid fire spread from floor to floor can occur via combustible exterior claddings creating a hazardous situation for building occupants.
2. In the case of very tall buildings, an exterior wall fire may create a problem for fire fighters, as the fire may not be accessible to fire-fighting appliances.
3. Fire spread on combustible cladding systems can also be life-threatening in some multi-storey buildings such as health-care units and detention centres as the occupants in these buildings may not be able to escape to a safe place without assistance.
4. Considerable property losses may arise by damage to the wall as a result of an exterior wall fire.
5. Fire spread on combustible cladding systems could also compromise sprinkler systems if the fire were to spread into several floors setting

sprinkler heads off at each level, and creating a greater demand for water than can be supplied.

Fire spread over combustible cladding has attracted the attention of researchers in recent years. Changes to building codes have also been initiated in countries such as New Zealand, in order to provide alternative classification schemes for the fire safety performance of exterior cladding systems. The design of buildings to control external fire spread on façades and between buildings varies between different countries with some giving no specific controls other than maintaining a non-combustible construction. Where fire engineering methods can be used, i.e., in a performance-based Building Code environment, little guidance is given and solutions rest with the specific fire engineer whose knowledge and experience is relied on to provide the complying solution.

A reason for performance-based codes is to allow for innovative design and to enable new materials to be used. Building Codes can take some time to catch up with new developments of building materials. This is true of composite materials where a potentially combustible core is laminated between non-combustible skins. Even though these products have non-combustible facings, they may be of low melting point which in a fire could easily melt and expose underlying combustible materials. It is this type of cladding that needs more specific testing and a fire engineering approach. There are two aspects of external claddings which are important to fire safety. These are:

1.   Ignition and flame spread of the cladding from an external source, and
2.   Ignition and flame spread caused by fires from the building itself.

4.1.1   External sources

This aspect relates to buildings separated by distance and ignition is mostly controlled by radiation with the possibility of flying brands causing piloted ignition. There are many means of calculating ignition of materials by this means and these are well covered by national building codes (DBH, 2005; DOE, 2000) and calculation methods (SFPE, 2002). The SP report by Emil Carlsson (Carlsson, 1999) gives a good overview of fire spread between buildings. There is only one specific test, NFPA 268 used to evaluate radiative heat transfer between buildings. Use is mostly made of standard ignitability tests such as ISO 5660-1 (ISO, 2002d), or data referenced in publications such as the *Ignition Handbook* (Babrauskas, 2003) and *Handbook of Fire Protection Engineering* (SFPE, 2002).

4.1.2   Fire within the building

The key mechanisms by which exterior cladding systems may ignite and contribute to vertical fire spread include:

- Flames projecting from broken windows in the room of fire origin, exposing the façade and any windows above. These upper windows may break allowing fire to enter the floors above.
- Inadequate fire stopping of the gap between the edge of the floor slab and the exterior wall allowing flames and hot gases to pass through to the floor above; particularly applies to curtain wall construction.
- Heat-induced distortion of low-melting-point metals or alloys, such as aluminium, exposing a combustible core or by causing fire stopping to become ineffective or fall out.

The first mechanism tends to be very spectacular, particularly when combustible materials on the upper floors are ignited by the window fire plumes. This is demonstrated by the large fires in Sao Paulo, Brazil (Willey, 1972; Sharry, 1974), the Las Vegas Hilton (Anon, 1982) and the First Interstate Bank, Los Angeles (Klem, 1989; Morris, 1990). The second mechanism of spread is not as spectacular as the first but can be a contributing factor in fire spread. Examples include: First Interstate Bank, Los Angeles (Klem, 1989; Morris, 1990) and the Avianca Building in Colombia (Sharry, 1974).

Examples of absent or ineffective fire-stopping include the President's Tower (Hartog, 1999) and the First Interstate Bank. Examples where combustible cladding directly contributed to vertical fire spread include an apartment building in Irvine Scotland (Anon., 1999a), a motel in Albury NSW (NSWFB, 1999) and an apartment complex in Seattle (FEMA, 1991a).

4.2   Building regulations

4.2.1   Building Code of Australia (BCA) 2004, (ABCB, 2004)

The BCA generally uses the combustibility test, AS1530.1 (SA, 1984) and the 'early fire hazard' test AS/NZS 1530.3, (SA, 1999) to control materials used for external walls and claddings. A requirement for an external wall to have a FRL does not apply to a curtain wall or panel wall which is of non-combustible construction and fully protected by automatic external wall-wetting sprinklers.

4.2.2   Asia

The regulations and test methods relevant to exterior claddings in countries such as Hong Kong, Malaysia and Singapore are generally decades old and are usually modified or directly taken from the USA, UK and Germany. The test methods that are in use in this region for determining external cladding materials performances include ASTM D2843 (ASTM, 1993a), BS 476 Part 4, 6, 7 and 11 (BSI, 1970, 1981, 1982, 1987) and DIN 4102 Part 1 (DIN, 1981). These test methods are generally not adequate for evaluation of the end-use performance of exterior claddings materials and systems. These countries usually look towards USA and Europe for relevant experiences and readily adopt controls and test methods developed and tested there (Yong, 1999).

4.2.3   Japan

In Japan, building and construction products are regulated by the Japanese Building Standard Law together with enforcement orders from Cabinet and Ministry notifications. Parts of buildings may be required to be of 'fireproof construction', 'quasi-fireproof construction' or 'fire preventive construction' (The Building Centre of Japan, 2004). All materials that require fire performance properties are classified as non-combustible, quasi-non-combustible or fire retardant. Different bench-scale test methods are used to determine the classification of materials into these groupings (Hedskog and Ryber, 1998) and include the use of the Cone Calorimeter test ISO 5660, (ISO, 2002d). And the Japanese version of the non-combustibility test (JSA, 1994).

4.2.4   United Kingdom (England and Wales)

The Building Regulations Approved Documents specify external surfaces of walls to be class '0', or in certain applications be of 'limited combustibility' which includes testing with the non-combustibility test BS 476:Part 11. The surface of materials (including cladding) are classified (i.e class '0') by reference to two test methods. These are a spread-of-flame test (BS 476 Part 7 (BSI, 1987)), which measures the distance flame will spread across the surface of a sample, and a fire propagation test (BS 476 Part 6 (BSI, 1981)), which assesses the contribution that the sample makes to fire development. The Building regulation 2000 also enables an exterior wall to be tested to BRE Fire Note 9 (Colwell and Smit, 1998), as an alternative to complying with class '0' or 'limited combustibility' requirements.

4.2.5   Canada

The National Building Code of Canada (NRCC, 1995) allows exterior non-load-bearing wall assemblies containing combustible components to be used in a building required to be of non-combustible construction, but requires the wall to be tested to CAN/ULC S134 'Standard Method of Fire Test of Exterior Wall Assemblies' (ULC, 1992a). The criteria are that flaming does not spread more than 5 m above the opening during or following the test, and the heat flux during flame exposure on the wall assembly is not more than 35 kW/m^2 measured 3.5 m above the opening.

4.2.6   USA – Uniform Building Code (UBC)

The Uniform Building Code (UBC, 1997a) requires external walls to be of non-combustible construction under certain circumstances. An exception is made to this for foamed plastics which may be used in all types of construction provided certain requirements are met. These include the requirement to pass the conditions of acceptance of UBC Standard 26-4 (UBC, 1997b) or UBC Standard 26-9 (UBC, 1997c). Jurisdictions may also accept a result to NFPA 285 (1998).

4.2.7   USA International Fire Code

This is very similar to the UBC but also calls up NFPA 268 (NFPA, 2001).

4.2.8   New Zealand

The test methods specified for external claddings in New Zealand are AS 1530 Part 1, (combustibility) and AS/NZS 3837: (cone calorimeter) (SA, 1998). An external cladding can have a non-combustible substrate with a combustible finish not greater than 1 mm thick, or be classified according to the results of the cone calorimeter test AS/NZ 3837 (SA, 1998). An option exists to evaluate an external wall to NFPA 285 (1998) or other full-scale faced test method acceptable to the regulatory authorities.

4.2.9   Sweden

Façades are required to be of non-combustible construction or have passed the full-scale SP FIRE 105 test.

4.3   Modelling external fire spread

Two aspects of modelling fire spread on façades are important:

1. The nature of the heating conditions, i.e., the radiative and convective components of the flames from the windows, which includes an estimate of height and projection.
2. Reaction to fire properties of the external claddings.

Yokoi (1960) was one of the first researchers to develop a calculation method based on extensive testing. These calculations show that the flame height from a window depends on many parameters including the size and shape of the window, the shape of the fire room and the fire load. Law also developed calculation procedures for predicting the shape of fire plumes projecting from window openings (Law, 1978; Anon., 1979).

The work published by Oleszkiewicz (1989b) on the heat transfer from a window flame to the façade above describes a mathematical model of heat flow (an extension of the previous work by Law (1978)) and discusses its applicability in comparison with the results obtained in the full-scale experiments. Calculated values of the total heat flux density, at a particular height above the window, were generally higher than those measured. A comparison of the heat transfer to the wall above the window for 'normal' fires produced calculated values which were mostly lower than those measured, with the values closer to the windows giving a better correlation. This model seems to be conservative but not excessively so.

Other researchers have undertaken studies of the fire dynamics of window fire plumes. These include:

- Klopovic and Turan (1998) who studied the effect of burn room ventilation and environmental conditions on venting plumes.
- Ohmiya *et al.* (1998) have also developed a model for predicting a room fire in the fully developed stage to estimate the heat release due to combustion of excess fuel in window fire plumes.
- Sugawa *et al.* (1997) conducted a reduced scale experimental study investigating the effect of wind on the flow behaviour of window plumes.

Galea *et al.* (1996) used CFD techniques in the analysis of fire plumes emerging from windows. They examined three window configurations; narrow, wide and wide with a 1 m deep external protrusion (apron). Bong (2000) gives a theoretical analysis of fire spread on external walls as part of a research study on exterior wall fires. Drysdale (1999) also gives some information on calculating flames out of windows. Cowles and Soja (1999) discussed a method of classifying external claddings based on the results of cone calorimeter testing.

4.4   Case studies of fires

Apartment Building, Munich, ~1996 (Mayr, 1996)

A five-level apartment building with a façade made of a composite thermal insulation (about 100 mm thick) comprising polystyrene and foam plastics slabs and a reinforced covering layer. A rubbish container fire on the exterior ignited the cladding and created extensive damage. Windows were broken and flames spread into rooms at upper levels.

Manchester, New Hampshire, 1985 (Bletzacker and Crowder, 1988; Crowder and Bletzacker, 1988)

A fire exposed an adjacent seven-storey office building located across a 6 m wide alley. The office building was clad with an EIFS. A flash fire on the exterior of the office building due to exposure to radiant heat was extinguished in a few minutes, however this short-duration fire was regarded as important because it revealed previously unrecognised fire performance characteristics of popular foam plastic exterior insulation and finish systems.

Lakeside Plaza, Virginia, 1983 (Belles, 1983)

This was a 12-storey apartment building clad with the Dryvit[TM] EIFS (External Insulation and Finish system), (EPS backing was 25 mm thick). The fire started in a waste chute on the fifth or sixth floor and there were openings into the chute at each floor level. Damage was limited to floors above the fire source and to areas where it had been directly exposed to heating from the flame (Belles, 1983). Thus extensive vertical spread did not occur in this case.

Apartment Building, Irvine, Scotland, 1999 (Anon., 1999a,b)

The fire, which started in a flat on the fifth floor of a 13-storey apartment block, broke out through the window and quickly spread vertically up the exterior face of the building engulfing the upper nine floors within minutes. The building was of concrete construction but had full height composite window units comprising a GRP panel below the window and PVC window frames. The fire ignited the GRP panels and spread vertically. There was one death in the incident and that was in the apartment of origin. Therefore the combustible window components were unlikely to have contributed to the death. This fire was a direct cause of external wall tests being included in the England and Wales Approved Documents for Fire Safety (DoE, 2000).

Country Comfort Motel, Albury, NSW, 1998 (NSWFB, 1999)

This was an eight-storey accommodation building generally of concrete/masonry construction but with fibreglass panels beneath the windows. A fire occurred at the third-floor level and following breakout through the window, flames rapidly spread to the top of the building via the fibreglass panels positioned vertically above the fire source. Fire penetrated into the third- and fourth-floor levels. There were no deaths associated with the vertical spread of fire.

Te Papa (Museum of New Zealand), Wellington (NZFS, 1997)

This was a large multi-level national museum building under construction. The exterior cladding used comprised a thin aluminium-faced panel with a polyethylene core, mounted over extruded foam polystyrene insulation board and building paper. A worker, heat welding a roofing membrane, ignited the building paper and this quickly spread up the exterior façade involving the polystyrene and cladding panel. There were no deaths or injuries associated with the fire.

Villa Plaza Apartment Complex, Seattle, Washington, 1991

This was a four-storey wood frame building with an exterior walkway façade and it had decorative lattice screens of 2″ × 6″ vertical cedar boards (with oil-based stain). The ceiling of the walkway was also exposed tongue and groove cedar. The space between the cedar boards was about 180 mm and the boards ran the full height of the building. This meant that the exit paths were enclosed in readily combustible material on three sides which contributed to rapid fire spread, including spread to other apartments facing the walkway. A fire coming out of a broken front bedroom window caused the cedar screen to ignite and it provided a path for the fire quickly to extend vertically up the entire 12 m height of the screen and also horizontally across the screen. There were no fatalities in this fire (FEMA, 1991b).

Knowsley Heights, Liverpool, 1991

This was an 11-floor apartment tower block. A deliberately lit fire, in the rubbish compound outside the building, spread up through a 90 mm gap between the tower's rubberised paint-covered concrete outer wall and a recently installed rain screen cladding. The rapid spread of fire was thought to have been caused by the lack of fire barriers in the cavity gap passing all 11 floors and providing a flue for hot gases to rise (Anon., 1993). The fire destroyed the rubbish compound and severely damaged the ground floor lobby and the outer walls and windows of all the upper floors. No smoke or fire penetrated into the flats and the building was reoccupied by tenants later the same day (Anon., 1992). The rain screen material was a Class O (limited combustibility) rated product using BS 476 parts 6 and 7 (BSI, 1981, 1987). Building regulations were changed as a result of this fire.

President Tower, Bangkok, 1997 (Hartog, 1999)

This was a 37-storey retail, commercial office and hotel development. A sprinkler system installed was not yet operational as the interior fit out was not fully completed. An explosion and fire started on level seven causing the destruction of an aluminium framed curtain wall system. The effectiveness of fire stopping at the floor edge was compromised by floor to floor cabling. Window and spandrel glass shattered and collapsed before the structural silicone sealant bonds between the glass and aluminium were destroyed. Many heat strengthened glass panels sustained elevated temperatures but fractured and collapsed as they cooled. The vertical fire spread extended up to level 10.

One Meridian Plaza Building, Philadelphia, 1991 (FEMA, 1991b; Klem, 1991)

This was a fire which occurred on the 22nd floor of a 38-storey bank building. The building frame was structural steel with concrete floors poured over metal decks and protected with spray-on fireproofing materials. The exterior of the building was covered by granite curtain wall panels with glass windows attached to perimeter floor girders and spandrels. At the time of construction only the below-ground service floors were fitted with sprinklers and since then they had been added to the 30th, 31st, 34th, and 35th floors and parts of floors 11 and 15. Fire broke through windows on the 22nd floor and the heat exposure from the window plumes ignited items on the floor above. The fire was stopped when it reached the 30th floor which was sprinklered.

First Interstate Bank Building, Los Angeles, California, 1988 (FEMA, 1988; Chapman, 1988; Klem, 1989; Morris, 1990)

The bank was a 62-storey tower with sprinkler protection only in the basement, garage and underground pedestrian tunnel. An automatic sprinkler system was being installed in the building at the time. The building had a structural steel frame with spray-on fire proofing with steel floor pans and lightweight concrete decking. The exterior curtain walls were glass and aluminium with a 100 mm gap between the curtain wall and the floor slab, where the fire stopping comprised 15 mm gypsum board and fibreglass caulking. The fire started on the 12th floor and extended to the floors above primarily via the outer walls of the building. Flames also penetrated behind the spandrel panels around the ends of the floor slab where there was sufficient deformation of the aluminium mullions to weaken the fire stopping, allowing the flame to pass through even before the windows and mullions had failed. The flames were estimated to be lapping 9 m up the face of the building and the curtain walls including windows, spandrel panels, and mullions, were almost completely destroyed by fire. The upward extension stopped at the 26th floor level.

Westchase Hilton Hotel, Houston, 1982 (Kim, 1990)

This was a 13-storey tower of reinforced concrete construction containing 306 guest rooms. The fire was located in a guest room on the fourth floor and spread internally to the adjacent corridor. There was also vertical exterior fire spread to three guest rooms on the fifth floor. Smoke spread throughout the fire floor and to varying degrees throughout the building. The door to the room of origin (a guest room) was not closed due to a malfunctioning or inoperative door closer. There were 12 deaths and three serious injuries. The exterior walls were of tempered glass in aluminium floor-to-ceiling frames. Aluminium plates spanned the gap between the edge of the floor slab and the exterior cladding. Sprinklers were installed in the linen chutes only. Fire was propagated to the floor above through the broken window by impinging on the exterior glass walls. Fire fighters prevented further upward spread.

Las Vegas Hilton Hotel, 1981

This was a 30-storey hotel of reinforced concrete construction. Glass windows between floors were separated vertically by a 1 m spandrel prefabricated of masonry, plaster and plasterboard on steel studs. Fire occurred on the eighth floor of the east tower lift lobby, where the fire involved curtains, carpeting on the walls, ceiling and floor, and furniture. The plate glass window to the exterior shattered allowing a flame front to extend upwards on the exterior of the building. It apparently took 20–25 minutes for exterior fire spread from the eighth floor to the top of the building (about 20 floors). Two mechanisms were identified for the vertical fire spread:

- Flames outside the upper windows radiated heat through the windows and ignited curtains, timber benches with polyurethane foam padding, which then ignited carpeting on room surfaces.
- Flames contacted the plate glass windows. It is believed that the triangular shape of the spandrels, and recessed plate glass caused additional turbulence which rolled the flames onto the windows resulting in early failure.

There were eight deaths in this fire. Three were in the eighth floor lift lobby, one by jumping/falling from the 12th floor, and four in hotel rooms in the east tower. The doors to the hotel rooms where the deaths occurred were open or had been opened during the fire. There were no fatalities in rooms where the door had been kept closed (Anon., 1982). This suggests that internal rather than external fire spread was the main reason for the deaths.

Joelma Building, Sao Paulo, Brazil, 1974

This was a 25-storey office building of reinforced concrete structure (beams, columns, floor slab) and an exterior curtain wall with hollow tiles rendered with cement plaster on both sides and windows with aluminium framing (Sharry, 1974). The floor slabs were poured in place and provided a 900 mm projection on the north wall and a 600 mm projection on the south wall. Fire started on the twelfth floor near a window. The fire spread externally up two of the façades to the top of the building, readily igniting combustible finishes inside the windows of the floors above which allowed the vertical spread to continue. There were 179 deaths.

Avianca Building, Bogota, Columbia, 1973

This was a 36-storey office building of reinforced concrete construction. The exterior walls consisted of glass and metal panels in metal frames set between concrete mullions (Sharry, 1974). Fire started on the thirteenth floor. Spaces between the cladding and the edge of the floor provided a path for the fire to spread from floor to floor. Fire spread was a combination of internal and external spread. In contrast to the Joelma building the rate of vertical fire spread was relatively slow. Apparently there were not enough combustibles directly inside the windows to sustain the vertical spread. There were four deaths in this fire.

Pima County Administration Building, 1973

This was an 11-storey office building. The fire occurred on the fourth floor. Flames extended from windows of this floor and entered windows on the fifth and sixth floors. There were no deaths and fire was extinguished after 45 minutes with about half the fourth floor burned out and minor damage to the fifth and sixth floors (Stone, 1974).

Andraus Building, Sao Paulo, Brazil, 1972

This was a 31-storey department store and office building. The fire developed on four floors of the department store and then spread externally up the side of the building, involving another 24 floors. Wind velocity and combustible interior finishes and contents were contributing factors to the fire spread. The building construction was reinforced concrete. The building façade had extensive floor to ceiling glazed areas, with a spandrel only 350 mm high and projecting 305 mm from the face of the building. After fire broke through the windows they formed a front exposing the three to four floors above the department store. Radiant heat then ignited combustible ceiling tiles and wood partitions on each floor. The flame front then increased in height as more floors became involved. At its peak the mass of flame over the external façade was 40 m wide, 100 m high and projecting at least 15 m into the street (Willey, 1972).

Miscellaneous, United Kingdom (DoETR, 1999)

- Mercantile credit building, Basingstoke, 1991. Fire on the eighth floor spread up the building. This was a 12-storey office block. Fire broke out on the eighth floor and spread externally behind glass curtain walling to the tenth floor, fanned by strong winds.
- Three-storey block in Milton Keynes, 1995. Room destroyed.
- Alpha House Coventry, 1997. Flames travelled up the outside of the block from the 13th to 17th floor. No fire penetration of the flats.

It is not known if combustible claddings were involved in these last two fires.

4.5   Full-scale or intermediate-scale test methods for façades

4.5.1   NFPA 268 (2001)

This method assesses the ability of an external cladding to ignite under an imposed radiant heat flux of 12.5 kW/m under pilot ignition (a spark igniter). The test lasts 20 minutes unless sustained flaming of greater than five seconds occurs. Figure 4.1 shows a schematic of the equipment. The distance between the radiant panel and the specimen is adjusted to give 12.5 kW/m at the specimen.

4.5.2   BRE/BSI test method (Colwell and Smit, 1998) (BSI, 2002)

The BRE test method assesses the fire performance of external cladding systems. It applies to non-load-bearing exterior wall assemblies, including external cladding systems, rainscreen overcladding systems, external insulation systems and curtain walling when exposed to an exterior fire. The method determines the comparative burning characteristics of exterior wall assemblies by evaluating fire spread over the external surface, fire spread internally within the system and mechanical response such as damage distortion and collapse. The facility consists of a 2.8 m wide vertical wall with a 1.5 m wide wing wall at right-angles to, and 250 mm to one side of, the opening in the main test face. The main face has a 2 m by 2 m opening for the combustion chamber. The height of the wall is at least 6 m above the opening.

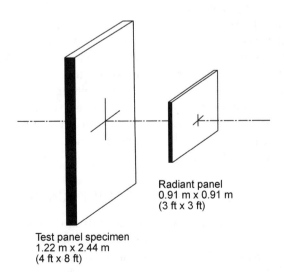

Radiant panel
0.91 m x 0.91 m
(3 ft x 3 ft)

Test panel specimen
1.22 m x 2.44 m
(4 ft x 8 ft)

4.1 Schematic of NFPA 268 Apparatus.

The test specimen is exposed to a heat flux of $90 \, kW/m^2$ at a distance of 1 m above the opening. Performance is evaluated against three criteria:

1. External fire spread – when the temperature of an external thermocouple (5 m above opening) exceeds 600 °C for a 30-second period.
2. Internal fire spread – when the temperature of an internal thermocouple (5 m above opening) exceeds 600 °C for a 30-second period.
3. Mechanical response – observations made of collapse causing a hazard.

4.5.3   ISO intermediate-scale façade test (ISO, 2002a)

The ISO 13785 Part 1 intermediate-scale test for façades is intended for the screening or evaluation of sub-components or families of materials whereas the large-scale test is intended to provide an end-use evaluation. The test specimen consists of sufficient cladding or façade panels to cover two areas of 1.2 m wide × 2.4 m high and 0.6 m wide × 2.4 m high (at right-angles to form a re-entrant corner). The bottom edge is closed by the method normally used for the inclusion of a window casement. Joints and fixings are used as they would be in practice. The apparatus consists of a specimen support frame and an ignition source. The ignition source is a propane burner having a right-angle top surface and a heat output of 100 kW. Temperature and heat flux measurements are made during the test and performance criteria are not included in the standard.

4.5.4   ISO full-scale façade test (ISO, 2002b)

The ISO 13785 Part 2 full-scale test for façades consists of a combustion chamber with volume in the range $20–100 \, m^3$, with an opening in the front wall (2 m wide by 1.2 m high). The height of the test facility is 4 m above the window opening, with a main façade 3 m wide. A vertically held wing façade, 1.2 m wide, is also required to form a re-entrant angle of 90° (see Fig. 4.2). Any fuel can be used to produce a window flame which exposes the test specimen to a heat flux of $55 \, kW/m^2$ at a height of 0.5 m above the opening, and $35 \, kW/m^2$ at a height of 1.5 m above the top opening. Heat fluxes are measured 3.5 m above the top of the window opening, and thermocouples are installed at the top of the test specimen and at the top of the window opening. Evaluation or performance criteria are not included in the standard.

4.5.5   Swedish full-scale façade test, SP FIRE 105 (SP, 1994)

This method (Fig. 4.3) comprises a façade specimen 4.18 m wide and 6 m high with an opening at the bottom edge of the specimen. The fire source is a pan filled with 60 litres of heptane. The test specimen is mounted on a lightweight concrete structure. The test specimen includes two 50 mm deep indentations on

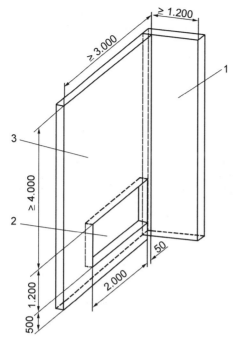

4.2 ISO 13785 Part 2 Façade, combustion chamber not shown.

the façade surface to represent (imaginary) window upper level openings. A heat flux meter is included at the centre of the second upper window. Thermocouples are also located at the top of the specimen and beneath a non-combustible eaves detail (Hermodsson and Månsson, 1992; Babrauskas, 1996).

4.5.6   Canadian full-scale façade test (CAN/ULC S-134)

The Canadian test method, developed by IRC National Research Council of Canada, was standardised in 1992 as CAN/ULC S-134, 'Standard Method of Fire Test of Exterior Wall Assemblies' (ULC, 1992a). This requires a three-storey facility with burn room and propane burners. The window opening is 2.6 m wide by 1.37 m high. The test specimen measures 5 m wide by about 10.3 m high. The exposure conditions are intended to provide a heat flux density of $45 \pm 5 \, kW/m^2$ measured 0.5 m above the opening and $27 \pm 3 \, kW/m^2$ measured 1.5 m above the opening on a non-combustible wall. The acceptance criteria are given in the National Building Code of Canada.

4.5.7   VTT façade test

VTT Building and Transport, Fire Technology possess a large-scale vertical test apparatus simulating up to a three-storey building. The façade is 8100 high ×

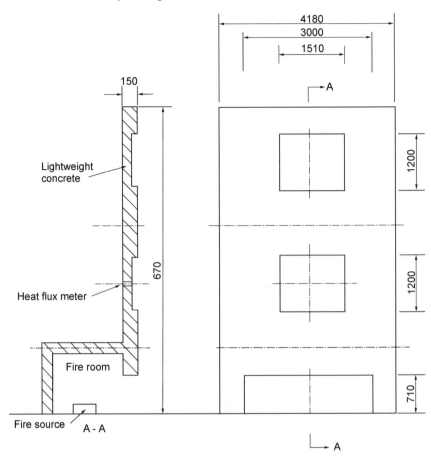

4.3 SP FIRE 105.

4200 wide, but can be reduced to 5600 mm high. It has no return wall. The ignition source is a propane gas burner giving either 100 kW or 200 kW output at the base. (Hakkarainen and Oksanen, 2002).

4.5.8   Vertical channel test (ASTM, 1992a)

The Vertical Channel Test apparatus (Fig. 4.4) consists of a combustion chamber (1.9 m high, 1.5 m deep, and 0.85 m wide) with two openings in the front wall, one at the bottom (440 mm high by 850 mm wide) and one at the top (630 mm high by 850 mm wide). The test specimen measures 850 mm wide by 7320 mm high. Two full-height vertical panels are installed either side of the specimen projecting 500 mm forward of the face of the specimen. Propane burners are used to produce window flames which expose the test specimen to a heat flux of 50 kW/m^2 at a height of 0.5 m above the top opening, and 27 kW/m^2 at a height of

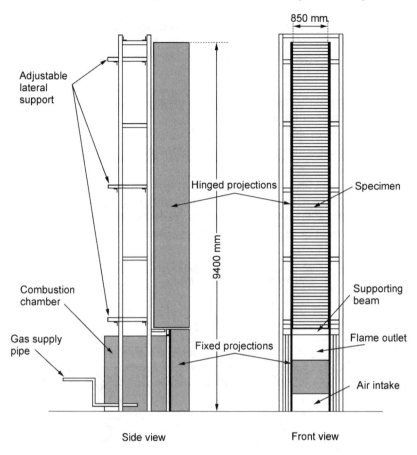

4.4 Vertical channel test diagram.

1.5 m above the top opening. The performance of the specimen is acceptable if the flame does not spread more than 5 m above the bottom of the specimen, and the heat flux density at a height 3.5 m above the top window opening does not exceed 35 kW/m².

The test was developed with the purpose of being less complex and less costly than the larger-scale test but yet still able to discriminate between the performance of wall assemblies in the same way as the larger-scale test. Development of this apparatus is being carried on at BRANZ, New Zealand and the current work is reported in a BRANZ Study Report (Whiting, 2005).

4.5.9 NFPA 285 (1998)

NFPA 285 Standard Method of Test for the Evaluation of Flammability Characteristics of Exterior Non-Load-Bearing Wall Assemblies Containing

Combustible Components Using the Intermediate-Scale, Multi-storey Test Apparatus (NFPA, 1998). This method is for determining the flammability characteristics of exterior non-load-bearing wall assemblies or panels. The performance is evaluated by:

- The capability of the test wall assembly to resist vertical spread of flame over the exterior face of the system.
- The capability of the test wall assembly to resist vertical spread of flame within the combustible core/component of the panel from one storey to the next.
- The capability of the test wall assembly to resist vertical spread of flame over the interior surface of the panel from one storey to the next.
- The capability of the test wall assembly to resist lateral spread of flame from the compartment of fire origin to adjacent spaces.

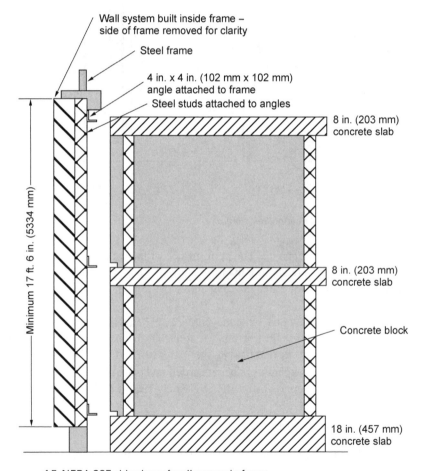

4.5 NFPA 285 side view of wall system in frame.

This test (see Figs 4.5 and 4.6) comprises a two-storey test structure having overall dimensions 4.6 m high by 4.1 m wide. Each room has dimensions 3.05 m by 3.05 m with a floor to ceiling height of 2.13 m. There is a simulated window opening in the lower storey 760 mm high by 1981 mm wide and sill height of 760 mm. The test wall assembly is required to measure at least 5.33 m high by 4.06 m wide. The performance of the test wall assembly is judged on the basis of visual observations in conjunction with temperature data.

UBC Standard 26-9 is very similar to NFPA 285 (NFPA, 1998) and uses a gas burner as the fire source. This is a newer method covering the same scope as UBC 26-4, which is also a two-storey test arrangement but with wood cribs as the fuel source. Both tests require extensive instrumentation. UBC Standard 26-4 (UBC, 1997b) is a method of test for the evaluation of flammability characteristics of exterior, non-load-bearing wall panel assemblies using foamed

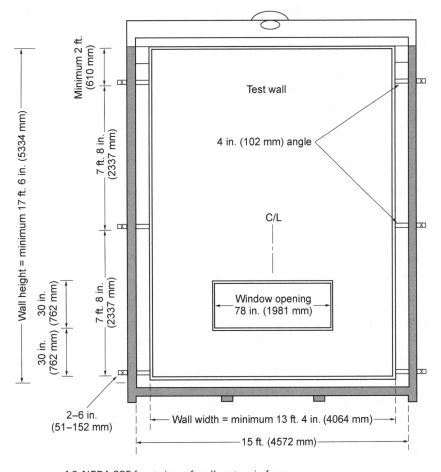

4.6 NFPA 285 front view of wall system in frame.

plastic insulation, and UBC Standard 26-9 (UBC, 1997c) is a method of test for the evaluation of flammability characteristics of exterior, non-load-bearing wall assemblies containing combustible components using the intermediate-scale multi-storey test apparatus.

4.5.10   CSIRO façade test

This test has similar features to the ISO full-scale test and BSI exterior wall test with a return wall.

4.6   Small- or bench-scale test methods used to classify external claddings

These methods are used within current building regulations and where 'deemed to satisfy' or 'approved documents' are invoked.

4.6.1   Non-combustibility tests

These are AS1530 Part 1 (SA, 1984) BS 476: Part 4 (BSI, 1970), BS476 Part 11 (BSI, 1981), ISO 1182 (ISO, 2002c), ASTM E136-82 (ASTM, 1982), and their equivalents throughout the world and are the general method for determining combustibility of building materials.

4.6.2   Spread of flame and rate of heat release tests

AS/NZS 1530 Part 3 (SA, 1999) exposes a vertical specimen to a gradually increasing heat flux up to a maximum of about $25\,kW/m^2$ (Dowling and Martin, 1985). The Fire Propagation BS 476 Part 6 (BSI, 1981) provides a comparative measure of the contribution to the growth of fire made by an essentially flat material, composite or assembly. The result is given as a fire propagation index. The test specimens measure 225 mm square and can be up to 50 mm thick. The apparatus comprises a combustion chamber attached to a chimney and cowl (with thermocouples). The test specimens are subjected to a prescribed heating regime for a duration of 20 minutes and the index obtained is derived from the flue gas temperature compared to that obtained for a non-combustible material.

Surface Spread of Flame BS 476 Part 7 (BSI, 1987) is used to determine the tendency of materials to support lateral spread of flame. The test specimen is 925 mm long × 280 mm wide with thickness up to 50 mm. The vertical specimen is exposed, at an angle of 90 degrees, to a 900 mm square gas-fired radiant panel. Depending on the extent of lateral flame spread along the specimen, the product is classified as Class 1, 2, 3 or 4 with Class 1 representing the best performance.

Bomb Calorimeter ISO 1716 test method (ISO, 1990) determines the gross calorific potential under constant volume. The apparatus consists of a calorimetric bomb, calorimeter (jacket, vessel, stirrer), ignition source and temperature measuring device. The gross calorific potential is calculated on the basis of the measured temperature rise in the test. Equivalent methods include ASTM D2015 (ASTM, 1993b) and NFPA 259 (NFPA, 1991).

Cone Calorimeter ISO 5660 test apparatus consists of a conical electric heater, ignition source and gas collection system. The test specimen measures 100 mm × 100 mm with a thickness between 6 mm and 50 mm. The test specimen is exposed to a heat flux from the electric heater (in the range 0–100 kW/m^2). When the mixture of gases above the specimen surface becomes flammable it is ignited by a spark igniter. Measurements are made of the heat release rate, mass loss, time to ignition, carbon monoxide and carbon dioxide production and light-obscuring smoke (ISO, 2002d). Also similar to ASTM E1354 (ASTM, 1992b) and CAN/ ULC-S135 (ULC, 1992b) and AS/NZS 3837 (SA, 1998).

The ICAL intermediate-scale calorimeter (ASTM, 1994) was developed to measure the heat release rate from wall assemblies, particularly those composites and assemblies that cannot be tested in a cone calorimeter in a representative manner. The apparatus consists of a vertical radiant panel of approximate height 1.3 m and width 1.5 m. The test specimen measures 1 m × 1 m and is positioned parallel to the radiant panel. The maximum radiant flux received by the sample is set at 60 kW/m^2 by adjusting the distance to the panel. The products of pyrolysis are ignited with hot wires located at the top and bottom of the specimen (but not in contact). The specimen is placed on a load cell to measure the mass loss during the test and both specimen and radiant panel are placed beneath an ISO 9705 hood for measurement of the heat release rate on the basis of oxygen consumption (Janssens, 1995).

4.7   Performance in selected test methods

This section provides data on the performance of various external wall claddings and assemblies when subjected to a range of fire test methods. The data presented is taken from the general literature. While it is acknowledged that much additional data and resource exists around the world, it is often of a proprietary nature and therefore often not generally available. More examples are given in Wade and Clampett (2000).

4.7.1 Performance of claddings in full-scale tests

Canada – CAN/ULC-S134

Full-scale tests were carried out on various cladding systems (Oleskiewicz, 1989a, 1990), using a three-storey facility with a test specimen measuring 5 m

Table 4.1 Full-scale experiments at National Research Council of Canada

Assembly	Flame distance (m)*	Heat flux density (kW/m²) @ 3.5 m	@ 5.5 m
Calibration			
Non-combustible board over concrete block wall	2.0	16	10
Assemblies not showing flame spread above the exposing flame			
Gypsum sheathing on glass fibre insulated wood frame wall	3.0	15	10
Assemblies showing flame spread which stops or recedes before end of test			
Vinyl siding on gypsum sheathing on glass fibre insulated wood frame wall	3.0	23	17
Aluminium siding on waferboard on glass fibre insulated wood frame wall	4.5	70	20
12.7 mm flame-retardant treated plywood on untreated wood studs, with phenolic foam insulation in cavities	3.0	29	20
Aluminium sheet (0.75 mm) on flame-retardant treated wood studs, with phenolic foam insulation in cavities	3.2	20	12
76 mm polystyrene foam, glass fibre mesh, 7 mm synthetic plaster on gypsum sheathing, glass fibre insulated steel stud wall	4.5	31	8
Composite panels (6 mm FRP membranes, 127 mm polyurethane foam core) attached to concrete block wall	4.0	24	10
102 mm expanded polystyrene insulation bonded to gypsum sheathing, covered with glass fibre mesh embedded in 4 mm synthetic plaster	4.5	48	37
76 mm expanded polystyrene insulation bonded to gypsum sheathing, covered with glass fibre mesh embedded in 4 mm synthetic plaster	2.0	27	11
Assemblies showing flame spread to top of wall			
8 mm waferboard on glass fibre insulated wood frame wall	7.5	61	79
Vinyl siding on 8 mm waferboard on glass fibre insulated wood frame wall	7.5	82	111
Aluminium siding on 25 mm strapping, 25 mm expanded polystyrene, 19 mm plywood, glass insulated wood frame wall	7.5	30	31

* The flame distance refers to the distance between the top of the window opening and the highest observable instance of flaming on the wall – determined from video recordings.

wide by about 10.3 m high. The performances of various assemblies were grouped according to the flame spread distance recorded. These performances can be categorised as three distinct behaviours:

1. Flame spread extending to the top of the wall (significant fire hazard).
2. Flame spread beyond the extent of the external flame but which stopped or receded before the end of the test (some fire incremental hazard).
3. No flame spread beyond the extent of the external flame (negligible incremental fire hazard).

The results obtained for a range of different cladding assemblies are given in Table 4.1.

The following proprietary wall cladding systems have passed the CAN/ULC-S134 (ULC, 1992a) test as reported by the Canadian Construction Materials Centre Registry of Product Evaluations (IRC, 2004).

*   ALPOLIC®/fr. This is an aluminium composite material (ACM) 4 mm or 6 mm thick comprising aluminium facings with an organic/inorganic core. It is manufactured by Mitsubishi Chemical Functional Products, Inc
*   'Dryvit Outsulation' manufactured by Dryvit Systems Canada. This is an EIFS system.
*   'Carea Exterior Wall Cladding System' manufactured by Productions CAREA Inc, Quebec. This is a prefabricated cladding system consisting of panels of polyester-glass fibre composite, highly filled with mineral particles.
*   'Sto Systems' manufactured by Sto Finish Systems Canada. This is an EIFS system.

Canada – Vertical Channel Test

The results for a range of wall assemblies are given in Table 4.2.

On the basis on the two acceptance criteria (flame spread distance and heat flux density), the Vertical Channel Test correctly classifies all of the assemblies tested except one. The exception was the composite panel which did not meet the acceptance criteria in the Vertical Channel Test, yet was satisfactory in the larger-scale test. However, since the Vertical Channel Test erred on the conservative side, this anomalous result should not be of significant concern.

Building Research Association of New Zealand (BRANZ) (Bong 2000; Saunders and Wade, 2002)

BRANZ used a test facility (see Fig. 4.7) consisting of a vertical wall of overall height 6 m and width 3.6 m to test four cladding materials. This was an *ad-hoc* construction and is not in general use. The main test face of the facility extended at least 5 m over the top of the opening 0.6 ± 0.1 m (high) and 2.1 ± 0.1 m (wide). The total heat flux at a location 1.0 ± 0.1 m above the opening on the centre line

Table 4.2 Vertical channel experiments at National Research Council of Canada

Assembly	Flame distance (m)*	Heat flux density (kW/m²) @ 3.5 m	@ 5.5 m
Acceptable			
Non-combustible board over concrete block wall	2.0	12	7
Vinyl siding on gypsum sheathing on glass fibre insulated wood frame wall	2.8	—	19
12.7 mm flame-retardant treated plywood on untreated wood studs, with phenolic foam insulation in cavities	2.8	14	9
Aluminium sheet (0.75 mm) on flame-retardant treated wood studs, with phenolic foam insulation in cavities	2.3	16	14
76 mm polystyrene foam, glass fibre mesh, 7 mm synthetic plaster on gypsum sheathing, glass fibre insulated steel stud wall	2.0	18	10
Not acceptable			
Aluminium siding on waferboard on glass fibre insulated wood frame wall	4.5	45	24
Composite panels (6 mm FRP membranes, 127 mm polyurethane foam core) attached to concrete block wall	7.3	42	18
8 mm waferboard on glass fibre insulated wood frame wall	7.3	70	65
Vinyl siding on 8 mm waferboard on glass fibre insulated wood frame wall	7.3	67	60

of the opening and in the plane of the façade of the system being tested was $70 \pm 20 \, kW/m^2$ over the period from 5 minutes to 15 minutes. The average value of total heat flux at the same location was intended to be $70 \pm 5 \, kW/m^2$ over the period from 5 minutes to 10 minutes. The heat source used was two fuel trays 1.0 m (length) by 0.25 m (width) and 0.1 m (depth) supported 0.1 m above the floor level and capable of holding approximate 40 l of liquid fuel. The liquid fuel was Pegasol AA by Mobil with a heat of combustion greater than 42 MJ/kg. The wall systems and their performance are described in Table 4.3.

Sweden – Lund University

Tests were carried out at Lund University (Ondrus, 1985) using a three-storey test building similar to the SP Fire 105 apparatus. Several external insulation wall assemblies were examined and a selection are described in Table 4.4. Performance in the full-scale tests were evaluated according to three criteria.

1. No collapse of major sections of the external additional thermal insulation system.

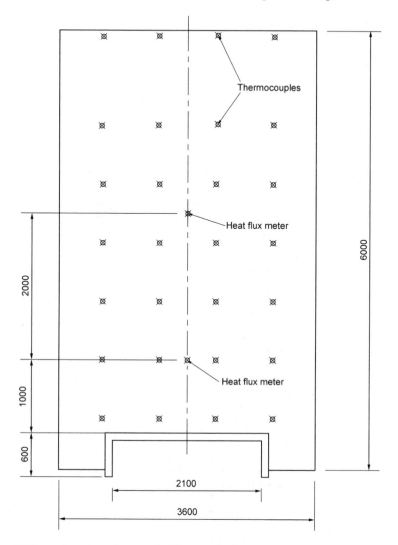

4.7 Front elevation full-scale cladding test rig (BRANZ).

2. The surface spread of flame (a) and the fire spread within the insulation (b) should be limited to the bottom part of the window on the 3rd floor. External flame which can ignite eaves is not permissible.

3. There must be no spread of fire to the 2nd floor through windows – deemed to be verified if the total heat flow toward the centre of windows was 80 kW/m^2.

None of the systems containing polystyrene passed the test criteria (Table 4.5).

Table 4.3 Wall claddings tested by BRANZ

Description	Performance and comments
Cellulose fibre cement sheet, 7.5 mm thick	Flame spread did not occur, localised damage in the region directly exposed to the flame
Extruded foamed uPVC weatherboard	Flame spread reached top of specimen, extensive melting and damage to wall
Plywood, 5-ply radiata pine,	Flame spread reached top of specimen
Reconstituted timber weatherboard	Flame spread reached top of specimen

Table 4.4 External insulation wall assemblies tested at Lund University

System no.	Description
1.1	Profiled steel sheet, asphalt felting, with 95 mm glass wool insulation, timber studs
1.2	Profiled aluminium sheet, wind-protective paper glued to mineral wool on fabric, with 95 mm glass wool insulation, timber studs
3.2	100 mm expanded polystyrene (20 kg/m^3), fastened to the wall by bolts, $19 \times 19 \times 1.05$ mm metal reinforcing mesh fastened to the polystyrene (also in the window splays), 6 mm glass fibre reinforced cement mortar, coloured plaster
3.4	Adhesive mortar with organic agents, 60 mm expanded polystyrene (20 kg/m^3), glass fibre fabric 4×4 mm, 4 mm ground coating of adhesive mortar, 3–4 mm acrylic co-polymer surface coating
3.5	Adhesive mortar with organic agents, 60 mm expanded polystyrene (15 kg/m^3), glass fibre fabric 4×4 mm, 13 mm mineral light plaster with filling of Perlit (expanded volcanic material)
3.8	55 mm polyurethane foam ($30–35 \text{ kg/m}^3$), fastened to the wall by a combination of adhesive and plastic plugs, 6 mm ground coating of vinyl co-polymer with 30% cement, glass fibre fabric 4×4 mm coated with 45% PVC, 2 mm decorating synthetic resin of vinyl (no cement), a gusset of steel at the upper edge of the window

Table 4.5 Lund University – results

System no.	Criterion 1 (collapse)	Criterion 2(a) (surface spread)	Criterion 2(b) (spread within insulation)	Criterion 3 (heat flux)
1.1	failed	failed	passed	failed
1.2	passed	passed	passed	passed
3.2	passed	passed	passed	failed
3.4	failed	failed	failed	failed
3.5	passed	passed	passed	failed
3.8	failed	passed	failed	failed
wood panel facing	failed	failed	failed	failed

Modified Swedish façade test

The SP Fire 105 test method was slightly modified by adding a re-entrant corner, 1 m wide, and mounted at the edge of the straight façade segment – its face was located 0.59 m from the edge of the opening (Babrauskas, 1996). The façade test rig was also situated underneath a large-scale products calorimeter allowing the heat release rate and other properties, such as production of smoke to be quantified. Three different EIFS specimens and a control were tested. Each of the three specimens was identical except for the type of insulation material.

The insulation material, rock wool, EPS and PUR for each test, 80 mm thick, was fixed to the concrete substrate. A reinforcing net was trowelled onto the surface on the insulation. A surface plaster coat was sprayed onto the surface with a total thickness of 8 mm. Performance in the full-scale tests were evaluated according to three criteria.

1. Flame spread and fire damage may not reach above the bottom of the second-storey window.
2. Large pieces of the façade may not fall down during the test.
3. The temperatures at the eaves may not exceed 500 °C for more than 120 seconds or 450 °C for more than 600 seconds. The heat flux meter readings may not exceed 80 kW/m^2 (for hospital occupancies only).

The performance of the wall assemblies is shown in Table 4.6. All the assemblies met the criteria for heat flux and eaves temperature, but specimens F2 and F3 did not meet the maximum damage criteria.

NFPA 285 – SWRI multi-storey apparatus

NFPA 285 is a standard method of test for the evaluation of flammability characteristics of exterior non-load-bearing wall assemblies containing combustible components using the intermediate-scale, multi-storey test apparatus (NFPA, 1998). The performance is evaluated by:

Table 4.6 Performance in modified Swedish façade test

	Maximum window heat flux (kW/m^2)	Maximum eaves temperature (°C)	Maximum damage
F0 – blank/control	43	260	None
F1 – rock wool	42	292	None
F2 – EPS	26	382	Top of 3rd-storey window
F3 – PUR	60	299	Bottom of 3rd-storey window

- The capability of the test wall assembly to resist vertical spread of flame over the exterior face of the system.
- The capability of the test wall assembly to resist vertical spread of flame within the combustible core/component of the panel from one storey to the next.
- The capability of the test wall assembly to resist vertical spread of flame over the interior surface of the panel from one storey to the next.
- The capability of the test wall assembly to resist lateral spread of flame from the compartment of fire origin to adjacent spaces.

The performance of the test wall assembly is judged on the basis of visual observations in conjunction with temperature data.

Six full-scale tests on exterior wall systems were as described in Table 4.7. The overall objective was to evaluate the performance characteristics of foamed-plastics-insulated non-load-bearing wall systems. Conclusions from the tests were:

- The wood crib fire source adequately produced a fire intensity similar to the ASTM E119 time temperature conditions for a period of 30 minutes.
- In all tests there was no flame penetration into the second storey during the 30-minute test period.
- In all tests there was total destruction of the core insulation to an elevation of about 12 feet.
- In all tests there was heat damage to the core insulation above the second floor line on the wall with the window opening.
- In all tests there was no significant flame propagation over the exterior face of the panels.

Table 4.7 Wall systems tested using the SWRI multi-storey apparatus

System no.	Description
A	Timber frame stud wall 100 × 50 mm, with 3 layers of 16 mm Type X gypsum wallboard on the fire side and 2 layers of 16 mm Type X gypsum wallboard on the non-fire side
B	Steel panels (22 g) insulated with 112 mm of glass fibre
C	Steel panels (22 g) insulated with 50 mm of urethane foam, and a thermal barrier consisting of one layer of 12.5 mm Type X gypsum wallboard on the interior side
D	Steel panels (22 g) insulated with 50 mm of urethane foam, no thermal barrier (Manufacturer A)
E	Steel panels (22 g) insulated with 50 mm of urethane foam, no thermal barrier (Manufacturer B)
F	Steel stud, 12.5 mm regular gypsum wallboard inside, exterior grade gypsum wallboard on the outside, 100 mm polystyrene external insulation system (Manufacturer C)

4.8   Conclusions

The flammability of external claddings has been controlled mostly by the application of prescriptive regulations using the concept of non-combustibility as the main parameter. This satisfied the building industry where concrete, glass and steel were the main construction materials where flammability was an important factor in the fire safety of a building.

With the trend towards performance-based codes, and innovation in materials and construction, the previous approach was restrictive and so a search was made for alternative means of assessing external claddings for flammability. It would seem obvious that carrying out essentially a full-scale test would be a solution and is an ideal for many applications and materials. This is the approach that most standards organisations have adopted (BSI, 2002; ISO, 2002a,b; NFPA 1998; ULC, 1992a; UBC, 1997a,b,c; SP, 1994). For small-economy countries this is not an appropriate approach as the costs of carrying out such tests are very prohibitive and would not justify their expense.

Using a fire engineering and mathematical approach is also not an option as the technology is not well developed and models for flame spread on façades, not in general use. An attempt has been made (Cowles and Soja, 1999) where the cone calorimeter test (ISO 2002d) is used but more work needs to be carried out to make use of bench-scale tests. An attempt to develop a medium-scale but representative test, the Vertical Channel Test (Wade and Clampett, 2000; ASTM, 1992a) has not been developed, to the author's knowledge, other than in New Zealand (Whiting, 2005).

We need more information on the behaviour of materials under façade fire conditions that can be used in modelling, and this includes making use of bench-scale tests to generate such data. Only then can we apply the available models to evaluate fire spread on exterior claddings.

4.9   References

American Society for Testing and Materials (ASTM). 1982. 'Standard Test Method for Behaviour of Materials in a Vertical Tube Furnace at 750C'. *ASTM E136-82*. American Society for Testing and Materials, Philadelphia.

American Society for Testing and Materials (ASTM). 1992a. 'Draft Proposed Standard Test Method for Surface Flammability of Combustible Cladding and Exterior Wall Assemblies'. *ASTM Task Group E5.22.07 Vertical Channel Test*, December 1992.

American Society for Testing and Materials (ASTM). 1992b. 'Standard Test Method for Heat and Visible Smoke Release Rates for Materials and Products Using an Oxygen Consumption Calorimeter'. *ASTM E1354-92*. American Society for Testing and Materials. Philadelphia.

American Society for Testing and Materials (ASTM). 1993a. 'Standard Test Method for Density of Smoke from the Burning or Decomposition of Plastics'. *ASTM D2843-93*. American Society for Testing and Materials. Philadelphia.

American Society for Testing and Materials (ASTM). 1993b. 'Standard Test Method for Gross Calorific Value of Coal and Coke by Adiabatic Bomb Calorimeter'. *ASTM D2015-93*. American Society for Testing and Materials. Philadelphia.

American Society for Testing and Materials (ASTM). 1994. 'Standard Test Method for Determination of Fire and Thermal Parameters of Materials, Products, and Systems Using an Intermediate Scale Calorimeter (ICAL)'. *ASTM E1623-94*. American Society for Testing and Materials. Philadelphia.

Anon. 1979. *Fire-Safe Structural Steel, A Design Guide*, American Iron and Steel Institute, Washington DC.

Anon. 1982. 'Eight Die in Las Vegas Hilton Hotel Fire'. *Fire Prevention* No. 150 pp. 33–36.

Anon. 1992. 'High-rise Flats'. *Fire Prevention* No. 252 p. 56.

Anon. 1993. 'BRE Plans Full Scale Tests of Building Cladding Systems'. *Fire Prevention* No. 264 p. 6.

Anon. 1999a. 'Parliamentary Inquiry into Cladding Fires'. *Fire and Flammability Bulletin*. July.

Anon. 1999b. 'Select Committee on Environment, Transport and Regional Affairs Minutes of Evidence', July 1999 (http://www.publications.parliament.uk/pa/cm199899/cmselect/).

Australian Building Codes Board (ABCB), 2004. *BCA 2004 Building Code of Australia*. Canberra.

Babrauskas, V. 1996. 'Façade Fire Tests: Towards an International Test Standard'. *Fire Technology*. pp. 219–230.

Babrauskas, V. 2003. *Ignition Handbook*, Fire Science Publishers/Society of Fire Protection Engineers, Issaquah WA 2003.

Belles, D.W. 1983. Fire Investigation Project 4d-129, Lakeside Plaza, 3800 Powells Lane, Falls Church, Virginia, 31 January 1983.

Bletzacker, R.W. and Crowder, J.L. 1988. 'A fire in New England exposes an exterior insulation and finish system'. *Part A. Building Standards* Vol. 57 No. 4, pp. 26–30.

Bong F.N.P. 2000. 'Fire Spread on Exterior Walls', *Fire Engineering Research Report 2000/1*, Department of Civil Engineering, University of Canterbury, New Zealand.

British Standards Institution (BSI). 1970. *BS 476 Part 4 Fire Tests on Building Materials and Structures. Non-combustibility test for materials.*

British Standards Institution (BSI). 1981. *BS 476 Part 6 Fire Tests on Building Materials and Structures. Method of Test for Fire Propagation for Products.*

British Standards Institution (BSI). 1982. *BS 476 Part 11 Fire tests on building materials and structures. Method for assessing the heat emission from building materials.*

British Standards Institution (BSI). 1987. *BS 476 Part 7 Fire Tests on Building Materials and Structures. Method for Classification of the Surface Spread of Flame of Products.*

British Standards Institution (BSI). 2002. 'Fire performance of external cladding systems. Test methods for non-loadbearing external cladding systems applied to the face of a building', *BS 8414-1:2002*, BSI, London.

Carlsson E. 1999. 'External Fire Spread to Adjoining Buildings – A Review of Fire Safety Design Guidance and Related Research,' *Report 5051*, Department of Fire Safety Engineering, Lund University, Sweden.

Chapman, E.F. 1988. 'High-Rise – An Analysis'. *Fire Engineering*. August.

Colwell, S.A. and Smit, D.J., 1998. 'Assessing the fire performance of external cladding systems: a test method', *Fire Note 9*, Building Research Establishment, Garston, UK

Cowles, G.S. and Soja, E. 1999. 'Flame Spread Classification Method for Exterior Wall Claddings'. *Proceedings 8th International Fire Science and Engineering Conference, INTERFLAM 1999*, Edinburgh, Scotland.

Crowder, J.L. and Bletzacker, R.W. 1988. 'A fire in New England exposes an exterior insulation and finish system. Part B : Critique of the tests used to evaluate exterior insulation and finish systems'. *Building Standards* Vol. 57 No. 5, pp. 12–15.

Department of Building and Housing (DBH). 2005. *Compliance Document for New Zealand Building Code Clauses C1, C2, C3, C4, Fire Safety.* Approved Document C/ASI: Part 7 Control of External Fire Spread, New Zealand.

DIN. 1981. *DIN 4102 Part 1. Brandverhalten von Baustoffen und Bauteilen.*

DoE. 2000. *The Building Regulations 2000* Fire Safety: Approved Documents B1, B2, B3, B4 and B5, Department of the Environment and The Welsh Office. England.

DoETR. 1999. Select Committee on Environment, Transport and Regional Affairs Minutes of Evidence, Memorandum by the Department of the Environment, Transport and the Regions (ROF31). *http://www.publications.parliament.uk/pa/cm199899/cmselect/.*

Dowling, V.P. and Martin, K.G. 1985. 'Radiant Panel Fire Tests on Cellular Plastics Insulation'. *Journal of Thermal Insulation Vol. 8 April.*

Drysdale D.D. 1999. *An Introduction to Fire Dynamics* 2nd edn, Wiley.

FEMA. 1988. Interstate Bank Building Fire Los Angeles, California. United States Fire Administration Technical Report Series. Report 022 of the Major Fires Investigation Project conducted by Tridata Corporation. Federal Emergency Management Agency, USA.

FEMA. 1991a. Apartment Complex Fire, 66 Units Destroyed Seattle, Washington. United States Fire Administration Technical Report Series. Report 059 of the Major Fires Investigation Project conducted by Tridata Corporation. Federal Emergency Management Agency, USA.

FEMA. 1991b. High-rise Office Building Fire, One Meridian Plaza Philadelphia, Pennsylvania. United States Fire Administration Technical Report Series. Report 049 of the Major Fires Investigation Project conducted by Tridata Corporation. Federal Emergency Management Agency, USA.

Galea, E.R., Berhane D. and Hoffmann N.A. 1996. 'CFD Analysis of Fire Plumes Emerging from Windows with External Protrusions in High-Rise Buildings'. *Proceedings of the 7th International Fire Science and Engineering Conference: Interflam '96*, pp. 835–839, Interscience Communications Ltd, London, UK, 1996.

Hakkarainen. T. and Oksanen T. 2002. 'Fire Safety Assessment of Wooden Facades', *Fire and materials* Vol. 26, pp. 7–27.

Hartog, P. 1999. 'Performance of the Lightweight Curtain Wall in the President Hotel Fire'. *Interflam 99, 8th International Fire Science and Engineering Conference.*

Hedskog, B. and Ryber, F. 1998. *The Classification Systems for Surface Lining Materials used in Buildings in Europe and Japan.* Report 5023. Department of Fire Safety Engineering, Lund University.

Hermodsson, T. and Månsson, L. 1992. *Facades: Fire Testing of Materials and Constructions – A First Report for a Test Method* (SP AR 1991: 64). Swedish National Testing and Research Institute.

International Standards Organisation (ISO). 1990. *ISO 1716 Fire Tests – Building Materials – Determination of Calorific Potential.* International Organisation for Standardisation, Geneva

International Standards Organisation (ISO). 2002a. 'Reaction-to-fire tests for façades – Part 1: Intermediate-scale test', *ISO 13785-1:2002*, Switzerland.

International Standards Organisation (ISO). 2002b. 'Reaction-to-fire tests for façades – Part 2: Large-scale test', *ISO 13785-2:2002*, Switzerland.

International Standards Organisation (ISO). 2002c. 'Reaction to fire tests for building products – Non-combustibility test', *ISO 1182:2002*, Switzerland.

International Standards Organisation (ISO). 2002d. 'Reaction-to-fire tests – Heat Release, Smoke Production and Mass Loss Rate – Part 1: Heat Release Rate', (Cone Calorimeter Method). *ISO 5660-1*, Switzerland.

IRC. 2004. *Registry of Product Evaluations,* Canadian Construction Materials Centre. 2004.

Janssens, M.L. 1995. 'Innovative Methods for Evaluating the Combustibility of Materials and Assemblies'. Paper presented at Asiaflam 1995.

Japanese Standards Association (JSA) 1994. 'Method of fire test for non-combustible structural parts of buildings', *JIS A 1302:1994*, Japan.

Kim, W.K. 1990. *Exterior Fire Propagation in a High Rise Building.* Master of Science Thesis, Worcester Polytechnic Institute, MA.

Klem, T.J. 1989. 'Los Angeles High Rise Bank Fire'. *Fire Journal.* May/June.

Klem, T.J. 1991. 'High-Rise Fire Claims Three Philadelphia Fire Fighters'. *Fire Journal.* Sep/Oct.

Klopovic, S. and Turan, O. F. 1998. 'Flames Venting Externally During Full-Scale Flashover Fires: two sample ventilation cases'. *Fire Safety Journal* Vol. 31 pp. 117–142.

Law, M. 1978. 'Fire Safety of External Building Elements The Design Approach', *Engineering Journal/American Institute of Steel Construction*, second quarter pp. 59–74.

Mayr, J. 1996. 'Burning rubbish container threatens residential building'. *Schaden Spiegel – Losses and Loss Prevention*, A publication of the Munich Reinsurance Company compiled to inform their business friends on losses and loss prevention, 39th year, 1996, No. 1, p. 12.

Morris, J. 1990. 'The First Interstate Bank Fire – what went wrong?' *Fire Prevention* No. 226 pp. 20–26. January/February.

NFPA. 1991. *NFPA 259 Standard Method of Test for the Potential Heat of Building Materials.* National Fire Protection Association.

NFPA. 1998. *NFPA 285 Standard Method of Test for the Evaluation of Flammability Characteristics of Exterior Non-Load-Bearing Wall Assemblies Containing Combustible Components Using the Intermediate-Scale, Multistorey Test Apparatus.* National Fire Protection Association.

NFPA. 2001. *NFPA 268 Standard Test Method for Determining Ignitability of exterior Wall Assemblies Using a Radiant Heat Energy Source.* National Fire Protection Association.

NRCC. 1995. *National Building Code of Canada.* National Research Council Canada.

NSWFB. 1999. Personal Communication. New South Wales Fire Brigade.

NZFS. 1997. *The Way Forward.* New Zealand Fire Service Commission.

Ohmiya, Y., Tanaka, T. and Wakamatsu, T. 1998. 'A Room Fire Model for Predicting Fire Spread by External Flames'. *Fire Science & Technology* Vol. 18 No. 1 pp. 11–21.

Oleszkiewicz, I. 1989a. 'Fire Spread on Building Facades'. Presented at the International Fire Protection Engineering Institute – V. National Research Council of Canada.

Oleszkiewicz, I. 1989b. 'Heat Transfer from a Window Plume to a Building Façade'. *Proceedings of the Winter Annual Meeting of the American Society of Mechanical Engineers*, San Francisco, HDT-Vol. 123, pp. 163–170.

Oleszkiewicz, I. 1990. 'Fire and Combustible Cladding'. *Construction Canada* Jul–Aug, pp. 16–18, 20–21.

Ondrus, J. 1985. *Fire Hazards of Facades with Externally Applied Additional Thermal Insulation*. Report LUTVDG/TVBB-3021 Lund Institute of Technology, Sweden.

Saunders N.M. and Wade, C.A. 2002. 'Fire Performance Of External Wall Claddings'. Unpublished study report BRANZ, 2002.

Sharry, J.A. 1974. 'South America Burning', *Fire Journal* Vol. 68 No. 4, pp. 23–34.

Society of Fire Protection Engineering, (SFPE) 2002. *Handbook of Fire Protection Engineering* 3rd edn, Chapter 4, NFPA, Quincy, Mass.

Standards Australia (SA). 1984. *Combustibility test for materials. AS1530 : Part 1*.

Standards Australia (SA). 1998. 'Method of test for heat and smoke release rates for materials and products using an oxygen consumption calorimeter'. *AS/NZS 3837:1998*.

Standards Australia (SA). 1999. 'Methods for fire tests on building materials, components and structures – Simultaneous determination of ignitability, flame propagation, heat release and smoke release'. *AS/NZS 1530.3:1999*.

Stone, W.R. 1974. 'Office Building'. *Fire Journal* January.

Sugawa, O., Momita, D. and Takazhashi, W. 1997. 'Flow Behaviour of Ejected fire Flame/Plume from an Opening Effected by External Side Wind'. *Proceedings (5th) International Symposium. March 3–7. Association for Fire Safety Science*.

Swedish National Testing and Research Institute (SP). 1994. 'External Wall Assemblies and Façade Claddings, Reaction to Fire', *SP Fire 105*, Issue 5, Borås, Sweden.

The Building Centre of Japan (ed.). 2004. *The Building Standard Law of Japan June 2004*.

UBC. 1997a. *Uniform Building Code* Vol. 1. International Conference of Building Officials. USA.

UBC. 1997b. Uniform Building Code Standard 26-4 Method of Test for the Evaluation of Flammability Characteristics of Exterior, Nonload-bearing Wall Panel Assemblies Using Foamed Plastic Insulation. Uniform Building Code Volume 3. International Conference of Building Officials. USA.

UBC. 1997c. Uniform Building Code Standard 26-9 Method of Test for the Evaluation of Flammability Characteristics of Exterior, Nonload-bearing Wall Assemblies Containing Combustible Components Using the Intermediate-Scale, Multistorey Test Apparatus. Uniform Building Code Volume 3. International Conference of Building Officials. USA.

ULC. 1992a. *Standard Method of Fire Test of Exterior Wall Assemblies*. National Standard of Canada CAN/ULC S134-92, Underwriters Laboratories of Canada, Scarborough, Ontario.

ULC. 1992b. *Standard Method of Test for Determining Degrees of Combustibility of Building Materials Using an Oxygen Consumption Calorimeter (Cone Calorimeter)*. 1st edn, National Standard of Canada CAN/ULC-S135-92, Underwriters Laboratories of Canada, Scarborough, Ontario.

Wade, C.A. and Clampett, J.C. 2000. 'Fire Performance of Exterior Claddings', Report of

FCRC Project 2B-2, *BRANZ*, April 2000.

Whiting, P.N. 2005. 'Vertical Channel Test Method', unpublished Study Report, BRANZ, Judgeford, New Zealand.

Willey, A.E. 1972. 'High-Rise Building Fire'. *Fire Journal* Vol. 66 No. 5.

Yokoi, S. 1960. *Study on the Prevention of Fire Spread Caused by Hot Upward Current. Report No. 34.* Building Research Institute, Ministry of Construction, Tokyo, Japan.

Yong, A. 1999. Personal communication.

5

Fire hazard assessment of wall and ceiling fire spread in rooms

C W A D E , BRANZ Ltd, New Zealand

5.1   Introduction

Over the years there have been a number of tragic fires where combustible room linings have contributed to tragic outcomes. Some of the better-known incidents include the Summerland Isle of Man in 1973, Stardust Club Dublin and Hilton Las Vegas both in 1981, and more recently, the Station Nightclub Rhode Island in 2003. Although in most cases, the major fire load in a room is due to the contents of the room, combustible linings can also provide considerable fire load. This chapter presents the latest research on the methods used to evaluate the fire hazard of combustible linings in a room. The approach researchers are working on involves the use of mathematical models of fire growth in conjunction with material flammability data measured and inferred from small-scale tests such as the cone calorimeter. This can provide an initial estimate of the fire hazard, and can minimise the need to conduct expensive large scale room-corner fire tests.

Traditional small-scale fire test methods used to rank reaction-to-fire behaviour have been subjected to criticisms that lead to a burst of research activity in the late 1980s and 1990s. In some cases, the same material or product was ranked highly in one country and poorly in another, based on test results from different apparatus that were intended to discriminate the fire spread performance and hazard of the materials in use. The need for both the test methods and classification systems to be reviewed and rationalised and given a greater scientific basis was apparent. Some of the following research activity that ensued has now resulted in regulatory changes in Europe, Japan and Australia, changes now based on a more scientific approach to fire tests and classification systems.

The EUREFIC project (EUropean REaction to FIre Classification) started in 1989 and primarily involved the Nordic countries Denmark, Finland, Norway and Sweden (ICL, 1991). This project was intended to develop evaluation methods based on the cone calorimeter (ISO, 1991) and the ISO 9705 room/ corner test (ISO, 1993) for wall and ceiling linings and to develop appropriate classification criteria based on these methods. The cone calorimeter apparatus has become an internationally accepted small-scale test for measuring flammability parameters including rate of heat release and smoke production (ISO, 1991). The ISO 9705 room corner test comprises a room 3.6 m × 2.4 m × 2.4 m with a 2 m × 0.8 m door opening (ISO, 1993). The test material is fixed to the interior walls and ceiling and ignited with a gas burner in a corner of the room. Smoke and gases leaving the room are collected in a hood and analysed using oxygen calorimetry techniques, allowing rate of heat release and other measurements to be made. The output of the gas burner is set to 100 kW for the first ten minutes and then the burner output is increased to 300 kW for a further ten minutes. The rate of heat release and smoke production from burner and linings is measured over the 20 minute test duration or until flashover occurs.

The project suggested that materials could be classified into groups based on the measured time to flashover in the ISO 9705 room/corner test. Although not always used directly for regulatory control, this test has nevertheless become the cornerstone behind a number of regulatory changes in recent years particularly those in Europe, Japan and Australia. The research also advanced progress in developing calculation models using cone calorimeter data intended to predict the outcome of the larger and more expensive room-corner test.

Classification systems in Europe, Japan and Australia

Regulatory authorities in the European Community did not adopt the recommendations of the EUREFIC project. However, in 1994 member countries in the European Community did agree to use common fire test methods and a classification system for surface lining materials. The classification system was primarily based on a Fire Growth Rate (FIGRA) index determined from the Single Burning Item (SBI) test EN 13823 (CEN, 2000). This is a comprehensive intermediate-scale calorimetry test simulating an item burning in the corner of a room. Assessment of heat release rate, lateral flame spread and smoke production are included in the results.

The SBI test has been reported as being the most extensively and thoroughly evaluated fire test method ever developed (Deakin, 2000) and a link with the full-scale reference scenario for rate of heat release behaviour has been demonstrated (Tsantaridis and Östman, 1998). However, as for other small- or intermediate-scale tests, there are potential scaling problems with combustible-core metal clad sandwich panels or similar types of product because their fire performance is closely associated with the panel installation method. Inevitably larger-scale tests are needed for these types of product.

The Euroclasses consist of seven categories; A1 (best), A2, B, C, D, E, F (worst). Four fire test methods are used to determine the appropriate class for a given material. These tests are the single burning item test, non-combustibility test (ISO, 2002a), determination of calorific value test (ISO, 2002b), and small flame ignitability test (ISO, 2002c).

In Japan, a new fire classification system was introduced mainly based on the results from the cone calorimeter. Hakkarainen and Hayashi (2001) compared the new Japanese and European classification systems for surface linings and found that, for the majority of products, they were consistent for many materials. The differences in test methods used did account for some variations. The new system included adopting international standards for non-combustibility (ISO, 2002a), cone calorimeter (ISO, 1991) and the reduced-scale model box (ISO, 2004). The latter test is a one-third scale model of the ISO 9705 room. Depending on the classification group required, the non-combustibility test and the reduced-scale model box are alternatives for the cone calorimeter test. In Japan, the ISO 9705 room corner test is the reference scenario, it is not used for actual product classification.

Research carried out in Australia (FCRC, 1998) also agreed that time to flashover in the ISO 9705 room was the appropriate parameter to use for regulatory purposes. Following the EUREFIC methodology, time to flashover was taken as the time for the heat release rate to reach 1 MW from the room. The research recommended that materials be grouped into one of four categories based on time to flashover as follows:

Group 4   materials that result in flashover in less than 120 seconds
Group 3   materials that result in flashover in more than 120 seconds but less than 600 seconds
Group 2   materials that result in flashover in more than 600 seconds
Group 1   materials that do not result in flashover during the 20 minute test duration.

Subsequent changes to the Building Code of Australia (ABCB, 2004) also permitted small-scale test results from AS/NZS 3837 (1998) (cone calorimeter method) to be used in conjunction with the mathematical model developed by Kokkala et al. (1993) in order to predict the time to flashover in the ISO room fire test and hence classify the material, as an alternative to carrying out the full-scale ISO 9705 test. Depending on the location in the building, occupancy and fire protection features present, surface lining materials are required by the Building Code of Australia to meet a designated group classification as given above. Parts of buildings such as fire-protected exitways require surface linings meeting Group 1 while Group 4 linings are generally not permitted in any locations.

Such an approach is consistent with an engineered basis for regulation of surface lining materials. However it could also readily be extended into performance-based design, where a fire model could be used to predict time to flashover for room configurations and fire scenarios that are case specific which vary from the ISO 9705 room corner reference case.

The Kokkala model, as referenced in the Building Code of Australia, is an example of an empirical correlation developed specifically for the ISO 9705 test conditions. It requires input in the form of time versus rate of heat release data from a cone calorimeter test measured at an irradiance level of $50\,kW/m^2$. The method requires an ignitability index (I_{ig}) and two rates of heat release indices (I_{Q1}, I_{Q2}) to be calculated as follows:

$$I_{ig} = \frac{60}{t_{ig}} \qquad\qquad 5.1$$

where t_{ig} = time for heat release to exceed $50\,kW/m^2$ in minutes.

$$I_{Q1} = \int_{t_{ig}}^{t_f} \left[\frac{q''(t)}{(t - t_{ig})^{0.34}} \right] \qquad I_{Q2} = \int_{t_{ig}}^{t_f} \left[\frac{q''(t)}{(t - t_{ig})^{0.93}} \right] \qquad 5.2$$

where t = time (seconds) and is the time at the end of the test (as defined in AS/NZS 3837). $q''(t)$ = rate of heat release (kW/m^2) at time t.

The following three integral limits are then calculated and the materials classified as shown.

$$I_{Q,10\,min} = 6800 - 540 I_{ig} \qquad\qquad 5.3$$
$$I_{Q,2\,min} = 2475 - 165 I_{ig} \qquad\qquad 5.4$$
$$I_{Q,12\,min} = 1650 - 165 I_{ig} \qquad\qquad 5.5$$

Group 1: $I_{Q1} > I_{Q,10\,min}$ and $I_{Q2} > I_{Q,2\,min}$
Group 2: $I_{Q1} > I_{Q,10\,min}$ and $I_{Q2} \leq I_{Q,2\,min}$
Group 3: $I_{Q1} \leq I_{Q,10\,min}$ and $I_{Q2} > I_{Q,12\,min}$
Group 4: $I_{Q1} \leq I_{Q,10\,min}$ and $I_{Q2} \leq I_{Q,12\,min}$

Research on flammability of lining materials

Other models have also been developed, some empirical and others based on fundamental laws of physics. Some models are intended to apply only to ISO 9705 room/corner geometry and procedure, while others are generic and able to be applied using different geometries and ignition sources, although they may not have been fully validated for these alternative configurations. Some of these models are listed below.

- 'Conetools' – an empirical flame spread model for prediction of room/corner and SBI results (Wickström and Göransson, 1992). It uses cone calorimeter data obtained at one heat flux only.
- Quintiere (1993) developed a thermal flame spread model for predicting room/corner fires. This model was subsequently extended by Wade and Barnett (1997) and used in the combined flame spread and zone model described comprehensively in this chapter.
- Karlsson (1992) used an analytical solution to the flame spread equations for a model predicting fire growth in the room corner test.

- Model by Lattimer *et al.* (2003) to predict material performance in the ISO 9705 test using small-scale test data from ASTM E1354 cone calorimeter. The model includes a flame spread model linked to a two-zone compartment model.
- BRANZFIRE – combined thermal flame spread and multi-room zone model (Wade and Barnett, 1997; Wade, 2004; Apte *et al.*, 2004).
- CFD flame spread codes, e.g., FDS (McGrattan, 2005) take a more fundamental approach for the prediction of fire spread and may include either pyrolysis models or cone calorimeter data as input. In particular, much greater resolution in the flame region means many simplifications applying to a thermal flame spread model are not required. However, like the simpler thermal models, results tend to be very sensitive to assumptions concerning the thermal and combustion properties of the materials involved (Apte *et al.*, 2004).

In the remainder of this chapter, a combined thermal flame spread and zone model developed by the Author (BRANZFIRE) is described and used as an example of how scientific methodology has been applied to the analysis of fire hazard from the ignition and burning of room wall and ceiling lining materials.

5.2   BRANZFIRE model basis and principles

The full-scale room corner fire test ISO 9705 (1993) has gained international acceptance as a suitable means of assessing and comparing the fire spread hazard for different surface lining materials, and has been established as a reference scenario for a small room with results correlated with cone calorimeter material data. The room corner test is modelled using a combined material flame spread and room zone model. Figure 5.1 illustrates the relationship between the zone model and thermal flame spread model with the key connections being the inclusion of radiant feedback effects to the room surfaces and its impact on the flame spread calculation, and also the direct input of the calculated heat release rate from the surface linings into the zone model.

A zone model description of a room fire assumes two homogeneous control volumes representing a hot upper layer and a cooler lower layer. Mass and energy flows enter and leave the control volumes due to buoyancy and hydrostatic forces (see Fig. 5.2). The fire plume is the main mechanism used to transport combustion products and entrained air from the lower to the upper layer. Applying laws of conservation of mass and energy leads to a set of first order differential equations which allow the upper layer volume, upper and lower layer temperatures and the pressure equation to be solved. The form of the equations are as follows and are the same as presented by Peacock *et al.* for the CFAST zone model (1993).

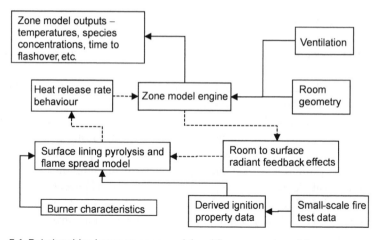

5.1 Relationships between zone model and flame spread model.

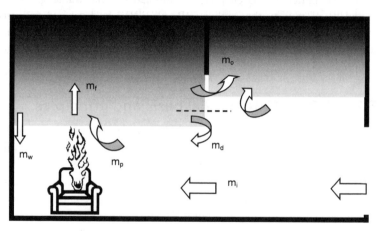

5.2 Mass flows for zone modelling.

Conservation of energy

$$\frac{dP}{dt} = \frac{\gamma - 1}{V_R}(\dot{h}_L + \dot{h}_U) \qquad\qquad 5.6$$

$$\frac{dV_U}{dt} = \frac{1}{\gamma P}\left[(\gamma - 1)\dot{h}_U - V_U\frac{dP}{dt}\right] \qquad\qquad 5.7$$

$$\frac{dT_U}{dt} = \frac{1}{c_P\rho_U V_U}\left[(\dot{h}_U - c_P\dot{M}_U T_U) + V_U\frac{dP}{dt}\right] \qquad\qquad 5.8$$

$$\frac{dT_L}{dt} = \frac{1}{c_P\rho_L V_L}\left[(\dot{h}_L - c_P\dot{M}_L T_L) + V_L\frac{dP}{dt}\right] \qquad\qquad 5.9$$

Conservation of mass

$$\frac{dM_U}{dt} = \dot{m}_p + \dot{m}_f - \dot{m}_d - \dot{m}_o + \dot{m}_w \qquad 5.10$$

$$\frac{dM_L}{dt} = \dot{m}_l + \dot{m}_d - \dot{m}_p - \dot{m}_w \qquad 5.11$$

The flame spread model is based on a thermal upward flame spread approach developed by Quintiere (1993) and modified by Wade (2004). It accounts for both wind-aided and lateral flame spread and while it was developed with the room/corner fire scenario in mind, it is also generic enough to be applied to other configurations. The key parts of the model include: estimating the heat flux from the burner (or other ignition source) to the wall/corner and the subsequent ignition of the wall; calculating the pyrolysing area and the extent of flame spread as the fire spreads up the wall and beneath the ceiling. The total heat release rate with time is then determined by multiplying and summing the pyrolysis area and the heat release rate per unit area measured in the cone calorimeter.

5.3   Material properties

The fire properties of the surface lining materials are characterised using small-scale fire test data, principally from a cone calorimeter, but also from the LIFT apparatus (ISO, 1996) if lateral spread is to be considered in addition to upward or wind-aided flame spread. For upward flame spread calculations, the model requires cone calorimeter data (time to ignition, rate of heat release, heat of combustion) measured at a range of different external heat fluxes for the surface lining material of interest.

The time to ignition data is correlated using a method described by Silcock and Shields (1995) which is referred to here as the Flux Time Product (FTP) method. Other methods of correlating the ignition data could also be used but this one was found to be convenient.

$$(\dot{q}_e'' - \dot{q}_{cr}'')^n t_{ig} = FTP_n \quad \text{for } n \geq 1 \qquad 5.12$$

Rearranging this equation gives:

$$\dot{q}_e'' = \dot{q}_{cr}'' + \frac{FTP_n^{1/n}}{t_{ig}^{1/n}} \qquad 5.13$$

This represents a straight line when plotting inverse time to ignition raised to the power $1/n$ against the externally applied heat flux. The value of n that results in the best fit is determined within the range 1 to 2 and the FTP constant is then determined from the slope of the line. The critical heat flux for the material is determined from the X-axis intercept.

$$FTP_n = (slope)^n \qquad 5.14$$

To calculate the time to ignition of the wall lining in the room-corner test simulation, eqn 5.12 is used where $\dot{q}_e''$ represents the total applied heat flux from the burner flame, hot layer and room surfaces calculated at each time step. Ignition is assumed to occur when $FTP_{sum} > FTP_n$. FTP_{sum} is given by:

$$FTP_{sum} = \sum_0^t (\dot{q}_e''(t) - \dot{q}_{cr}'')^n \Delta t \qquad 5.15$$

The heat release rate (kW/m^2) time histories measured in the cone calorimeter along with the calculated pyrolysis area at each time step are used to compute the total energy release (kW), which in turn becomes the input for the zone model. Cone calorimeter data for the surface lining is required at discrete external heat flux values, with the model linearly interpolating this data, based on the value of the calculated heat flux to the wall or ceiling at each time step, in order to estimate the heat release rate per unit area to use in the calculation. This approach requires the cone calorimeter heat release data at three or preferably more irradiance levels. For lateral flame spread, an ignition correlation developed by Grenier and Janssens (1997) is used to estimate effective thermal inertia $k\rho c$ and T_{ig}. These properties, along with the lateral flame spread parameter ϕ, and $T_{s,min}$ are used for the solution of the lateral flame spread ordinary differential equation.

5.4   Determining flame spread and fire growth

The ignition source is represented by a gas burner positioned in the corner of the room with the burner flame directly impinging on the wall/corner surfaces. The peak heat flux from the burner flame to the corner wall surface, Q_p, is approximated using a correlation from Lattimer et al. (2003):

$$Q_p = 120(1 - e^{-4.0b_w}) \quad [kW/m^2] \qquad 5.16$$

This results in a calculated peak incident heat flux on the corner wall surface of $83.9\,kW/m^2$ for a 300 mm square burner or $59.2\,kW/m^2$ for a 170 mm square burner (the two most common burner sizes used in room/corner testing). This incident flux from the burner flame is assumed independent of the burner output.

The heat flux over the lower 40% of the flame height is assumed to be uniform and equal to the peak flux calculated above. The side dimension of the burner and 40% of the flame height defines the region or area first ignited by the burner flame. The flame height is based on a correlation for a corner configuration using Lattimer's correlation (Lattimer et al., 2003).

$$L_f = 5.9b_w\dot{Q}_b'' \qquad 5.17$$

At each time step, the pyrolysis area on the wall and ceiling is calculated by solving the differential eqns 5.19 and 5.20 (for upward and lateral flame spread).

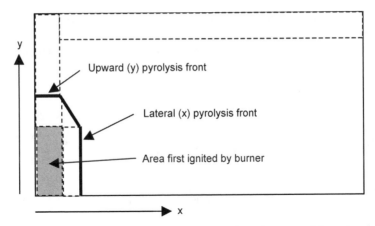

5.3 Wall elevation showing pyrolysis area when the upward front has not reached the ceiling (adapted from Quintiere, 1993).

The assumed geometry of fire spread on the wall surfaces is shown in Figs 5.3 and 5.4. The time step at which each incremental area first ignites is therefore known. The total heat release rate is the sum of the incremental pyrolysis areas multiplied by the time-dependent heat release rate for each incremental area.

The total energy released (at each time step) is the sum of that from the burner, the walls and the ceiling and this is given by:

$$\dot{Q}(t) = \dot{Q}_b + \sum (\dot{Q}''(t)\Delta A_p(t))$$ 5.18

where $\dot{Q}''$ is the energy release per unit area for each incremental area (i.e. the change in area from the previous time step) and depends on the elapsed time of burning for each incremental area. This is determined from an input set of cone

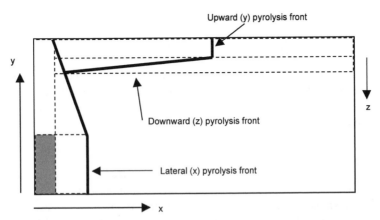

5.4 Wall elevation showing pyrolysis area when the upward front has reached the ceiling (adapted from Quintiere, 1993).

calorimeter heat release rate curves for the material measured at a range of applied heat fluxes.

The governing equation for upward flame spread is given by Quintiere (1993) as:

$$\frac{dy_p}{dt} = \frac{y_f - y_p}{t_{ig}} \quad \text{where } t_{ig} = \frac{FTP_n}{(\dot{q}''_{ff} - \dot{q}''_{cr})^n} \qquad 5.19$$

$\dot{q}''_{ff}$ is the heat flux from the flame to the surface ahead of the pyrolysis front and is taken as $30\,kW/m^2$ as proposed by Quintiere (1993).

The lateral pyrolysis front is given by:

$$\frac{dx_p}{dt} = \frac{\Phi}{k\rho c(T_{ig} - T_s)^2} \quad \text{where } T_s \geq T_{s,\min} \qquad 5.20$$

Φ is a flame spread parameter and $T_{s,\min}$ is the minimum surface temperature for spread with both parameters determined from the LIFT test (ISO, 1996).

5.5   Smoke development

The production of smoke and other combustion products from fires is important for determining the tenability and availability of escape routes. Soot or smoke production from the fire affects visibility and is determined by both the material chemistry and the combustion environment. The model was modified to include a functional dependency on global equivalence ratio for the soot yield. Smoke (or soot) is modelled as a 'gas' species uniformly distributed throughout each gas layer. The applicable mass conservation equations for the upper and lower layers respectively are (Wade, 2004):

$$\frac{dY_{s,u}}{dt} = \frac{1}{M_u}\left[\dot{m}_p Y_{s,l} + \dot{m}_f \Psi_s - (\dot{m}_d + \dot{m}_w + \dot{m}_o)Y_{s,u} - Y_{s,u}\frac{dM_u}{dt}\right] \qquad 5.21$$

$$\frac{dY_{s,l}}{dt} = \frac{1}{M_l}\left[\dot{m}_i Y_{s,\infty} - \dot{m}_p Y_{s,l} + (\dot{m}_d + \dot{m}_w)Y_{s,u} - Y_{s,l}\frac{dM_l}{dt}\right] \qquad 5.22$$

In the case of the ISO 9705 room corner test, the smoke generation term comprises contributions from the ignition burner and the wall and ceiling lining materials, where the subscripts b, w, c represent the burner, wall lining and ceiling lining respectively.

$$\dot{m}\Psi_s = \dot{m}_{f,b}\Psi_{s,b} + \dot{m}_{f,w}\Psi_{s,w} + \dot{m}_{f,c}\Psi_{s,c} \quad \text{(kg-smoke/sec)} \qquad 5.23$$

The burning rate of the surface linings is calculated using the heat release rate from eqn 5.18 and the lining material heat of combustion.

The smoke production rate, SPR at any point in time, using a specific extinction coefficient (σ_s) of $8.79\,m^2/g$-smoke from Mulholland and Choi (1998) is given by:

$$SPR = \sigma_s \dot{m} \Psi_s \quad (\text{m}^2/\text{s}) \qquad\qquad 5.24$$

The light extinction coefficient is calculated from the smoke mass concentration ($C_s = \rho_u Y_{s,u}$) in the gas layer and the specific extinction coefficient, i.e.,

$$k = \sigma_s C_s = \sigma_s \rho_u Y_{s,u} \qquad\qquad 5.25$$

The yield of smoke is dependent on the fuel and the combustion environment. A correlation for the dependence of the soot yield on the global equivalence ratio was developed by Tewarson, Jiang and Morikawa (1993) and this was included in the model thus effectively increasing the smoke yield from its assumed pre-flashover or well-ventilated value, as the fire becomes increasingly ventilation limited. This is not relevant for an ISO 9705 simulation since the test compartment is well ventilated.

5.6   Model verification

The EUREFIC research programme included experiments on different surface lining products using the ISO 9705 full-scale room test apparatus and bench-scale cone calorimeter. BRANZFIRE modelling relies on use of the cone calorimeter data for input to the model. Heat release rate predictions of twenty-one experiments are shown in Figs 5.5 to 5.25. The experiments include both ISO 9705 and ASTM room-corner methods, and with materials attached to either the wall only or ceiling only or both wall and ceiling as indicated. Figures

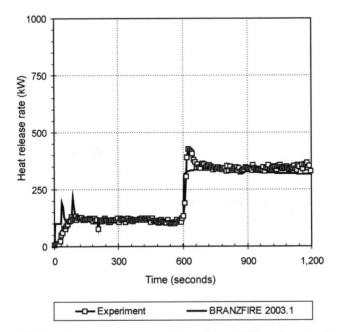

5.5 Painted gypsum paper-faced plasterboard (ISO, walls and ceiling).

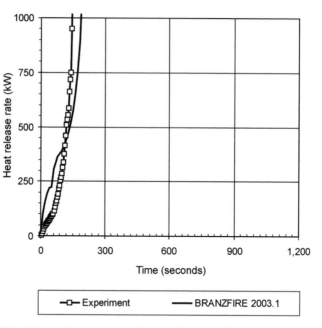

5.6 Ordinary birch plywood (ISO, walls and ceiling).

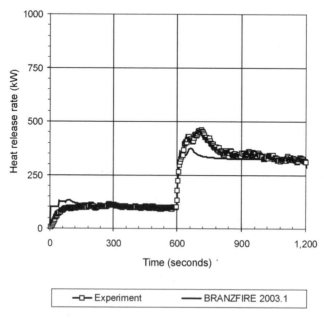

5.7 Melamine-faced high density noncombustible board (ISO, walls and ceiling).

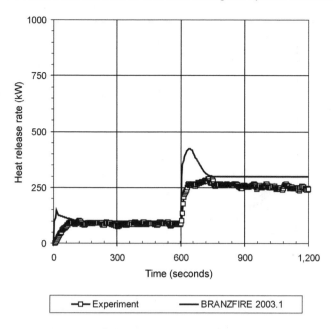

5.8 Plastic-faced steel sheet on mineral wool (ISO, walls and ceiling).

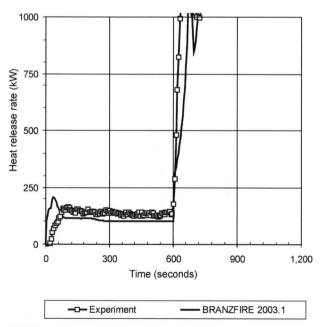

5.9 FR particleboard (ISO, walls and ceiling).

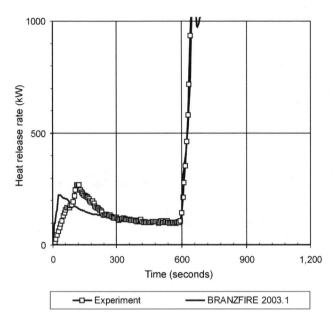

5.10 PVC wallcarpet on gypsum paper-faced plasterboard (ISO, walls and ceiling).

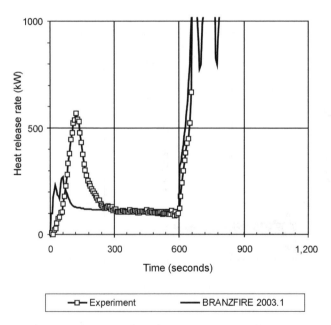

5.11 Textile wallcovering on gypsum paper-faced plasterboard (ISO, walls and ceiling).

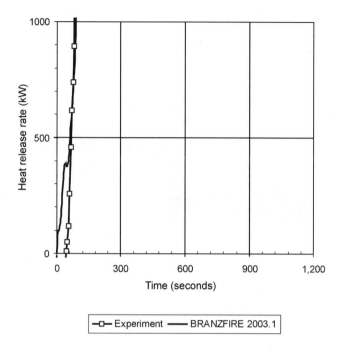

5.12 FR polystyrene foam (ISO, walls and ceiling).

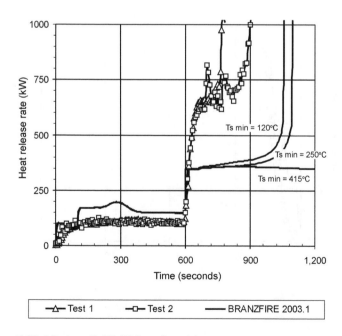

5.13 GR phenolic 70 (ISO, walls only).

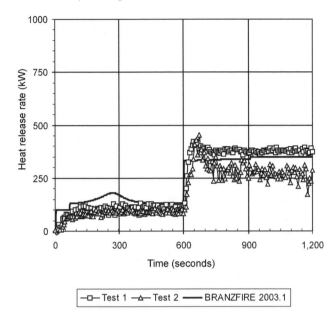

5.14 GR phenolic 71 (ISO, walls only).

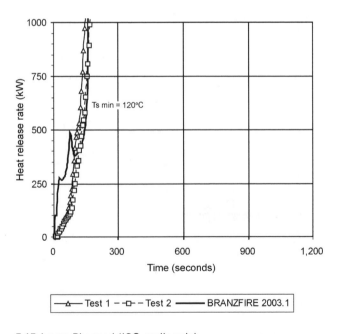

5.15 Lauan Plywood (ISO, walls only).

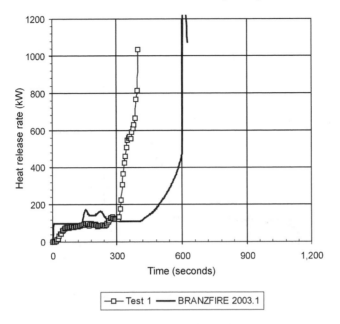

5.16 Lauan Plywood (ISO, ceiling only).

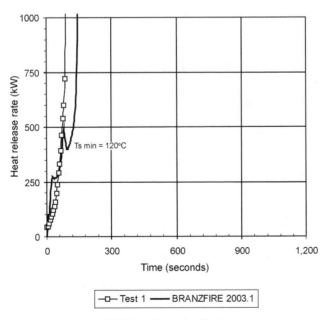

5.17 Lauan Plywood (ISO, walls and ceiling).

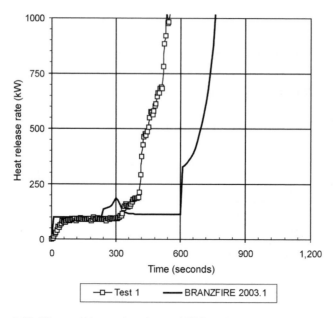

5.18 FR treated hoop pine plywood (ISO, ceiling only).

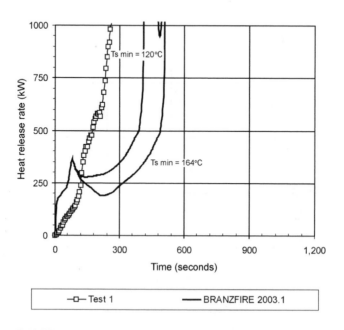

5.19 FR treated hoop pine plywood (ISO, walls only).

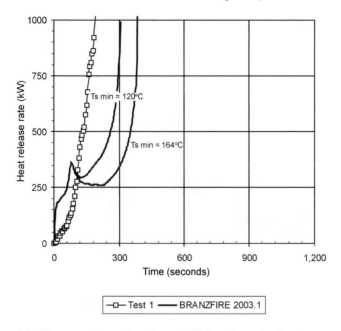

5.20 FR treated hoop pine plywood (ISO, walls and ceiling).

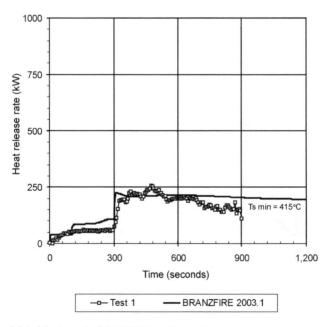

5.21 GR phenolic 70 (ASTM, walls only).

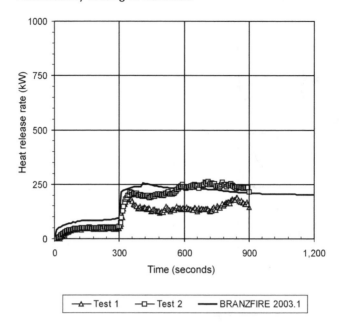

5.22 Plywood (ASTM, walls only).

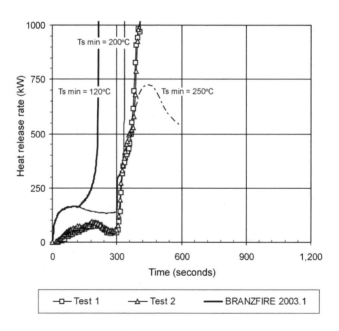

5.23 Plywood FR (ASTM, walls only).

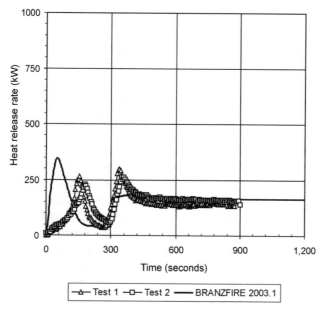

5.24 FR polystyrene foam (ASTM, walls only).

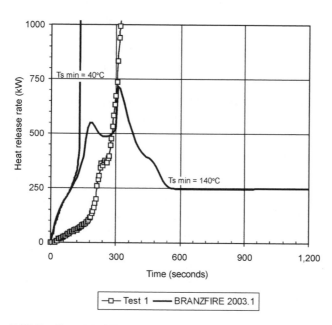

5.25 Hardboard (ASTM, walls only).

5.5 to 5.12 are materials from the EUREFIC project tested using the 100/300 kW burner regime from ISO 9705, while Figs 5.13 to 5.25 are materials tested by CSIRO, Australia (Apte *et al.*, 2004). The experiments shown in Figs 5.13 to 5.20 use the 100/300 kW burner regime from ISO 9705, while those in Figs 5.21 to 5.25 use a 40/160 kW regime. The factors that contribute to uncertainty in the model predictions include parameters derived from cone calorimeter ignition data (*FTP* and $\dot{q}''_{cr}$) and lateral flame spread variables from LIFT data (Φ and $T_{s,min}$).

Materials that exhibited only limited burning in the cone calorimeter, are generally well predicted by the model and do not exhibit significant flame-spread for the burner output regimes that were modelled as shown, for example, in Figs 5.5, 5.7, 5.8, 5.14, 5.21, 5.22 and 5.24. It is observed that the model predictions give a higher heat release at the start of the simulation since the initial compartment smoke-filling time delays the recorded rise in the heat release rate in the experiments since the instrumentation is in the exhaust duct and not within the actual burn room. This effect does not have much influence on the results in terms of the subsequent flame spread prediction. Lateral flame spread parameters are more important for the 'wall only' cases (see Figs 5.14, 5.15, 5.19 and 5.21–5.25). In some cases they will determine the prediction of whether flashover occurs or not as shown in Figs 5.23 and 5.25.

Some materials/products are not well suited to the type of thermal flame spread modelling that has been described here. For example, thermoplastic foams will shrink and melt during the fire (e.g. Figs 5.12 and 5.24), and for metal-clad sandwich panels with combustible cores (not modelled), the performance is very much dependent on the installation method including the jointing details and fixings. Table 5.1 compares the time for the total heat release to reach 1 MW (flashover) from the experiment with the prediction from the model. In most cases the model predicts the correct BCA classification group – in 3 of the 21 cases it differed, while in one further case (Fig. 5.13) it depended on the selection of the assumed minimum surface temperature for spread.

5.7   Conclusions

The current suite of reaction-to-fire tests including cone calorimeter (ISO 5660) and the room-corner (ISO 9705) are more robust and scientific than their predecessors. They have greater applicability to realistic end-use applications and fire scenarios. When used in conjunction with mathematical models of fire growth and flame spread, particularly for characterising flammability properties of materials and for validation of models, then the models could potentially be used to better predict fire growth and flame spread under different non-standard conditions, room geometries or ventilation arrangements, for which it is not currently practical to conduct full-scale fire tests. This provides the fire safety system designer with new possibilities for the analysis of fire hazards in

Table 5.1 Calculated and measured classification groups based on time to flashover (1 MW)

Product	Time to flashover group (calculated using BRANZFIRE model)	Time to flashover group (measured)	Figure reference
Painted gypsum paper-faced plasterboard	1	1	5.5
Ordinary birch plywood	3	3	5.6
Melamine-faced high-density non-combustible board	1	1	5.7
Plastic-faced steel sheet on mineral wool	1	1	5.8
FR particleboard type B1	2	2	5.9
PVC wallcarpet on gypsum paper-faced plasterboard	2	2	5.10
Textile wallcovering on gypsum paper-faced plasterboard	2	2	5.11
FR polystyrene foam	4	4	5.12
GR phenolic 70 (w)	1/2	2	5.13
GR phenolic 71 (w)	1	1	5.14
Plywood (w)	3	3	5.15
Plywood (c)	2	3	5.16
Plywood	3	3	5.17
FR plywood (c)	2	3	5.18
FR plywood (w)	3	3	5.19
FR plywood	3	3	5.20
*GR phenolic 70	1	1	5.21
*Plywood FR US	1	1	5.22
*Plywood US (w)	3	3	5.23
*Polystyrene foam FR (w)	1	1	5.24
*Hardboard (w)	1	3	5.25

* -- These are ASTM room tests not ISO 9705.
(w) = materials fixed to walls only
(c) = materials fixed to ceilings only

buildings resulting from the use of combustible surface-lining materials as part of a larger risk assessment process.

5.8   Nomenclature

P	room pressure (Pa)
t	time (s)
V_R	room volume (m^3)
h_u and h_l	respective enthalpies (kJ) in the upper and lower layers
V_u and V_l	respective gas volumes of the upper and lower layers (m^3)
T_u and T_l	respective gas temperature of the upper and lower layers (K)

$Y_{s,u}$ and $Y_{s,l}$ respective mass fractions of smoke/soot in the upper and lower layers

M_u and M_l respective masses of the upper and lower layer (kg)

$\dot{m}_p$ rate of air entrained in the fire plume (kg/s)

$\dot{m}_f$ fuel mass loss rate (kg/s)

$\dot{m}_d$ vent/door mixing flow rate (kg/s)

$\dot{m}_w$ natural convective wall flow (kg/s)

$\dot{m}_o$ and $\dot{m}_i$ air flow rates leaving and entering through the vent respectively (kg/s)

FTP flux time product

I_{ig} ignitability index (min^{-1})

I_{Q1}, I_{Q2} rate of heat release indices

$I_{Q,i\,\text{min}}$ rate of heat release integral limits (for $i = 2$, 10 or 12 minutes)

$\dot{q}_e''$ external heat flux incident on surface (kW/m^2)

$\dot{q}_{cr}''$ critical heat flux (kW/m^2)

$\dot{q}''(t)$ rate of heat release at time t (kW/m^2)

n power index

b_w burner width (m)

L_f flame height (m)

Q_p peak heat release rate (kW/m^2)

Q_b burner output (kW)

Ψ_s mass generation rate of smoke per unit mass of fuel burned (kg/kg)

x_p position of the lateral pyrolysis front (m)

y_p position of the upward pyrolysis front (m)

y_f position of the flame front (m)

t_{ig} time to ignition (s)

T_{ig} surface temperature at ignition (K)

T_s surface temperature (K)

$T_{s,\text{min}}$ minimum surface temperature for spread (K)

$\dot{q}_{ff}''$ heat flux to the surface ahead of the flame (kW/m^2)

$k\rho c$ thermal inertia

$\dot{Q}''(t)$ heat release rate per unit area from cone calorimeter (kW/m^2)

A_p pyrolysis area (m)

k extinction coefficient (1/m)

SPR smoke production rate (m^2/s)

γ ratio of specific heats

Φ lateral flame spread parameter

σ_s specific extinction coefficient (m^2/s)

5.9   References

Apte V.B., Bui, A., Paroz, B., Wade, C.A., Webb, A.K. and Dowling V.P. (2004). 'An Assessment of Fire Growth Models – BRANZFIRE and FDS Against CSIRO Fire

Tests on Combustible Linings in a Room'. *Proceedings Interflam 2004*, Edinburgh. Interscience Communications Ltd.

AS/NZS 3837:1998: Method of test for heat and smoke release rates for materials and products using an oxygen consumption calorimeter. Standards Australia.

Australian Building Codes Board (ABCB). 2004. BCA 2004 Building Code of Australia. Canberra.

CEN. 2000. *EN 13823 Reaction to fire tests for building products – Building products excluding floorings – exposed to the thermal attack by a single burning item, CEN*.

Deakin, G. 2000. http://www.wfrc.co.uk/useful_publications/harmonised_fire_safety.htm

Fire Code Reform Centre (FCRC). 1998. 'Fire Performance of Wall and Ceiling Lining Materials. Final Report – With Supplement'. Project Report FCRCPR98-02.

Grenier, A.T. and Janssens, M.L. 1997. 'An Improved Methods for Analyzing Ignition Data of Composites'. *Proceedings of the International Conference on Fire Safety*, 23.

Hakkarainen, Tuula and Hayashi, Yoshihiko. 2001. 'Comparison of Japanese and European Fire Classification Systems for Surface Linings'. *Fire Science & Technology* Vol. 21 No. 1 pp. 19–42.

Interscience Communications Ltd (ICL). 1991. 'EUREFIC European Reaction to Fire Classification'. *Proceedings of the International EUREFIC Seminar*, Copenhagen Denmark, September 1991.

ISO. 1991. *ISO 5660 Fire Tests – Reaction to Fire – Rate of Heat Release from building products*. International Standards Organisation (ISO).

ISO. 1993. *ISO 9705 Room fire test in full scale for surface products*. International Standards Organisation (ISO).

ISO. 1996. *ISO 5658-2 Room Reaction to fire tests – Spread of flame – Part 2: Lateral spread on building products in vertical configuration*. International Standards Organisation (ISO).

ISO. 2002a. *ISO 1182 Reaction to fire tests for building products – Non-combustibility test*. International Standards Organisation (ISO).

ISO. 2002b. *ISO 1716 Reaction to fire tests for building products – Determination of the heat of combustion*. International Standards Organisation (ISO).

ISO. 2002c. *ISO 11925-2 Reaction to fire tests – Ignitability of building products subjected to direct impingement of flame – Part 2: Single-flame source test*. International Standards Organisation (ISO).

ISO. 2004. *ISO/DIS 17431 Fire test – Reduced-scale model box test*. International Standards Organisation (ISO).

Karlsson, B. 1992. 'Modelling Fire Growth on Combustible Lining Materials in Enclosures'. *Report TVBB-1009*, Lund University, Lund, Sweden.

Kokkala, M.H., Thomas, P.H. and Karlsson, B.1993. 'Rate of Heat Release and Ignitability Indices for Surface Linings'. *Fire and Materials* 17, pp. 209–216.

Lattimer, Brian Y. *et al.* 2003. 'Corner Fire Growth in a Room with a Combustible Lining. Fire Safety Science', *Proceedings of the Seventh International Symposium*. Worcester, USA June 2002. International Association for Fire Safety Science.

McGrattan, K. (ed.). 2005. *Fire Dynamics Simulator (Version 4) Technical Reference Guide*. NIST Special Publication 1018. US Department of Commerce. National Institute of Standards and Technology. Washington.

Mulholland, George W. and Choi, Mun H. 1998. *Measurement of the Specific Extinction Coefficient for Acetylene and Ethene Smoke Using the Large Agglomerate Optics*

Facility. Twenty-Seventh Symposium (International) on Combustion. The Combustion Institute, pp. 1515–1522.

Peacock, R.D., Forney, G., Reneke, P.A., Portier, R. and Jones, W.W. 1993. 'CFAST, the Consolidated Model of Fire and Smoke Transport'. *NIST Technical Note 1299*, National Institute of Standards and Technology, USA.

Quintiere, J.G. 1993. 'A simulation model for fire growth on materials subject to a room corner test'. *Fire Safety Journal* 20, pp. 313–339.

Silcock, G.W.H. and Shields, T.J. 1995. 'A Protocol for Analysis of Time-to-Ignition Data from Bench-Scale Tests'. *Fire Safety Journal* 24, pp. 75–95.

Tewarson, A., Jiang, F.H. and Morikawa, T. 1993. 'Ventilation Controlled Combustion of Polymers'. *Combustion and Flame* 95, pp. 151–169.

Tsantaridis, L. and Östman, B. 1998. 'Cone Calorimeter Data and Comparisons for the SBI RR Products'. *Trätek Report 9812090*. Swedish Institute for Wood Technology Research. Sweden.

Wade, C.A. 2004. *BRANZFIRE Technical Reference Guide. Study Report 92* (revised). Building Research Association of New Zealand. Porirua.

Wade, C.A. and Barnett, J.R. 1997. 'A Room-Corner Fire Model Including Fire Growth on Linings and Enclosure Smoke-Filling'. *Journal of Fire Protection Engineering* Vol. 8 No. 4, pp. 27–36.

Wickström, U. and Göransson, U. 1992. 'Full-Scale/Bench-Scale Correlations of Wall and Ceiling Linings'. *Journal of Fire and Materials* Vol. 16.

6

Fire behaviour of sandwich panels

P V A N   H E E S , SP Fire Technology, Sweden

6.1   Introduction: description of sandwich panels

One of the most recent definitions for a sandwich panel is the one used in the proposed European product standard (draft) prEN 14509. This product standard defines a sandwich panel as a building product consisting of two metal faces positioned on either side of a core that is a rigid thermally insulating material, which is firmly bonded to both faces so that the three components act compositely when under load. In many cases sandwich panels can be self-supporting which means that the panels are capable of supporting, by virtue of their materials and shape, their own load and in the case of panels fixed to spaced structural supports all applied loadings (e.g. snow, wind, internal air pressure), and transmitting these loadings to the supports.

6.2   Typical fire problems related to sandwich panels

The number of fires involving sandwich panels with a combustible core has increased substantially over the last few years and they are considered to be a growing fire problem in the fire safety of buildings especially in food factories and in cold storage rooms, warehouses, etc. Fires in so-called food factories in the UK have caused much concern for insurance companies and fire fighters. For example, the total loss from fires in the UK food industry where panels are often used was more than €30m in 1995. Indeed, this was even sufficient to force one UK insurance company into liquidation. Since that time, correct specification and good fire safety management procedures have significantly reduced the number of sandwich panels related fires. The legacy of large losses is, however, that the insurance industry has targeted applications involving sandwich panels with combustible cores and has resulted in a requirement to use only certified products that have been subjected to large-scale fire tests.

As mentioned above, sandwich panels are built up as composite construction products, i.e., simply expressed, they consist of two metal sheets with an inner core insulation. This core insulation can be either combustible or noncombustible. The metal can cause a delay in ignition but still transfer heat to the core. Once on fire, such panels with a combustible core are known to be difficult to extinguish since the combustible material is behind a metal facing. In the following paragraphs some typical fire problems related to sandwich panels are mentioned.

6.2.1   Reaction to fire

The reaction of a product to fire involves characteristics such as ignition, flame spread, heat release rate, smoke and gas production, occurrence of burning droplets and parts. Due to the presence of the metal sheet, delay in ignition occurs but ignition will also only take place through openings between the panels, i.e., joints or other penetrations made through the panels for electrical applications, etc. Once ignited the flame or fire spread will depend on how easily the joints will open and how the pyrolysis gases escape from between the two metal sheets. It is clear that overall fire behaviour is thus greatly dependent on the whole system. So that characteristics such as the design of the panel joints and details of the junctions are important. Another very important factor is how the panels are fixed to the supporting structure if they are not self-supporting. It is therefore obvious that small-scale tests are not the most appropriate way to determine the reaction-to-fire behaviour of panels and that full-scale tests are much more appropriate. For this reason this chapter will deal only with major full-scale tests in sections 6.3 and 6.4.

6.2.2   Fire resistance

Fire resistance testing of sandwich panels provides a measure of the ability of the panel to isolate the fire to the area of ignition when exposed to a fully developed fire for a specific time. Fire resistance testing assesses integrity, insulation and stability of the construction. Similar to the reaction to fire the design of the joints and the fixing of the panels to the surrounding of the structure is extremely important. The testing of the sandwich panel system should therefore reflect the end-use details of the whole structure. Testing is most often done by means of fire resistance tests such as referred to in ISO 834, EN 13501-2, EN 1363-1 and ASTM E112 and the reader is referred to these procedures for further information.

6.2.3   Fire protection

Specific problems occurring with sandwich panels when using fire protection measures such as, e.g., a sprinkler or other extinguishing system is the fact that the combustible material is behind a metal sheet which reduces the ability of the extinguishing agent to access the burning material or to reduce the temperature of the fuel. When designing the fire protection system this should be taken into account.

6.3   Overview of major recent developments in fire test methods

Before going into the details of the full-scale test for sandwich panels it is worth mentioning that at European level there is the development of a product standard with respect to sandwich panels. The product standard (prEN 14509) covers fire safety and items such as structural stability, air permeability, acoustical properties, etc. This standard prescribes both reaction to fire (based on EN 13501-1) , fire resistance and external roof (pr EN 13501-5) classes. For reaction to fire this means that the major test method proposed at this time is the SBI method, despite some poor correlation to full-scale scenarios, see section 6.5.

When assessing the fire risk of sandwich panels, most small-scale tests are inadequate because they cannot predict the behaviour of joints and mechanical fastenings during fire, so we will not go into the detail of this. ISO TC92 SC1 understood this and produced two large-scale tests for sandwich panels which were published in 2002 as ISO 13784 part 1 and 2. ISO 13784 part 1 can especially be considered as a specific version of ISO 9705 [2,5] suitable for a full end-use mounting of sandwich panels. In this chapter the main focus will be an assessment of the fire behaviour of sandwich panels by means of these full-scale tests.

6.3.1   ISO 13784-1 Reaction to fire tests – Scale tests for industrial sandwich panels – Part 1: Intermediate scale test

This standard is very similar to the so-called Room/Corner Test (ISO 09705), see Fig. 6.1. First we will discuss the concepts of the ISO 9705 method. The Room/Corner Test was first published by ASTM in 1982 and then by NORDTEST in 1986. The international standard, ISO 9705, was published in 1993. The Room/Corner Test is a large-scale test method for measurement of the burning behaviour of building products, mainly linings in a room scenario with a single burning item (waste paper basket, furniture, etc.) as an ignition source. The principal output is the occurrence and time to flashover. Also a direct measure of fire growth (Heat Release Rate, HRR) and light obscuring smoke (Smoke Production Rate, SPR) are results from a test. The product is mounted on three walls and on the ceiling of a small compartment. A door opening ventilates the room.

Experience with testing products has been gained during more than ten years of work with the Room/Corner Test. A considerable amount of information on

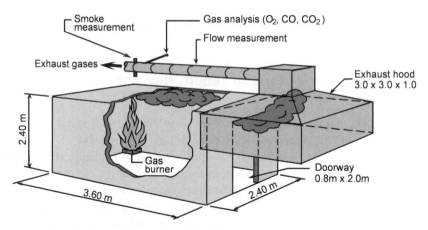

6.1 The Room/Corner Test, ISO 9705.

product burning behaviour by this method is available and the thermal conditions during a test fire has been carefully mapped and guidance is given, for example, in ISO 9705 part 2. ISO 9705-2 is a valuable source of background data and theory for the test. The complete heat balance of the test room is described. Equations for mass flow are given as well as a mapping of the heat flux from the different burner levels. ISO 9705-2 also contains a large biography of technical papers related to the Room/Corner Test. However, when performing tests on sandwich panels in ISO 9705 the panels are positioned inside the room. This has some drawbacks:

1.   Due to the fact that the dimensions are fixed, the room size will depend on the thickness of the panels. When panels for, e.g., cold rooms have to be tested actual thickness can very easily be up to more than 30 cm meaning that the volume of the room is decreased. This can introduce a 'thickness' dependency.
2.   In many cases the joints have an important effect on the fire behaviour of the complete system. That is why complex joint systems are developed within the sandwich panel industry and why these joints have to be exactly the same in the test as in reality. With the actual mounting technique inside the test room it is impossible to do this since the outer walls of the room corner structure prevents adequate finishing of the joints located at the rear of the panel (i.e. those between the sandwich panel construction and the walls of the room corner itself).
3.   Testing panels inside the ISO 9705 room does not allow observations to be made of the outside of the construction, e.g., flaming or excessive deformation.

These drawbacks can be solved only by testing either the construction as a 'stand alone' without walls and ceiling from a surrounding structure or as a

6.2 An ISO 13784 part 1 test.

construction where an enclosure is built around the sandwich panel construction at a sufficient distance allowing for finishing the joints and performing the observations. Such set-ups will result in a correct evaluation of the product. For this reason ISO TC 92 SC 1 developed ISO 13784 part 1 which represents a small room test scenario. A typical set-up for such a test can be seen in Fig. 6.2.

ISO/CD 13784-1 is a variant where the Room/Corner Test is used for the testing of sandwich panels. These panels are generally well insulated and self-supporting or fixed to a structure. Therefore they are themselves used as construction materials for the test room or will be the whole wall system fixed to a supporting system. In all other aspects ISO/CD 13784-1 testing is performed very similarly to an ordinary ISO 9705 test, see Table 6.1.

Table 6.1 The ISO 13784 part 1 test specifications

Specimens	Set-up of a room with inner dimensions of 3.6 × 2.4 × 2.4 m and with a door opening of 0.8 by 2 m (width by height) Sample material covers all four walls and ceiling of the test set-up
Specimen position	The specimen is either mounted as a self-supporting structure or as a wall/ceiling panel fixed to a supporting structure which is either inside or outside the room. The structure should be fire protected if this is done in practice.
Ignition source	Gas burner placed in one of the room corners. If there is a supporting structure in the corner, the burner is positioned at the back wall at the nearest joint (see standard for more instructions). The burner heat output is 100 kW for the first ten minutes and then 300 kW for another ten minutes.
Test duration	20 min or until flashover.
Measurements and conclusions	A number of parameters relating to a room fire such as temperatures of the gas layers, flame spread and heat fluxes can be measured. However, the most important outputs are HRR, SPR and time to or occurrence of flashover. The HRR and SPR can be measured by two methods. In method 1 the room is connected to a standard hood described in ISO 9705. With this option only the gases coming from the door opening are collected. In method 2 an enlarged hood or structure is used which can collect all gases escaping from the structure.

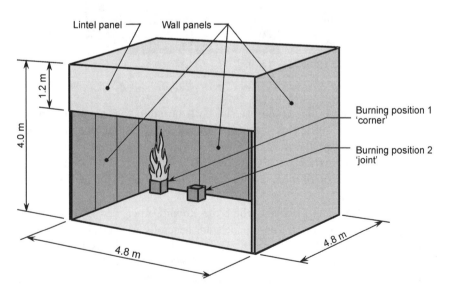

6.3 ISO 13784-2, large-scale test for sandwich panels.

Table 6.2 ISO 13784-2 test specifications

Specimens for constructing the test room	The test room is made up from the sandwich panels and shall consist of four walls at right-angles and a ceiling and shall be located on a rigid, non-combustible floor surface. The room shall have the following inner dimensions: (a) length: (4.8 ± 0.05) m (b) width: (4.8 ± 0.05) m (c) height: (4.0 ± 0.05) m A doorway shall be provided in the centre of one wall, and no other wall shall have any openings that allow ventilation. The doorway shall have the following dimensions: (a) width: (0.8 ± 0.05) m (b) height: (2.8 ± 0.05) m The room may be located indoors or outdoors.
Specimen position	The specimens are either mounted as a self-supporting structure or as a wall/ceiling panel fixed to a supporting structure which is either inside or outside the room. The structure should be fire protected if this is done in practice.
Ignition source	The burner is placed on the floor in a corner directly opposite the doorway wall in contact with the specimens. If there is a supporting structure in the corner, the burner is positioned at the back wall at the nearest joint (see standard for more instructions). The heat output from the burner is 100 kW for the first 5 minutes of the test, 300 kW for the subsequent 5 minutes, increasing to 600 kW for the remaining 5 minutes test duration if ignition and sustained burning of the test specimens has not occurred.
Test duration	Until flashover, or if the structure collapses, or max 15 minutes.
Conclusions	A number of parameters relating to a room fire such as temperature, flame spread and structural failure. HRR and SPR measurements are not mandatory but can be done as an option. Larger hoods than the ISO 9705 hood are then, however, necessary.

6.3.2   ISO/CD 13784-2 Reaction to the fire tests – Scale test for industrial sandwich panels – Part 2: Large-scale test

Constructions made of sandwich panels may reach a considerable size. Therefore a large-scale test is also needed which can represent fire scenarios in larger industrial buildings. ISO/CD 13784-2 describes such a test set-up, see Fig. 6.3. The test specifications are given in Table 6.2.

6.4   Overview of major test methods and approaches used by insurance companies

Many insurance companies have established procedures to approve the use of sandwich panels in, for example, industrial buildings. This is as a consequence of the large losses occurred in many applications such as food factories, etc. Two of the most widely used procedures are those developed by the LPC (Loss Prevention Council, UK) and FM global (USA) and will shortly be commented on. Also other full-scale test methods have been introduced by insurance companies or users of sandwich panels. It is not intended to go into details of the different procedures but to give a short overview of them.

6.4.1   FM approvals protocol for sandwich panels

For all insulated wall panels other than thermoplastic products, FM uses the Fire Propagation Apparatus (FPA) to measure convective flame spread on the foam core. If the foam core passes the FPA test, then FM requires the use of the UBC or ISO 9705 room test to test flame propagation on the specific assembly. However, if the foam core fails the FPA test, then FM requires the use of the room test as well as the 25 foot corner test for approval of the assembly. For some specific product assemblies such as thermoplastic products FM relies on the 25-foot corner test and the UBC or ISO 9705 room test to test flame propagation on the assembly. The assembly must pass both tests for approval.

6.4.2   The LPC approach

LPC approvals use two full-scale tests (LPS 1181-1 for external walls and roofs, and LPS 1181-2 for internal walls and ceilings) together with a design guide for the use of sandwich panels in industrial buildings. The approval is based on a room 10 × 4.5 × 3 m (length × width × height) which is constructed from the specific sandwich panel system. The ignition source is a timber crib that provides a heat source of approximately 1 MW. The standard requires that the extent of flame spread is limited and that no flashover occurs within the room.

6.4.3   Activities of the International Association of Cold Storage Constructors (IACSC)

Due to a number of serious fires, the IACSC has developed its own test method for free-standing internal structures. This test method is basically a stability test for the structure. This test is different from those developed by the insurance industry and assesses different characteristics and is therefore not entirely compatible but quite often used in cold-storage applications.

6.4.4   Other approaches

The French insurance industry has also recently established a test procedure (APSAD test, document T14) and published guidelines containing necessary technical provisions for fire safe constructions.

6.5   Recent research related to sandwich panels

This section gives a brief overview of a recently performed research project which has looked into the behaviour of sandwich panels using large-scale tests (ISO 9705 and ISO 13784 part 1) and intermediate scale tests (EN 13823). The project was funded by Nordtest, recently incorporated in the Nordic Innovation Centre.

6.5.1   Testing programme

In order to obtain the objectives of this project the following work programme was set up:

1. Selection of products for the tests
2. Room tests according to ISO 9705
3. Room tests with a free-standing set-up under a large hood
4. Comparison of the two test set-ups
5. SBI tests on the selected products
6. Comparison of the SBI with the full-scale tests results.

6.5.2   Selection and description of the specimens

From discussions with industrial partners the following selection as given in Table 6.3 was made. Table 6.3 gives a generic description of the products. As this project did not intend to compare products but only to investigate the different test procedures, the description is kept to a minimum. In order to avoid any problems of different room sizes in both set-ups or between different products it was decided to keep the thickness of the product identical in all tests.

6.5.3   Mounting of the specimens

Similar to the description of the specimens, no detailed construction information is given. It is important to note that most products have been installed by the suppliers and constructors of the products (Products B, C and D) with assistance of SP. In the case of product A the instructions of the constructor were followed

Table 6.3 Overview of the products

Product	Insulation material	Thickness of the panel (mm)	Application
A	Stone wool	100	Industrial or cooling houses
B	Polystyrene	100	Cold stores
C	PIR	100	Cold stores
D	PUR	100	Cold stores

by SP. In every case the mounting was thoroughly discussed with the supplier and the possible changes between the mounting of the specimens inside the room and the free-standing set-up were discussed in detail.

ISO 9705 room tests

All panels were mounted in an identical manner inside the ISO 9705 room and according to the instructions as given in ISO 9705. Figure 6.1 shows a schematic drawing of the ISO 9705 room. The mounting was as follows. First the wall panels were mounted. On top of these wall panels the roof panel was mounted. Specific corner profiles and/or sealants were used depending on the practice of the supplier of the product. The ends of the panels close to the door wall were sealed with ceramic or mineral wool to avoid ignition of exposed core material.

Free-standing set-up

In the free-standing set-up the product formed the walls and ceiling of the test rig. This meant that the wall with the door opening was also constructed with the same product as the other walls. In certain cases supporting structures were necessary as this is accepted in practice. The outer dimensions were identical to the inner dimensions of the ISO 9705 room, which resulted in a constant volume in both test procedures. In Table 6.4 an overview is given of which products were used with or without supporting structure. The actual way of mounting was decided by the supplier of the product.

The supporting steel frame was constructed of U profiles, constructed at the outside of the room and in which the products could be fixed by self-drilling and tapping screws or by means of bolts. For each product those corner profiles and sealants were used as instructed by the supplier of the product. In Fig. 6.2 an example of mounting such a system can be seen. The picture was taken during a test and shows the free-standing sandwich panel construction. The fire and smoke plume entering the large hood can also be observed.

Table 6.4 Overview of mounting free-standing set-up

Product	Use of supporting frame	Fixed to supporting frame	Type of supporting frame
A	Yes	Yes	Steel
B	Yes	No	Steel
C	No	No	Not applicable
D	Yes	Yes	Steel

6.5.4   Full-scale test results

An overview of the test results is given in Table 6.5 and Fig. 6.4. For clarification it should be noted that in Fig. 6.4 material B and C reach flashover almost at the same time in the free-standing test set-up. This might not be clear from the graphs.

6.5.5   SBI test results

In order to get a first impression of the link between the two test methods and the SBI method (EN 13823) a number of SBI tests were performed. The results are given in Table 6.6. If the additional tests of ISO 1182 and ISO1716 were performed, material A would be A1 or A2.

6.5.6   Comparison of SBI test results with full-scale tests

Table 6.7 gives a comparison between the EUROCLASS (excluding smoke and droplets) determined by means of the different tests.

6.5.7   Observations

1. The free-standing set-up allows an identical mounting as in practice both with and without supporting structure. For this project a large calorimeter allowed for HRR and SPR measurements. If such a calorimeter is not present extended hoods can be used and this was applied successfully at SP

Table 6.5 Comparison of test results

	Product A	Product B	Product C	Product D
Flashover time (time to 1000 kW) (min:s)				
ISO 9705	No FO	6:54	No FO	14:42
Free-standing	No FO	12:08	11:44	7:04
Max HRR (kW) 0–20 min (excl. burner)				
ISO 9705	74	>900*	317	>700*
Free-standing	195	>700*	>700*	>900*
Euroclass according to ISO 9705[2]				
ISO 9705	≥B	D	≥B	C
Free-standing	≥B	C	C	D

FO = Flashover.
* = HRR higher than 1 MW including burner (flashover cases).

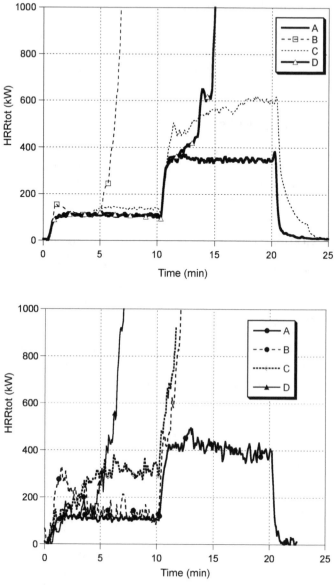

6.4 Heat release rate incl. burner (non-smoothed data) of all products in the ISO 9705 tests (top) and in the free-standing test set-up (bottom).

within another project. Such extended hoods are also described in ISO 13784 part 1.

2. The free-standing test procedure can be pointed out as being more severe than the ISO 9705 test procedure with the exception of one product. This product differed a great deal in mounting between the two test set-ups

Table 6.6 Test results according to the SBI method

Product	FIGRA	THR 600s	SMOGRA	TSP 600s	Droplets	Expected Euroclass*
A**	0	0.4	4	46	No	Bs1d0
B	92	1.8	38	87	No	Bs2d0
C	60	5.5	30	377	No	Bs3d0
D	36	2.9	16	164	No	Bs2d0

*Not taking into account the results of the small flame test (EN ISO 11925-2).
** average of three test results. For one test the TSP600s was 51 m^2.

(improvement of the joints in the free-standing test set-up). One should take into account that no exactly similar mounting is possible in both set-ups. It is clear that for most products an earlier flashover is observed.

3. In both test procedures varied fire behaviour was observed from early flashovers to no flashover at all.

4. Finishing of the joints between different sandwich panels and mounting of the panels are important factors influencing the fire behaviour of the panels. For this reason, the free-standing set-up is more suitable as it allows an identical mounting and finishing of the joint as in practice.

5. It is also observed that mounting a free-standing set-up was easier than in the ISO 9705 room. Similarly, dismounting was much easier than in the ISO 9705.

6. A first comparison with the SBI test results shows that there is little correlation between the test results of both full-scale test procedures and the SBI test results. This can be explained by the limited capabilities of the SBI method to mimic the influence of the mechanical behaviour of panels and joints. All products show limited burning in the SBI test as the joints remain closed and no deformation occurs. All products are classified in a higher class than in the ISO 9705 test. In one case the result is the same while in another case the SBI test results is two classes better compared to the reference scenario. Hence it is questionable if sandwich panels should be classified solely based on the SBI and the small flame test. It would no doubt also be wrong to change the mounting conditions in the SBI method so that they are no longer reflecting end-use conditions. Examples of such

Table 6.7 Euroclass according to the different test methods

	Product A	Product B	Product C	Product D
ISO 9705	≥B	D	≥B	C
Free-standing	≥B	C	C	D
SBI	Bs1d0	Bs2d0	Bs3d0	Bs2d0

changes would leave the top ends open or not using joint profiles. Such solutions would be unfair to systems which obtain good fire performance by means of advanced joint and mounting systems.

6.5.8 Conclusions

Simulating end-use conditions in a realistic way is very important when evaluating the burning behaviour of sandwich panels. Based on the results of a series of tests conducted both in free-standing test rooms, the ISO 9705 room test configuration and according to the SBI test protocol, a number of conclusions can be drawn.

- Mounting technique, especially joint behaviour, is an important factor for the fire behaviour. Any test used for evaluation of the fire risk of sandwich panels should be able to cover the so-called end-use conditions.
- A first comparison with the SBI test shows poor correlation with both full-scale set-ups. The products behave in most cases much better in the intermediate-scale SBI test. Use of the SBI method for sandwich panels is therefore questionable.
- The free-standing set-up allows for correct mounting in end-use conditions of panels and allows easier mounting and dismounting.
- The free-standing test set-up used for the project allows for the measurement of the heat release rate (HRR) and smoke production rate (SPR). This means that the procedure for HRR and SPR measurements as mentioned in ISO 13784 part 1 can be used.
- The free-standing test set-up is more severe than the ISO 9705 room tests.

6.6   Areas for further progress in testing of sandwich panels

The areas of further progress in testing of sandwich panels can be listed as follows:

1. Harmonisation and reduction of the number of full-scale tests in order to reduce the burden on industry for type approval both for prescriptive regulations and for insurance approvals.
2. Development of screening methods in order to get a good prediction of the full-scale behaviour of the panels in a smaller scale.
3. Development of reliable methods to allow for substitution of components in a sandwich panels system without the need for re-running full-scale tests.
4. Development of methods for advanced modelling of the fire behaviour of sandwich panels by means of CFD and FEM calculations which would be able to be used within fire performance based engineering.

6.7   References and bibliography

ASTM E119-00a Standard Test Methods for Fire Tests of Building Construction and Materials, ASTM.

BRE Report on verification of reaction to fire performance of sandwich panels used in the food industry, BRE, January 2001.

Development of a test procedure for sandwich panels using ISO 9705 philosophy, Nordtest project nr 1432-99, Patrick Van Hees, Patrik Johansson, SP Rapport 2000:26, Fire Technology, Borås 2001.

DRAFT prEN 14509 Self-supporting double skin metal faced insulating sandwich panels – Factory made products – Specification, CEN November 2004.

EN 1363-1 Fire Resistance test part 1 – General requirements, CEN.

EN 13501-1, Fire classification of construction products and building elements – Part 1: Classification using test data from reaction to fire tests, CEN, 2002.

EN 13501-2, Fire classification of construction products and building elements – Part 1: Classification using test data from fire resistance test, excluding ventilation services, CEN.

EN 13823 Reaction to fire tests for building products – Building products excluding floorings – exposed to the thermal attack by a single burning item, CEN, 2002.

Full scale test report of a sandwich panel according to ISO 13784 Part 1, test conducted at the General Assembly of Panama International, Borås June 2001.

Full scale tests on free standing industrial sandwich panels and comparison with ISO 9705 test, Patrick Van Hees, Patrik Johansson, Proceedings of Fire and Materials 2001 conference, San Francisco, Interscience Communications, 2001.

ISO 9705:1993(E), Fire Tests – Full-scale room test for surface products, ISO 1993.

ISO 13784 Part 1, Fire tests Reaction to Fire Part 1: Small-scale room test for sandwich panel products for industrial buildings, 2002.

ISO 13784 Part 2, Fire tests Reaction to Fire Part 1: Large-scale room test for sandwich panel products for industrial buildings, 2002.

NT FIRE 025 Edition 2 Surface Products – Room fire tests in full-scale, Nordtest 1991.

Serious fires in food industry premises during 1995, Fire prevention 302, September 1997.

Smith, D., Marshall, N., Shaw, K. and Colwell, S., Correlating large-scale fire performance with the single burning item test, *Interflam 2001 Proceedings*, Interscience Communications, London, 2001.

Sundström, B., Thureson, P. and Van Hees, P. Results and Analysis from Fire Tests of Building Products in ISO 9705, the Room/Corner Tester, report from the SBI Research Programme, SP Report 1998:11, Borås 1998.

Van Hees, Patrick and Johansson, Patrik, The need for full-scale testing of sandwich panels, comparison of full-scale tests and intermediate-scale tests, *Interflam 2001 Proceedings*, Interscience Communications, London, 2001.

7

Flammability tests for upholstered furniture and mattresses

C M F L E I S C H M A N N , University of Canterbury, New Zealand

7.1   Introduction

When a fire occurs within a residential dwelling, there are few objects that have the potential to bring about untenable conditions as swiftly as upholstered furniture. In the worst case scenario, the Heat Release Rate (HRR) of an upholstered furniture item can reach values of 3 MW in a very short period of time (3–5 minutes) following ignition.[1] Furthermore, it is not only the heat given off during the growth period, but also the toxic combustion products (primarily CO) produced that overwhelm the occupants. These coupled effects are directly linked to the materials used in the construction of the furniture, and in particular the soft combustibles (covering and padding materials).

This chapter describes the current level of understanding of the fire behaviour of upholstered furniture and mattresses. The first section will briefly describe the US fire statistics which clearly show why we need to research the combustion behaviour of upholstered furniture. We will proceed with a description of how upholstered chairs burn once the fire is greater than about 20 kW. This description is intended to give the reader a better understanding of how a burning chair evolves and how changing foam and fabric can make a significant difference in the combustion behaviour. Currently, there are several mandatory and voluntary standard test methods available for furniture and mattresses. The most relevant methods will be briefly described with the appropriate references given.

Recently a great deal of research has been completed on upholstered furniture and mattresses conducted with support from the European Commission. This study, Combustion Behaviour of Upholstered Furniture (CBUF) incorporated both extensive experimental research as well as enhanced fire modelling. The results of this research are the basis for the discussion on modelling upholstered furniture fires in this chapter. Also included is a brief section highlighting where additional information can be found.

7.2   The role of upholstered furniture and mattresses in fires

In the United States between 1994 and 1998 inclusive, there were 14,850 fire deaths in 1,822,700 residential structure fires from non-incendiary and non-suspicious fires.[2] This is an average death rate of 2970 fire deaths per year; in approximately 2/3 of these fatal fires, the first material ignited was reported. Figure 7.1 shows a breakdown of the average number of fire deaths per year versus the first material ignited. In these fatal fires, for 32% of the fires, the first material ignited was upholstered furniture and for 25% of the fires, the first material ignited was mattress/bedding. This clearly shows the importance of upholstered furniture and mattress/bedding in fatal fires.

In addition to the total number of fires, the individual ignition scenarios are also of interest when assessing the hazards associated with upholstered furniture and mattress/bedding fires. Research has shown that the fire hazard of upholstered furniture and mattress/bedding fires is a function of the ignition source. Figure 7.2 shows that smoking material is clearly the most common ignition source when upholstered furniture or mattress/bedding is the first item ignited although open flames and other sources cannot be ignored. Clearly, upholstered furniture and mattress/bedding fires play a significant role in the

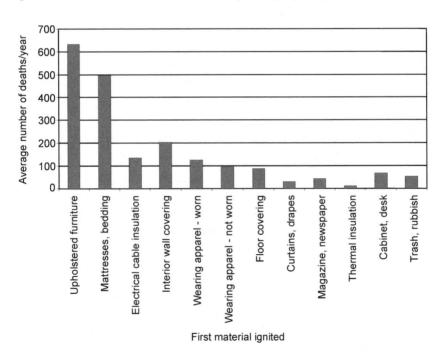

7.1 Average number of fatalities in residential fires broken down by first material ignited.

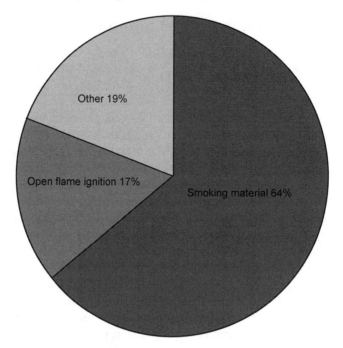

7.2 Source of ignition for fatal fires in which upholstered furniture or bedding/mattress was the first item ignited.

fatal fire statistics and further research and potentially more regulation is warranted to reduce the hazard posed by upholstered furniture and mattress/bedding fires.

7.3   Description of the burning behaviour of upholstered furniture[3]

This section discusses the results from a series of experiments concluded on domestic furniture. In this study, 55 one- and two-seater chairs modelled after the CBUF standard chairs were burned. The following description is based on the ignition of an upholstered chair from a 30 kW ignition burner centred between the armrests, 50 mm from the back cushion and placed 25 mm above the seat cushion in accordance with CAL TB 133.[4] The burner is left on for two minutes and then removed. Previous research has shown that the location and intensity of the ignition source has little impact on the peak heat release rate, however, the ignitability and time to peak were greatly affected.[5] The application of the description given here is appropriate once radiation becomes the dominant form of heat transfer governing the burning rate. Thus the ignition scenario is not included in this description. A sophisticated load cell system was used to isolate the mass of the burning chair from the mass of the burning foam/

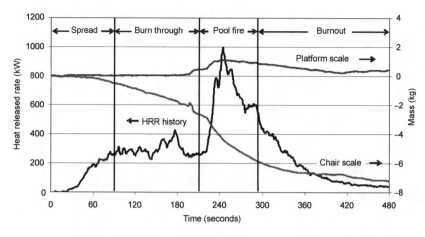

7.3 Mass and HRR histories for the wool fabric covering domestic furniture foam – showing the four general phases of burning: (1) spread; (2) burn through; (3) pool fire, and (4) burnout.

fabric pool that collects on the platform below the chair. The burning chair was supported on an insulated steel frame, suspended 25 mm above the platform. The frame was physically isolated from the platform and supported on two load cells to continuously record the mass of the chair. The platform below the chair was supported on four load cells, one on each corner, to record the weight of the foam and fabric that fell from the chair. The heat release rate from the burning chair was measured using oxygen depletion calorimetery.

After flaming ignition has occurred, the general burning behaviour of an upholstered furniture item, ignited on the seat, can be broken down into four main phases: spread; burn through; pool fire; and burnout. Figure 7.3 shows the mass and Heat Release Rate (HRR) histories for a chair with domestic polyurethane foam covered by wool fabric. The four phases of burning have been highlighted on the plot.

The 'spread' phase is the time taken for the flame to spread over the surfaces of the seat cushion, up the back cushion and onto the armrest assemblies. Figure 7.4 is a series of photographs taken during the same experiment shown in Fig. 7.3, i.e., domestic foam covered with wool fabric. The clock seen in the images is reported in minutes:seconds and includes the 3-minute baseline, such that ignition was at 3:00 (clock time). Figure 7.4(a) shows a standard chair configuration 60 seconds after ignition. As the seating fire becomes established it is then able to propagate to the arm assemblies, and quickly engulfs all of the soft combustible components in the seating area, as seen in Fig. 7.4(b). However, the fire is restricted to just the seating region of the chair, with the lower half of the chair still unaffected by the fire. The foam and fabric from the seat back cushion that does not immediately burn will melt down onto the seat cushion where it starts to form a pool within the seat. For the more heat-resistant fabrics,

7.4 Standard CBUF Series 2 chair with wool fabric covering domestic furniture polyurethane foam. Photos (a) and (b) show the characteristics of the spread phase, photo (c) shows the characteristics of the burn-through phase. Photos (d) and (e) show pool-fire phase. Photo (f) shows the burnout phase.

the seat cushion, underside fabric and sides, form a fabric pan that contains the molten material. In addition to the seat back foam and fabric, the seat cushion is also contributing to the 'seat pool fire'. All of the soft combustibles in the seating region that do not immediately burn, melt and accumulate on the seat cushion where they pool before eventually burning. During the spread phase there was little measurable mass of molten fuel on the platform below the chair, but the HRR tends to grow steadily as the fire spreads over the surface of the seating area.

The 'burn-through' phase begins once the flame has spread over all of the surfaces including seat, back, and arm rests. Burn through ends when the fabric on the underside of the seat cushion and/or webbing fails spilling the molten foam and fabric onto the platform beneath the chair. The seat fire is present during this phase, as shown in Fig. 7.4(c). As this phase proceeds, the seat cushion starts to show the initial signs of failing through the gradual increase in mass measured on the platform load cell as seen in Fig. 7.3. During the burn-through phase, the fire enters into a relatively steady burning region as seen in Fig. 7.3 between 80 and 200 seconds. At 120 s the ignition burner is switched off sometimes causing a decline in the HRR. The length of time before the seat pool fire burns though the seat fabric and webbing is strongly dependent on the fabric material. Thermosetting fabrics, that char, were supported by the webbing and were very effective at containing the molten foam fire in the seat region.

From the experimental observations, a combination of features has to occur before the pool-fire phase could proceed. Firstly, the seat cushion and/or webbing must have failed, contributing significantly to the initial stages of the pool fire, and allowing the pool fire to become established below the chair. Secondly, some or all of the fabric on the front, back and side walls must have failed, allowing a plentiful supply of air to the pool fire. Figure 7.4(d) shows the transition from the burn-through to pool-fire phases. The photo shown in Fig. 7.4(d) was taken just after the molten fuel spilled onto the platform below the chair. This is evident by close examination of the photograph; note the remaining fabric on the front lower half of the chair is being backlit from the pool fire on the platform. Figure 7.4(e) shows the chair 225 seconds after ignition as the fire is growing to the peak heat release rate. The rapid growth can be seen in both the HRR and platform curves in Fig. 7.3. As shown in Fig. 7.3, this phase was predominately short lived, of the order of seconds not minutes, and the HRR and platform load cells quickly climbed to peak values before starting the burnout phase.

The burnout phase takes over, the HRR and platform load cell indicate the spilled molten mass starts to decline, once the bulk of the soft combustibles are consumed. Typically, there are no soft combustibles present on the burning frame, and any that remain are pooled on the platform below the furniture item. Figure 7.4(f) shows the chair 360 seconds after ignition where only the remains of the fuel on the platform are burning. The burnout phase can be seen in Fig. 7.3 as the characteristic gradual decay in the HRR curve and also the platform scale. This represents the depletion of the pool fire and the slow burning of the timber frame.

7.3.1 Alternative scenarios

The physical description of the burning phases given above is considered typical for furniture. However, for certain foam-fabric combinations, the description requires modification. The alternatives are the result of either readily combustible materials or highly resistant materials.

Readily combustible materials

For the more readily combustible combinations, i.e., thermoplastic fabrics covering non-combustion modified foams, there may be bypassing of the burn-through phase as a distinct phase because the chair quickly transitions into the pool-fire phase. The flame quickly propagates around the seating area of the chair due to the fabrics poor thermal resistance. The fabric rapidly melts and exposes the foam directly to the flame. The low ignition resistance of the materials means the seat pool fire quickly burns through the seat fabric and webbing, discharging the pool directly onto the platform. Once the pool fire is established on the platform scale, the fabric on the lower half of the chair is compromised, allowing free flow of air into the pool fire. The HRR for the chair shows a steep rise to peak values before the ignition burner is turned off. Figure 7.5 shows the heat release rate, chair mass, and platform mass histories for a standard chair with polypropylene fabric over public auditorium polyurethane foam and includes the highlighted burning phases. For this case, there is no distinguishable burn-through phase shown in Fig. 7.5.

Figure 7.6 is a series of photographs taken during the experiment shown in Fig. 7.5. The photo in Fig. 7.6(a), taken 45 seconds after ignition, shows the fire spreading over the chair and along the top of the armrests during the spread phase. The spread phase in Fig. 7.5 shows almost continuous growth up to the pool-fire phase. The fire developed so rapidly that the integrity of the seat failed quickly (before the burner was extinguished) spilling the contents of the seat

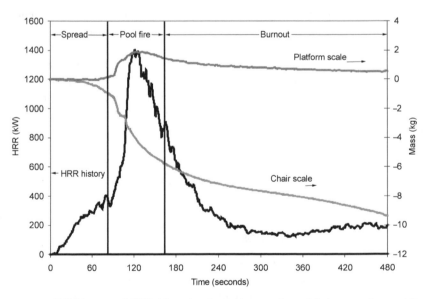

7.5 Mass and HRR histories for polypropylene fabric covering public auditorium foam – showing spread, pool fire, and burnout, however, burn through is effectively non-existent.

7.6 Photographs showing different phases of burning for a polypropylene fabric covering public auditorium foam. (a) spread phase, (b) pool-fire phase, (c) burnout phase.

onto the platform below. Once the integrity of the seat was compromised the fire entered into the pool-fire phase and quickly reached its peak heat release rate of 1400 kW at 120 seconds. Coincidentally, the burner gas flow was shut off around the same time. Figure 7.6(b) shows the fire 120 seconds after ignition during its peak heat release rate. The pool-fire phase was short lived and the fire dropped into the burnout phase as seen in the HRR plot in Fig. 7.5 and the photo taken 240 seconds after ignition in Fig. 7.6(c).

Fire-resistive materials

A chair with excellent fire-resistive properties may not be able to support flaming once the gas flow to the burner is shut off and may self-extinguish during the burn-through phase. This behaviour was observed for two out of three identical chairs with wool fabric covering (combustion modified) aviation foam. However, when a third wool fabric chair coupled with aviation foam was tested, the fire did eventually burn through and ignite the pool fire below. The chair reached its peak HRR at 594 s after burning for more then 300 s with a heat release rate less than 5 kW.

Figure 7.7 shows the heat release rate and mass histories for a chair constructed with combustion modified foam covered with wool fabric along with the three burning phases observed in this experiment. In this case, the chair had the characteristic spread, burn-through, and burnout phases. The spread phase lasted beyond 120 seconds when the gas flow to the burner was turned off. Once the gas flow to the burner was turned off, the HRR continued to decline as seen in Fig. 7.7. At approximately 150 seconds a portion of the fabric on the underside of the seat cushion failed spilling molten fuel onto the platform.

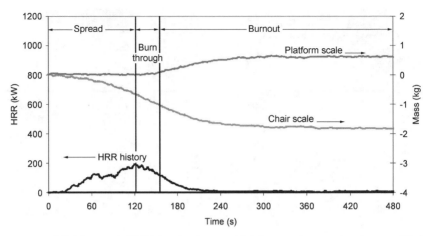

7.7 Mass and HRR histories for wool fabric covering combustion modified foam showing spread, burn through and burnout, however, pool fire was effectively non-existent because the pool never ignited.

However, the pool fire phase never eventuated as the molten material spilled onto the platform but never ignited. The chair self-extinguished around 360 s.

Figure 7.8 shows a series of photographs taken during each phase of the experiment presented in Fig. 7.7. Figure 7.8(a), taken 45 seconds after ignition, shows the fire spread across the seat, up the back cushion, and onto the arm rest in a similar manner to the most combustible chairs, however, the development is noticeably slower. Figure 7.8(b) is taken 120 seconds after ignition when the burner is shut off. It took another 30 seconds before any measurable mass is detected on the platform scale. Figure 7.8(c) shows the chair 180 seconds after ignition where the fire is clearly into its burnout phase. Figure 7.8(d) is the post-fire photo showing large portions of the seat and back foam still in place although the wool fabric has burned through the fabric on the underside of the seat. The fabric on the back of the chair is still intact although somewhat scorched.

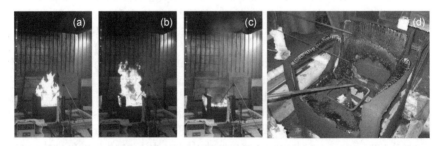

7.8 Series of photos showing a wool fabric covering combustion modified aviation foam. (a) spread phase, (b) burn-through phase, (c) burnout phase, (d) post-fire showing large portions of the seat and back foam still in place.

7.4   Standard test methods for upholstered furniture and mattresses

There is a broad spectrum of experimental methods for upholstered furniture and mattresses covering a wide range of fire scenarios. Test standards can be broadly divided into three areas based on the item used in the test. For example, some standards such as TB 117 *Requirements, Test Procedures and Apparatus for Testing the Flame Retardance of Resilient Filling Materials Used in Upholstered Furniture*[6] use only individual components, namely, the padding material and fabric. The test is a small bench-scale test that exposes the sample to a small laboratory flame. In this test, individual components are exposed to the burner flame, but there is no radiant feedback as in a real test. Therefore, this test is not representative of a real fire. A separate procedure is also included in the standard for exposing the padding material to a burning cigarette. However, this style of test has proved to be a poor predictor of the fire performance of the finished furniture item and is currently under review. Other standards such as BS 5852:1990 *Methods of Test for Assessment of Ignitability of Upholstered Seating by Smouldering and Flaming Ignition Source*[7] relies on mock-ups constructed using the actual covering and padding materials together and use simple pass/fail criteria based on ignition of the mock-up. More detail about BS 5852 is given below.

The more applicable test standards involve testing of actual furniture items and quantifying the actual fire hazard with oxygen depletion calorimetry (ODC). TB 133 *Flammability Test Procedure for Seating Furniture for Use in Public Occupancies*[4] is an example of a regulatory standard used in California, USA that uses ODC as part of the pass/fail criteria. The state of California Bureau of Home Furnishing and Thermal Insulation (CBHFTI) has been a vanguard in terms of developing regulations for controlling the flammability of upholstered furniture and mattresses. At the time of writing this chapter, the Consumer Product Safety Commission (CPSC) is involved in a major research effort to determine the most appropriate way to reduce fire deaths from upholstered furniture and mattress fires on a national level within the USA. The CPSC is working with the CBHFTI using many of their standards as a starting point for the development of new national standards.

Table 7.1 gives a summary of many of the more common experimental methods used for testing upholstered furniture. The table summarises the sample type, ignition sources, and pass/fail criteria. In several cases, an identical or slightly modified version of the standards is available from other standards-writing authorities. Table 7.1 is intended to give an overview of the standards and not a comprehensive comparison. Specific details about the standards can be obtained from the actual standard which is given in references. More detailed summaries of the more relevant standards are given below. In Table 7.1 only TB 133 and TB 129 measure HRR and use it as a criterion in line with the current accepted fire test methods.

Table 7.1 Summary of upholstered furniture and mattress standard test methods comparing the type of sample, ignition sources used and acceptance criteria

Test Method	Sample type			Ignition source			Acceptance criteria						
	Component	Mock-up	Finished product	Cigarette	Small flame	Large flame	Ignition	Progressive smouldering	Char length	Flame spread	Heat release rate	Total heat release	Other
Upholstered furniture tests													
BS 5852		X	X	X	X	X	X	X		X			
TB 116			X	X			X	X	X				
TB 117	X				X				X	X			
TB 133			X			X					X	X	X[3]
Mattress tests													
BS 6807		X	X	X	X	X	X	X		X			
TB 106			X	X			X	X					
TB 603									X				
TB 121[1]			X			X							X[4]
TB 129[2]			X			X					X	X	X[5]

1. Technical Bulletin 121, *Flammability Test Procedure for Mattresses for Use in High Risk Occupancies*, California Bureau of Home Furnishings and Thermal Insulation, North Highlands, CA, USA, 1980.
2. Technical Bulletin 129, *Flammability Test Procedure for Mattresses for Use in Public Buildings*, California Bureau of Home Furnishings and Thermal Insulation, North Highlands, CA, USA, 1992.
3. Pass/fail criteria include ceiling temperature, temperature 1.2 m above floor, smoke opacity, CO concentration and mass loss.
4. Test criterion also includes total mass loss, temperature above mattress, and CO concentration in the compartment.
5. Pass/fail criteria also include mass loss.

7.4.1 Standard test for assessing upholstered furniture flammability

BS 5852:1990 Methods of Test for Assessment of the Ignitability of Upholstered Seating by Smouldering and Flaming Ignition Sources[7]

The objective of the standard is to measure the ignitability of upholstered furniture composites and/or complete pieces of furniture. The standard incorporates eight ignition sources including a smouldering cigarette, three butane flames, and four burning wooden cribs. The strength of the flaming sources of ignition are designed to approximate a lighted match or to be as severe as the burning of four double sheets of a full-size newspaper. The standard allows for the composites of the furniture materials to be tested rather then the entire piece of furniture. Figure 7.9 shows the geometry defined in the standard. For the cigarette and the small open flame test, the horizontal cushion is 150 mm by 450 mm and the vertical cushion is 300 mm high and 450 mm wide. For the wooden crib experiments the horizontal cushion is 300 mm by 450 mm and the vertical is 450 mm high and 450 mm wide.

For smouldering ignition, the lit cigarette is placed at the junction between the vertical and horizontal cushions. The test lasts for 60 minutes in which there can be no ignition or progressive smouldering observed. If there is no observable ignition or progressive smouldering observed then the test is repeated at another location on the same sample. For the small butane flame test the flame height and exposure time range from approximately 35 mm for 20 seconds (source 1), 145 mm for 40 seconds (source 2) to 240 mm for 70 seconds (source 3). The flame is placed at the junction between the vertical and horizontal cushion. The same pass/fail criteria apply, i.e. no observable ignition or progressive

7.9 Photograph showing BS 5852 upholstered furniture mockup during experiments, ignition from small butane flame.

smouldering. If the sample fails to ignite or progressively smoulder then the test is repeated. If the sample fails to ignite or progressively smoulder in the repeated test then the sample is considered to have 'no ignition'. For the wooden cribs (ignition source 4 to 7), a progressively larger wood crib is placed on a new sample for each test. Source 4 (the smallest crib) is 40 mm square assembled with five stacked layers of 6.5 mm sticks. Source 5 is identical to source 4 but has ten layers of sticks. Source 6 is 80 mm square and assembled from four layers of 12.5 mm sticks. Source 7, the largest crib, is identical to source 6 but is nine layers high. The crib is placed on the horizontal cushion and against the vertical cushion. The crib tests are 60 minutes long in which there can be no observed flaming ten minutes after the crib ignition for sources 4 and 5 or 13 minutes after ignition for the larger cribs used for source 6 and 7. BS 5852 has been replaced with BSEN 1021 which is an adaptation of ISO 8191[8] but the basic principles remain the same.

TB 133 Flammability Test Procedure for Seating Furniture for Use in Public Occupancies[4]

The objective of TB 133 is to test seating for use in occupancies that are identified for public use, such as prisons, health care facilities, public auditoriums, and hotels. It is not intended for residential use. The test is conducted in a 3 m by 3.6 m by 3 m high compartment. The only ventilation is a 0.95 m wide by 2 m high door. The test can be conducted in a smaller compartment such as an ISO 9705[9] room when equivalent test results can be demonstrated. A full-scale furniture item is placed in the corner opposite the door. Ignition is by a 250 m square tube burner generating approximately 18 kW for 80 seconds. For furniture items 1 m or less the burner is centred on the seat area, 25 mm above the seat surface 50 mm from the rear vertical cushion. For items larger than 1 m, the tube burner is placed 125 mm from the left arm of the item. The test is allowed to continue until all combustion has ceased, one hour has elapsed, or flameover or flashover appears to be inevitable. The pass/fail criteria allows for either measurements within the room or from oxygen depletion calorimetry. The criteria vary slightly depending on the measurements used. The criteria are summarised here, however, further detail on the pass/fail criteria can be found in the actual standard.[4] The furniture item fails if any of the following criteria is exceeded within the room:

1. Temperature increase of 111 °C or greater at ceiling thermocouple
2. Temperature rise of 28 °C or greater at the 1.2 m thermocouple
3. Greater than 75% opacity at the 1.2 m high smoke opacity meter
4. CO concentration in the room of 1000 ppm or greater for five minutes
5. Weight loss due to combustion of 1.36 kg or greater in the first ten minutes.

The furniture item fails if any of the following criteria is exceeded within the room test using ODC:

1. Maximum heat release rate of 80 kW or greater
2. Total heat release rate of 25 MJ or greater in the first ten minutes
3. Greater than 75% opacity at the 1.2 m high smoke opacity meter
4. CO concentration in the room of 1000 ppm or greater for five minutes.

7.4.2   Standard test of mattress

CAL TB 106 Requirements, Test Procedures and Apparatus for Testing the Resistance of a Mattress or Mattress Pad to Combustion Which May Result from a Smouldering Cigarette (Federal Standard 16 CFR 1632)[10]

The objective of TB 106 is to assess the resistance to ignition of a mattress from a burning cigarette. The test procedure requires a finished product ready for sale to the consumer. The procedure requires that three burning cigarettes be placed on the smooth surface of the mattress, three cigarettes along the tape edge, three cigarettes on quilted locations and three burning cigarettes on tufted locations. Tests are carried out with the cigarettes on the bare mattress and sandwiched between two cotton sheets over the mattress. The mattress fails the test if any of the following conditions occur:

- obvious flaming combustion occurs
- smouldering combustion continues for more than five minutes after the cigarette has extinguished
- char develops more than 50 mm in any direction from the cigarette.

CAL TB 603 Requirements and Test Procedure for Resistance of a Mattress/Box Spring Set to a Large Open-Flame[11]

The objective of this standard is to determine the burning behaviour of mattress plus foundation. Derived from extensive research conducted by the National Institute of Standards and Technology (NIST), the mattress and the foundation are placed on top of a bed frame that sits over a catch surface. The mattress is exposed to two gas burners one on top and the other on the side, as seen in Fig. 7.10. The burners are designed to mimic the local heat flux imposed on a mattress/foundation from burning bedclothes. The top burner flows 19 kW for 70 s and the side burner flows 10 kW for 50 s. The test can be conducted in one of three configurations: open calorimeter, 2.4 m × 3.6 m × 2.4 m high room, or 3.05 m by 3.66 m by 2.44 m high room. A large extraction system with oxygen depletion calorimetery is required. The test can last 30 minutes unless all signs of burning have ceased or the fire develops to such a size that it must be extinguished. The mattress is considered to have failed if the peak heat released exceeds 200 kW or the total heat release rate exceeds 25 MJ in the first ten minutes.

7.10 Photograph showing the burner location for CAL TB 603 test standard for mattress/foundation exposed to open flame.

BS 6807 Assessment of the Ignitability of Mattresses, Divans, and Bed Bases with Primary and Secondary Sources of Ignition[12]

The objective of BS 6807 is to assess the ignitability of a mattress. The test procedure uses the same ignition sources: a cigarette, small gas flames and wood cribs as specified in BS 5852. The standard specifies procedures for primary ignition of the mattress, i.e., the ignition source placed directly onto the surface of the mattress and for secondary ignition where the ignition source is placed on the bedcovers. Bedcovers includes sheets, blankets, bedspreads, valances, quilts and mattress covers.

For smouldering ignition, the lit cigarette is placed on top of the mattress. For the flaming sources the standard requires testing with the source on top of the mattress and underneath the mattress. The precise location of the ignition source is mapped out in detail within the standard. The same pass/fail criteria used in BS 5852 applies, i.e., no observable ignition or progressive smouldering as defined in the standard.

7.5   Modelling of furniture fires

The ultimate goal of furniture/bedding fire modelling is to be able to predict the fire behaviour of the furnishings with a minimal amount of experimentally derived data. In an ideal world the user could simply go to a database and download the necessary properties for a desired furniture item and run their model to predict the fire growth rate, perform a forensic analysis, or carry out a hazard assessment based on the results from a Computational Fluid Dynamics model (CFD) or zone model. For details on compartment fire modelling see references 13, 14 and 15.

Unfortunately, the modelling of furniture and bedding fires is still in its infancy. The burning behaviour of upholstered furniture and bedding creates significant challenges for fire modellers beyond simply the solid-phase burning characteristics. Upholstered furniture and mattresses are often complex composites, constructed from at least three materials, padding, covering and framing. The problem is further complicated when the most commonly used soft materials and common construction techniques lead to complex burning phenomena including burn through, tunnelling, melting, flowing, and forming pool fires remote from their original locations. Modelling efforts have been developed using experimental correlations[16] and physics-based models to predict the burning behaviour of upholstered furniture.[17,18] In these models, the researchers have had to over-simplify the problem leaving out many of the important details of the burning item. However, useful information can still be obtained from these models and simple correlations that can be useful in developing less combustible furniture.

7.5.1   Combustion Behaviour of Upholstered Furniture (CBUF)[19]

The CBUF research programme was conducted by a consortium of research organisations one each from Belgium, Denmark, Finland, France, Germany, Italy, Sweden, and three from the UK. The programme is the most extensive study on upholstered furniture and bedding ever conducted. There were 71 room tests, 154 furniture calorimeter tests, and 1098 composites tested in the cone calorimeter. This does not include all of the preliminary experiments to establish the ignitability, effects of room size, ventilation, test procedures, repeatability and reproducibility. Details about the CBUF programme can be found in the comprehensive final report given in reference 19. The primary findings of the research are summarised below:

1.   The Heat Release Rate from the full-scale calorimeter experiments on full-scale furniture items was identified as the primary measure of the fire hazard from upholstered furniture. The test procedure for the furniture calorimeter was substantially similar to the TB 133. The TB 133 tube burner was used as the ignition source; however, the HRR of the burner was increased to 30 kW for 120 s. This thermal assault on the furniture item assured ignition of most items.

2.   Three models were developed for predicting the HRR from an item of upholstered furniture burned in the furniture calorimeter using the results from foam-fabric composites in the cone calorimeter.

    (a)   Model I is a set of correlations developed to predict the peak HRR, time to peak, and time to untenable conditions in an ISO 9705 room from an item of upholstered furniture burned in the furniture calorimeter. The correlations use the results from foam-fabric composites in the cone calorimeter. This correlation can help to reduce the number of full scale tests required.

(b) Model II is a combined physics/correlation model that uses an area-deconvolution model (based on an earlier flame spread model by Wickstrom and Goransson[20]) for predicting full-scale HRR time histories from results of the foam-fabric composite cone calorimeter tests and representative furniture calorimeter test.

(c) Model III is a physics-based model developed for predicting the HRR of burning mattresses only.

3. Existing room fire models can be used to predict the escape time and interface height using the full-scale HRR results. However, it is up to the regulator to determine the allowable escape time from the room of fire origin.

The CBUF Model I (described in detail in references 19 and 21) is a factor-based method that uses a series of statistically correlated factors to predict the peak HRR, total heat release, time to peak, and time to untenability. The model is an improvement on the earlier (1985) factor-based prediction from NIST.[22] The original model was examined for applicability to the CBUF items. It was found to apply only generally and tended to under-predict the more modern and varied European furniture. The study undertook further development and refinement of this model. They tested a series of differing furniture styles constructed from the same 'soft' combustible material combinations (soft being the foam, fabric, and interliner). An analysis of the results brought about several refinements from the 1985 NIST model to the CBUF Model I. Notably, the mass of soft combustibles replaced the mass of total combustibles, and the power was raised from 1 to 1.25. The time to ignition in the cone calorimeter test was seen as an important variable and was included.

The style factor also required significant change to account for the new European furniture. Incorporated in the calculation of the peak heat release rate and time to peak, the style factor accounts for the physical differences that cannot be resolved by the cone calorimeter test method including the ornate and intricate detail that can be found in some furniture. Table 7.2 provides the style factors needed in the predictive model.

7.5.2 Peak HRR

Equation 7.1 is the first correlating variable for the peak heat release rate. It was found that the partially correlating variable x_1 represented well the general trend with the exception of groupings of high peak HRR (over 1200 kW). Considering only these data points, the second correlating variable x_2 emerged as given in eqn 7.2 (the symbols together with their units are defined in section 7.7).

Table 7.2 Furniture style factors defined in the CBUF predictive model

Code	Style factor A	Style factor B	Type of furniture
1	1.0	1.0	Armchair, fully upholstered, average amount of padding
2	1.0	0.8	Sofa, two-seat
3	0.8	0.9	Sofa, three-seat
4	0.9	0.9	Armchair, fully upholstered, high amount of padding
5	1.2	0.8	Armchair, small amount of padding
6	1.0	2.50	Wingback chair
7	—	—	Office swivel chair, plastic arms (unpadded), plastic back shell
8	—	—	High-back office swivel chair, plastic arms (unpadded), plastic back, and bottom shell
9	—	—	Mattress, without inner spring
10	—	—	Mattress with inner spring
11	—	—	Mattress and box spring (divan base) set
12	0.6	0.75	Sofa-bed (convertible)
13	1.0	0.80	Armchair, fully upholstered, metal frame
14	1.0	0.75	Armless chair, seat and back cushions only
15	1.0	1.00	Two-seater, armless, seat and back cushions

$$x_1 = (m_{\text{soft}})^{1.25} \cdot (\text{style\_fac.A}) \cdot (\dot{q}''_{pk} + \dot{q}''_{300})^{0.7}(15 + t_{ig})^{-0.7} \qquad 7.1$$

$$x_2 = 880 + 500 \cdot (m_{\text{soft}})^{0.7}(\text{style\_fac.A}) \cdot \left(\frac{\Delta h_{c,eff}}{q''}\right)^{1.4} \qquad 7.2$$

The input data used in eqns 7.1 and 7.2 are from the cone calorimeter test conducted in accordance with the strict CBUF protocol. Selection rules are established, that are termed 'regimes', to determine when to use x_1 and x_2, with x_1 displaying a partial dependence. Regimes:

{1} If, $(x_1 > 115)$ or $(q'' > 70$ and $x_1 > 40)$ or (style = $\{3, 4\}$ and $x_1 > 70$) then,

$$\dot{Q}_{peak} = x_2 \qquad 7.3$$

{2} else If, $x_1 > 56$ then,

$$\dot{Q}_{peak} = 14.4 \cdot x_1 \qquad 7.4$$

{3} else,

$$\dot{Q}_{peak} = 600 + 3.77 \cdot x_1 \qquad 7.5$$

7.5.3 Total heat release

The total heat release is determined by the actual mass of the furniture item and small-scale effective heat of combustion. Differentiation is noted between the 'soft' and total combustible masses. Experimental observation reveals that the affect of a wooden frame is not seen until nearly all of the 'soft' materials are consumed. Equation 7.6 was found to represent the total heat release:

$$Q = 0.9 m_{soft} \cdot \Delta h_{e,eff} + 2.1 \left(m_{comb,total} - m_{soft} \right)^{1.5} \qquad 7.6$$

7.5.4 Time to peak

The time to peak heat release is as important as the peak heat release rate in hazard calculations. Equation 7.7 is developed to predict time to peak HRR from sustained burning (50 kW). It is recognised that often other hazard variables are maximised at or near the time of peak HRR.

$$t_{pk} = 30 + 4900 \cdot (\text{style\_fac.B}) \cdot (m_{soft})^{0.3} \cdot (\dot{q}''_{pk\,2})^{-0.5} \cdot$$
$$(\dot{q}''_{trough})^{-0.5} \cdot (t_{pk\,1} + 200)^{0.2} \qquad 7.7$$

Note that a different style factor is incorporated into the time to peak calculation.

7.5.5 Time to untenable conditions (untenability time)

Equation 7.8 was developed to predict time to untenable conditions in a standard room. Untenability time is defined as the time from 50 kW HRR to 100 °C temperature 1.1 to 1.2 m above floor level.

$$t_{UT} = 1.5 \times 10^5 (\text{style\_fac.B}) (m_{soft})^{-0.6} (\dot{q}''_{trough})^{-0.8} \cdot$$
$$(\dot{q}''_{pk\,2})^{-0.5} (t_{pk\,1} - 10)^{0.15} \qquad 7.8$$

A more complete discussion of the model and a discussion of the accuracy can be found in the original CBUF report.[19] A subsequent study in New Zealand[24] on eight different furniture items did not show as encouraging results as the original work.

7.5.6 Intrinsically safe upholstered furniture

When assessing the hazard from flaming combustion of furniture, the most important consideration is whether or not the fire will propagate over the surface of the item. For a propagating item, the fire will spread over the surfaces and consume most of the soft combustibles although often large pieces of the frame may survive. The propagating behaviour can be seen in Fig. 7.6. For non-propagating items, the fire behaviour is quite different. Once the ignition source is removed or burns out, the fire is not self-sustaining and will burn out, leaving most of the soft combustibles unburned. Based on the experimental observations in the CBUF study, non-propagating full-scale chair fires typically have a heat release rate between 20 and 100 kW. This is consistent with the pass/fail result of 80 kW used in TB 133. Comparing the results from the cone calorimeter experiments in the rigorous CBUF cone calorimeter protocol, it was found that a value of $\dot{q}''_{180} = 65 \, \text{kW/m}^2$ or less correlated well with the non-propagating items.

The total HRR and emission of toxic products for a non-propagating item is expected to give substantially similar results to those measured in the furniture calorimeter. This is based on experimental results which demonstrate that the heat and radiation levels of a non-propagating furniture item (i.e., producing less than 100 kW) are too low to cause serious injury to occupants of the room unless they are intimate with the fire. It can therefore be concluded that non-propagating furniture items are as intrinsically safe[1] as a combustible item can be.

However, for propagating items the results from the furniture calorimeter for the total HRR and toxic product emission is considered to be a lower limit of what can be expected in a real fire. This is due to the fact that a propagating item is capable of spreading to an adjacent combustible item that was not present in the test. In addition, independent analysis of the CBUF room experimental results showed that the room interaction effects can be found when the item's HRR is as low as 500 kW.[5] Room interaction can result in an increased HRR from the radiation feedback from the room and can increase the production of toxic species by reducing the available oxygen thus increasing the CO production.

7.6   Additional information

This chapter has served only as an introduction to a very complex topic with a large body of research having already been conducted but also an even greater amount of research still being required. There are some excellent books available on upholstered furniture and mattress fires; most notable is *Fire Behavior of Upholstered Furniture and Mattresses*[5] by Krasny *et al.*, which includes over 500 references. A more general reference on heat release rate from burning objects, which includes several chapters on upholstered furniture and mattresses is *Heat Release in Fires*[23] edited by Babrauskas and Grayson. The original CBUF Final Report[19] and subsequent research can be found in refs 5, 24, 25.

Some useful websites on the flammability of upholstered furniture and mattresses are:

- Combustion Behaviour of Upholstered Furniture (CBUF)
  http://www.sp.se/fire/Eng/Reaction/cbuf.htm

- Building Fire Research Laboratory – National Institute of Standards and Technology – includes copies of research reports and videos from a wide range of fire research including upholstered furniture and mattresses. http://www.bfrl.nist.gov/
- California Bureau of Furnishings and Thermal Insulation – includes copies of technical bulletins on testing furniture and mattresses. http://www.bhfti.ca.gov/index.html
- Consumer Product Safety Commission – contains up-to-date information on developments in possible regulations on the flammability testing of furniture and mattresses. http://www.cpsc.gov/
- Upholstered Furniture Action Council – furniture manufacturers association includes several voluntary standards for furniture flammability. http://www.homefurnish.com/UFAC/index.htm

7.7   Nomenclature

$\Delta h_{c,eff}$	effective heat of combustion of the bench-scale composite sample (MJ/kg)
m_{soft}	mass of the soft combustible material of the full-scale item (kg)
$m_{comb,total}$	mass of the total combustible material of the full-scale item (kg)
q''	total heat release per unit area of the bench-scale composite sample (MJ/m^2)
$\dot{Q}_{peak}$	peak HRR, measured or predicted, of the full-scale item (kW)
Q	total heat release rate of the full-scale item (MJ)
$\dot{q}''_{pk}$	peak HRR per unit area of the bench-scale composite sample (kW/m^2)
$\dot{q}''_{pk\,2}$	second peak HRR per unit area of the bench-scale composite sample (kW/m^2)
$\dot{q}''_{trough}$	trough between two peak HRR, per unit area of the bench-scale composite sample (kW/m^2)
$\dot{q}''_{300}$	HRR per unit area (bench scale) averaged over 300 s from ignition (kW/m^2)
t_{ig}	time to ignition of the bench scale composite sample (s)
t_{pk}	time peak in the full-scale item (s)
$t_{pk\,1}$	time to characteristic 'first' peak of the bench-scale composite sample (s)
t_{UT}	time to untenable conditions in a standard room (s)

7.8   References

1. Babrauskas, V., 'Upholstered Furniture and Mattresses', *Fire Protection Handbook*, 18th edn, Section 8 Chapter 17, National Fire Protection Association, Quincy, Massachusetts, 2002.

2. Mah, J., *1998 Residential Fire Loss Estimate – U.S. National Estimates of Fires, Deaths, Injuries and Property Losses from Non-Incendiary Non-Suspicious Fires*, Consumer Product Safety Commission, USA.

3. Fleischmann, C.M. and Hill G.R., *Burning Behaviour of Upholstered Furniture*, Interflam '04, Interscience Communications Co., London, 2004.

4. Technical Bulletin 133, *Flammability Test Procedure for Seating Furniture in for Use in Public Occupancies*, California Bureau of Home Furnishings and Thermal Insulation, North Highlands, CA, USA, 1991.

5. Krasny, J.F., Parker, W.J., Babrauskas, V., *Fire Behavior of Upholstered Furniture and Mattresses*, Noyes Publications, New Jersey, USA 2001.

6. Technical Bulletin 117 *Requirements, Test Procedures and Apparatus for Testing the Flame Retardance of Resilient Filling Materials Used in Upholstered Furniture*, California Bureau of Home Furnishings and Thermal Insulation, North Highlands, CA, USA, 2000.

7. BS 5852:1990 *Methods of Test for Assessment of the Ignitability of Upholstered Seating by Smouldering and Flaming Ignition Sources*, British Standard Institute, 1990.

8. ISO 8191 *Furniture Assessment of the Ignitability of Furniture, Part 1, Ignition Sources: Smouldering Cigarette, Part 2: Ignition Source: Match Equivalent Flame*, International Standards Organization, Geneva.

9. ISO 9705 *Fire Tests – Full-scale Room Test For Surface Products*, International Standard Organization, Geneva, 1993.

10. Technical Bulletin 106 (16 CFR 1632) *Requirements, Test Procedures and Apparatus for Testing the Resistance of a Mattress or Mattress Pad to Combustion Which May Result from A Smoldering Cigarette*, California Bureau of Home Furnishings and Thermal Insulation, North Highlands, CA, USA, 1986.

11. Technical Bulletin 603 *Requirements and Test Procedure For Resistance of a Mattress/Box Spring Set to a Large Open-Flame*, California Bureau of Home Furnishings and Thermal Insulation, North Highlands, CA, USA, 2004.

12. BS 6807:1990, *Assessment of the ignitability of mattresses, divans, and bed bases with primary and secondary sources of ignition*, British Standard Institute, 1990

13. Janssens, M.L., *An Introduction to Mathematical Fire Modelling*, Technomic Publishing, APA, USA, 2000.

14. Fire Dynamics Simulator (Version 4) Technical Reference Guide, NIST Special Publication 1018, U.S. Government Printing Office, Gaithersburg, USA, 2004.

15. Karlsson, B. and Quintiere, J.G., *Enclosure Fire Dynamics*, CRC Press LLC, London, 2000.

16. Babrauskas, V., Myllymäki, J., and Baroudi, D., 'Predicting Full Scale Furniture Burning Behavior from Small Scale Data', Ch 8, *CBUF: Fire safety of upholstered furniture - the final report on the CBUF research program*. Director-General Science, Research and Development (Measurements and Testing). European Commission. *Report EUR 16477 EN*, 1995.

17. Dietenberger, M.A., *Technical Reference and User's Guide for FAST/FFM* (Version 3), NIST-GCR-91-589, NIST, Gaithersburg, MD, 1991.

18. Pehrson, R., *Prediction of Fire Growth of Furniture Using CFD*, PhD Thesis, Worcester Polytechnic Institute, Worcester, MA, USA.

19. Sundstrom, B. (ed.) CBUF: Fire safety of upholstered furniture – the final report on the CBUF research program. Director-General Science, Research and Development

(Measurements and Testing). European Commission. *Report EUR 16477 EN,* 1995.

20. Wickstrom, U. and Goransson, U., 'Full-scale/Bench-scale Correlation's of Wall and ceiling Linings', Chapter 13 in: *Heat Release in Fires,* ed. V. Babrauskas and S. Grayson, Interscience Communications London, pp. 461–477 (1992).

21. Babrauskas, V. *et al.,* 'The Cone Calorimeter Used for Predictions of the Full-Scale Burning Behaviour of Upholstered Furniture', *Fire and Materials 21*: 95–105 (1997).

22. Babrauskas, V. and Krasny J. F., 'Fire Behaviour of Upholstered Furniture', *NBS Monograph 173*, US National Bureau of Standards, Gaithersburg (1985).

23. Babrauskas, V. and Grayson, S.J. (eds), *Heat Release in Fires*, Elsevier, London, 1992.

24. Enright, P.A. and Fleischmann, C.M., 'EC-CBUF Model I Applied to Exemplary New Zealand Furniture', *Proceedings of Sixth International Fire Safety Symposium,* 1999: 147–158.

25. Enright, P.A., Fleischmann, C.M. and Vandevelde, P., 'CBUF Model II Applied to Exemplary NZ Furniture (NZ-CBUF)', *Fire and Materials,* **25**, 2001: 203–207.

8

Flammability tests for cables

B S U N D S T R O M , SP Fire Technology, Sweden

8.1   Introduction: fire hazards related to cables

8.1.1   Fire hazard situations

A fire in a cable installation can lead to substantial damage in different ways. The burning cables will of course stop functioning. If a major power cable is damaged, the power supply to large areas may disappear. The fire in the Akalla tunnel in Stockholm in March 2001 led to power loss for substantial parts of the city and important functions within society stopped working. Communication cables that are damaged will not lead to loss of power but loss of control. Certain communication cables controlling processes, for example, in a nuclear power plant or a control system of an aircraft must not be allowed to burn. Therefore, substantial efforts are made to make these systems failsafe so that accidents are relatively few. On the other hand it is easy to imagine the consequences of a cable fire in the wrong situation.

Another hazard comes from the fire itself. A cable ladder full of cables can burn at a very high intensity. Depending on the type of cable, polymers and flame retardants involved, large amounts of smoke and toxic gases can be produced. In Fig. 8.1 a fire experiment on a cable installation with three cable ladders mounted on top of each other is shown. Every cable ladder carries one layer of cables. The top two ladders are burning at full intensity over a length of about 3.5 m the resulting fire gives a heat release rate of 2000 kW corresponding to a flashover fire of a living room. This is a large fire that would endanger people's lives and threaten their possibilities to escape. A fire of this size would also easily spread further through a building.

8.1.2   Burning behaviour of cables

Cables are often laid on ladders or on trays. They may be installed in bundles or as a loose disorganised web structure. The number of cables can be large and as they have a large content of polymers the heat released during a cable fire can be substantial. It is well established by fire-testing experts that the

8.1 A fire in a cable installation. Two cable ladders are burning and the third ladder has just ignited by burning droplets from the upper layers.

burning behaviour of cables is especially sensitive to the way they are mounted. If the mounting is like a relatively loose web structure the fire growth may be dramatically faster than if the cables are mounted in twisted bundles, see Fig. 8.2.

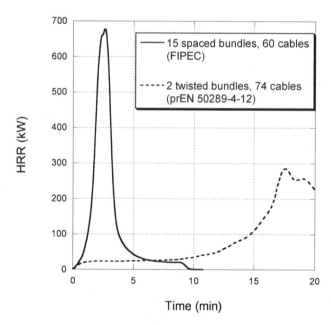

8.2 An example of differences in test results when a cable type is tested in a bundle or according to the FIPEC project[1] (long form) mounting practice.

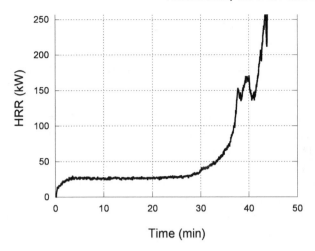

8.3 Experiment showing delayed ignition due to size and construction of the cable.

The spaced bundles result in a fire that by itself is large enough to create a flashover in a small room after only about three minutes. This is a short time for a substantial fire development. It is the same growth rate as we would expect from a bedroom lined with wood panel ignited by a waste-paper basket. The fire in the twisted bundles on the other hand peaks only after about 17 minutes and at a relatively low heat release rate. This fire is so small that it may not ignite other items and therefore may extinguish by itself.

Large cables can, in some circumstances, when exposed to an ignition source for a long a time, start to burn at high levels of heat release. This is due to the fact that only after a certain period of time the heat penetrates to more combustible segments of the cable and suddenly there is pyrolysis and burning occurs, see Fig. 8.3. We see that nothing happens for 30 minutes when this large power cable is exposed to an ignition source of 30 kW. Substantial fire growth starts around 30 minutes. The behaviour of this cable allows time in a fire situation. However, a fire test that is not severe enough may lead to a declaration that this product is perfectly safe, which not is the case.

The examples in Figs 8.2 and 8.3 demonstrate the importance of mounting and selection of test conditions when estimating the fire performance of cables. It can be seen that the configuration of cables affects their fire performance. Loosely spaced cables are more prone to fire growth than twisted bundles of cables. It is important that fire tests are linked to real installations and that the test conditions are selected to reflect actual hazard. As discussed in a later section, the concept of using reference scenarios to define fire performance is a powerful tool in defining the fire performance of cables.

8.2   International standards for the fire performance of cables

Most of the international standardisation in the electrical area takes places inside the International Electrotechnical Commission (IEC). IEC produces fire standards for all sorts of electrical equipment. Cables are handled in the IEC/ TC20 committee. Tests of wide use are the vertical small flame test, IEC 60332-1, for a single insulated cable and the large-scale vertical cable ladder test, IEC 60332-3. These two tests are well established and are also found in many national regulations. IEC 60332-1 measures the tendency for a single cable to be ignited and spread flames when exposed to a small Bunsen burner flame, comparable to a candle-like ignition. In IEC 60332-3 cables mounted on a vertical ladder are exposed to flames from a gas burner giving approximately 20 kW. The classification criterion is the damaged length and different classes are created on the basis of the amount of combustible material, i.e., the load of cables, and the application time of the burner.

The International Standardisation Organisation (ISO) TC 92 Fire Safety Committee is developing test methods for measurement of the heat release rate, ignitability, flame spread, smoke production and production of gas species from all sorts of products exposed to fire. In addition toxicity of fire gases is evaluated and tools for fire safety engineering are standardised. These tests and techniques are also applicable to cables.

European standardisation takes place in the European Committee for Electrotechnical Standardisation (CENELEC). Important tests are the EN 50265-2-1, the small flame test, and the prEN 50399-2-1 and prEN 50399-2-2 which are the large-scale vertical ladder tests. These tests are very similar to IEC 60332-1 and 60332-3 respectively. However, prEN 50399-2-1/50399-2-2 is equipped with a system to measure heat release rate and smoke production rate from the fire. This allows a classification system to be used where the quantitative parameters for fire growth can be used that can be compared with real-life fires. The CENELEC tests combined with the FIPEC findings takes us to the proposal for European harmonisation (see Fig. 8.7 and Table 8.1) which is explained in more detail in the next section.

8.3   European Union requirements for cables

8.3.1   Technical background

The Construction Products Directive (CPD) 89/106/EEG from the European Union lays down the essential requirements for building products. One of the essential requirements is 'safety in case of fire'. To allow free trade between the member states it is necessary to have the same test methods and to have a classification system that contains fire classes useful for, and matching the safety requirements of, the member states. In addition to this the market situation may change for the industries, products may have to be reformulated and so on. Therefore creating a testing and classification system that can be implemented may be a complicated process. The European Commission Decision 2000/147/EC sets up a European harmonised system for classification of reaction to fire performance of construction products, mainly linings and floorings. This has already been in operation for some years. However, it was recognised that not all construction products could be fire tested and classified in their end-use conditions in a fully satisfactory way. Cables represent such a case and only now a system for classifying cables useful in European legislation is under way.

Reference scenarios

An expert group created by the European Commission called the European Group of Fire Regulators (FRG) has been working with the fire aspects of the directive. The FRG had to decide on the scenarios that they needed to legislate for and what fire performance of the cable installation they wished to reach. The procedure shown in Fig. 8.4 was followed in principle.

The starting point is the real-life cable installations, simply because any fire regulation aims at addressing real-life fires. However, realistic cable installations cannot be used in a testing and classification system. The costs will be enormous as the number of different installations is almost infinite. The solution is therefore based on the assumption that certain large-scale reference scenarios can be representative of real-life hazards and that performance requirements of the cables can be identified in these reference scenarios. The term reference scenario is here used for an experimental set-up that is deemed to represent real life.

In exact terms the representation will never be true. However, a reference scenario is created in such a way that experimental fires in the scenario will be representative of a large number of real practical cases sufficiently accurately for a regulator. The burning behaviour of cables in the reference scenarios can then be linked to the burning behaviour in standardised test procedures. This is achieved by analysing fire parameters like heat release rate, flame spread and smoke production from experiments in the reference scenario and comparing them to the standard rate. When this link is established it is possible to use measurements in the standardised tests for classification. Thus the classification

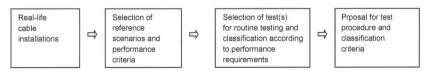

8.4 The process leading to a proposal for testing and classifying cables in Europe.

of a table in a standard test will reflect a certain burning behaviour in the reference scenario which in turn is linked to real-life hazard situations.

The European project Fire Performance of Electric Cables, FIPEC[1] included a survey of cable installations in Europe, a large number of fire experiments, evaluation of test procedures, fire modelling and so on. The availability of this work made it possible to propose a European testing and classification system. The cable installations used in Europe were mapped in a survey in the FIPEC[1] project. The most common installations found were characterised in a number of different scenarios. These installation scenarios were:

- open horizontal configuration (no walls or roof in the vicinity of the cable trays; just tray supports)
- semi-closed horizontal configuration (one wall and roof)
- closed horizontal configuration with two walls and a roof and the upper section of the end wall closed

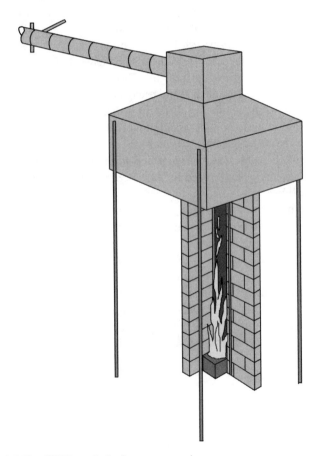

8.5 The FIPEC vertical reference scenario.

- closed horizontal configuration with two side walls and a roof and a fully open end (with and without forced ventilation)
- open vertical configuration
- semi-closed vertical configuration (corner configuration)
- closed vertical configuration (with and without ventilation)
- vertical void (small space) configuration
- horizontal void (small space) configuration.

These common European 'real-life' scenarios were then represented by two distinct reference test scenarios, the vertical scenario (see Fig. 8.5) and one horizontal scenario (see Fig. 8.6).

In the vertical scenario a cable ladder is mounted in a vertical corner. The heat output from the propane burner used for ignition is 40 kW for five minutes, then 100 kW for another ten minutes and finally 300 kW for the remaining ten minutes. Except for the 40 kW level, this burner programme in the Room Corner

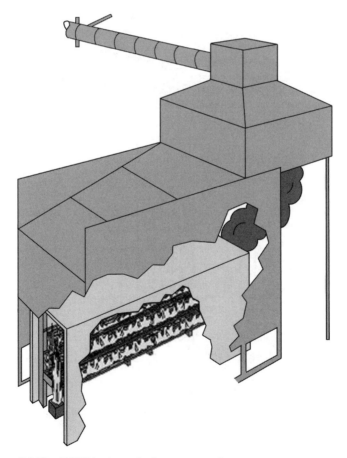

8.6 The FIPEC horizontal reference scenario.

Test (ISO, 9795), which is the reference scenario used for the European system for classification of linings. In the horizontal scenario three cable ladders are mounted horizontally and on top of each other. The bottom layer is ignited at one end with a burner following the same stepwise heat output programme as for the vertical scenario. The ignition source programme was also decided from the survey to represent realistic initial fires. No forced ventilation was used.

These two scenarios can be considered as reference scenarios representing a large variety of real-life cable installations in Europe. The FIPEC[1] work then established the standard test that can represent the two reference scenarios and which is used for classification. This test is called prEN 50399-2 and has two

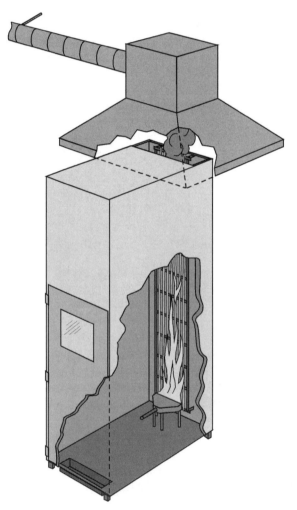

8.7 The test apparatus for prEN 50399-2-2 (FIPEC$_{20}$ Scenario 2) and prEN 50399-2-1 (FIPEC$_{20}$ Scenario 1) test procedures. The test chamber is 4 m high.

variants (see below). This test was modified to measure heat release rate and smoke production rate. In addition the mounting of the cables was changed after a sensitivity study allowing better identification between different fire performances.

The approach taken was successful and the Fire Regulators Group was able to decide the FIPEC[1] vertical and horizontal scenarios to be reference scenarios for the European regulations on cables. In addition the modified FIPEC[1] tests were suggested as the classification test procedure. The actual classification proposal contains seven classes. These are called A_{ca}, $B1_{ca}$ $B2_{ca}$, C_{ca}, D_{ca}, E_{ca} and F_{ca}. The fire performance is described by using data for the heat release rate and flame spread. Additional classes are defined for smoke production, burning droplets/particles and smoke production. The proposal is based on two test procedures using the same test equipment for the measurement of heat release, smoke production and droplet parameters, prEN 50399-2-2 for class $B1_{ca}$ and prEN 50399-2-1 for classes $B2_{ca}$, C_{ca} and D_{ca}. Therefore the test procedures are called FIPEC$_{20}$ Scenario 2 and FIPEC$_{20}$ Scenario 1 respectively, see Fig. 8.7.

The ignition source is a premixed propane gas burner giving a heat output of 20 kW for prEN 50399-2-1 and 30 kW for prEN 50399-2-2. The cables are mounted vertically in 3.5 m lengths on the front of a standard ladder (EN 50266-1). The lower part of the cables shall be 20 cm under the lower edge of the burner. The cables are positioned in the middle of the ladder (with respect to its width). The testing time is 20 minutes and during this period the heat release rate (HRR) using oxygen consumption calorimetry, and the smoke production rate, SPR, are measured. The occurrence of burning droplets/particles is also noted. From the measured parameters FIGRA, THR and TSP are derived. The small flame test EN 50265-2-1, is required for all classes $B1_{ca}$-E_{ca}. The additional classification for acidity is based on EN 50267-2-3.

8.3.2 The proposed classification system

Performance requirements that correspond to the limit values given in Table 8.1 for the different classes can be described as follows.

- Class A_{ca}. This class defines non-combustible products. Therefore the calorific potential is used as the sole criterion to define a product that cannot burn.
- Class $B1_{ca}$. Products that show a non-continuous flame spread and very limited HRR in the vertical and the horizontal reference scenario when exposed to the 40–100–300 kW ignition source. This also applies for the 30 kW test exposure in FIPEC$_{20}$ Scenario 2.
- Class $B2_{ca}$. Products that show a non-continuous flame spread when exposed to the 40 kW ignition source in the vertical reference scenario and the 40–100 kW ignition source in the horizontal reference scenario. They should also

Table 8.1 Classification proposal

Class	Test method(s)	Classification criteria	Additional classification
A_{ca}	EN ISO 1716	PCS $\leq$ 2.0 MJ/kg (1) and PCS $\leq$ 2.0 MJ/kg (2) and	
$B1_{ca}$	FIPEC20 Scen 2 (6) and	FS $\leq$ 1.75 m and $THR_{1200s} \leq$ 10 MJ and Peak HRR $\leq$ 20 kW and FIGRA $\leq$ 120 Ws^{-1}	Smoke production (3, 7) and Flaming droplets/ particles (4) and Acidity (5)
	EN 50265-2-1	H $\leq$ 425 mm	
$B2_{ca}$	$FIPEC_{20}$ Scen 1 (6) and	FS $\leq$ 1.5 m; and $THR_{1200s} \leq$ 15 MJ and Peak HRR $\leq$ 30 kW and FIGRA $\leq$ 150 Ws^{-1}	Smoke production (3, 8) and Flaming droplets/ particles (4) and Acidity (5)
	EN 50265-2-1	H $\leq$ 425 mm	
C_{ca}	$FIPEC_{20}$ Scen 1 (6) and	FS $\leq$ 2.0 m; and $THR_{1200s} \leq$ 30 MJ and Peak HRR $\leq$ 60 kW and FIGRA $\leq$ 300 Ws^{-1}	Smoke production (3, 8) and Flaming droplets/ particles (4) and Acidity (5)
	EN 50265-2-1	H $\leq$ 425 mm	
D_{ca}	$FIPEC_{20}$ Scen 1 (6) and	$THR_{1200s} \leq$ 70 MJ and Peak HRR $\leq$ 400 kW and FIGRA $\leq$ 1300 W/s	Smoke production (3, 8) and Flaming droplets/ particles (4) and Acidity (5)
	EN 50265-2-1	H $\leq$ 425 mm	
E_{ca}	EN 50265-2-1	H $\leq$ 425 mm	Acidity (5)
F_{ca}	No performance determined		

(1) For the product as a whole, excluding metallic materials.
(2) For any external component (i.e. sheath) of the product.
(3) Smoke classes to be determined from the FIPEC tests but also the needs for tunnel operators which require testing according to the so-called 3 m cube, EN 50268-2, is required.
(4) For $FIPEC_{20}$ Scenarios 1 and 2: **d0** = No flaming droplets/particles within 1200 s; **d1** = No flaming droplets/ particles persisting longer than 10 s within 1200 s; **d2** = not d0 or d1.
(5) EN 50267-2-3: **a1** = conductivity < 2.5 μS/mm and pH > 4.3; **a2** = conductivity < 10 μS/mm and pH > 4.3; **a3** = not a1 or a2. No declaration = No performance determined.
(6) Air flow into chamber shall be set to 8000 $\pm$ 800 l/min.
   $FIPEC_{20}$ Scenario 1 = prEN 50399-2-1 with mounting and fixing according to FIPEC (see Test Procedures).
   $FIPEC_{20}$ Scenario 2 = prEN 50399-2-2 with mounting and fixing according to FIPEC (see Test Procedures).
(7) The smoke class declared for class $B1_{ca}$ cables must originate from the $FIPEC_{20}$ Scen 2 test.
(8) The smoke class declared for class $B2_{ca}$, C_{ca}, D_{ca} cables must originate from the $FIPEC_{20}$ Scen 1 test.
Symbols used: PCS – gross calorific potential; FS – flame spread (damaged length); THR – total heat release; HRR – heat release rate; FIGRA – fire growth rate; TSP – total smoke production; SPR – smoke production rate; H – flame spread.

show a non-continuing flame spread, a limited fire growth rate, and a limited heat release when exposed to the 20 kW test procedure, $FIPEC_{20}$ Scenario 1.

- Class C_{ca}. Products that show a non-continuing flame spread when exposed to the 40–100 kW ignition source in the horizontal reference scenario and a non-continuing flame spread, a limited fire growth rate, and a limited heat release when exposed to the 20 kW test procedure, $FIPEC_{20}$ Scenario 1.
- Class D_{ca}. Products that show a fire performance better than ordinary, not flame-retardant treated polyethylene, and a performance approximately like wood when tested in the reference scenarios. When tested in $FIPEC_{20}$ Scenario 1 the products show a continuous flame spread but a moderate fire growth rate, and a moderate heat release rate.
- Class E_{ca}. Products that show a continuous flame spread when exposed to the 40 kW ignition source in the horizontal reference scenario. The small flame test (EN 50265-3-1) is used.

Based on these performance requirements, national regulations and proposals by industry, the following classes (see Table 8.1), were proposed in February 2004. They were fully accepted in principle by the Fire Regulators Group.

The analysis leading to the proposal comprised the whole of the FIPEC databases, additional tests done for the purpose of this work and an additional study of test data from industry and test institutes. Thus a large database was used to finalise a proposal of the evaluation system. The proposal is now being finally considered by European regulators and the European Commission. Amendments are possible, for example, the additional classification of acidity is considered, but a final decision is expected without any further substantial alterations.

8.4   Some results of research projects related to cables

A recent study[2] on cable fires in concealed spaces shows that ventilation may be a decisive factor for fire growth. Of special interest is to note that fire modelling can be very helpful to optimise the ventilation rate for minimising the fire hazard. The ability of cables to function when exposed to fire was studied in an experimental and theoretical study.[3] A first attempt was made to model the performance based on solving the heat transfer equations. If the temperature inside the cable can be calculated then its function when exposed to fire can be estimated.

At BRE,[4] work on communication cables used a test scenario where a very large fire in a room penetrates a suspended ceiling above which there is a cable installation. The room dimensions were approximately 7 m × 6 m × 4 m (height) and the fire source was more than 1 MW. These experiments were unique in scale and provide very valuable information on fire spread mechanisms that may occur for large installations of communication cables. Fire modelling of these experiments was also performed by BRE. Sandia National Laboratories, USA, has studied how cables fail to operate when exposed to fire.[5] They found that short-circuiting was the dominant failure mode. The cables did not fail due to open circuits as a result of the fire exposure.

The FIPEC[1] project is a very large study on cable fire performance, fire modelling, analysis of test procedures and classification systems and so on. This project was sponsored by the European Union and was performed by partners from Italy, Sweden, Belgium and the UK. As mentioned earlier, the FIPEC work is the basis for the proposal for a harmonised classification system in Europe.

Hirschler has done work on the comparison of electrical cables in both large and small scale tests.[6,7] The large scale tests were ASTM D9.21 and IEC 332-3. The small scale test was the Cone Calorimeter. A number of cables were tested. It was found that the heat release rate and the smoke production rate measured in the Cone Calorimeter could be used to predict large scale data. However, Hirschler also pointed out the difficulties in using simple correlations and that detailed modelling may be required.

At FM Global, Khan et al.[8] made correlation studies comparing the small scale test ASTM E-2058 (the Fire Propagation Apparatus) with the large scale tests UL-910 and NFPA 262. They found that ASTM E-2058 is useful tool for screening cables thus reducing testing costs and complexity.

8.5   Areas for further development

The acidity of the fire gases from burning cables can be measured by using a special test standard for cables called EN 50267-2-3. This test measures the conductivity of a fluid through which the fire gases are made to pass. The higher the conductivity the more acid the smoke is. Some cables are marketed as being 'halogen free' and the test is used to determine that. However, it needs to be further studied how that information relates to fire hazard. In a somewhat wider perspective also the question of toxicity of fire gases needs to be addressed. This is a large and difficult question and so far it has only been fragmentarily addressed on an international level. Reliable test methods are lacking. However, ISO TC 92 is working on these issues and it is hoped that advanced tools to assess toxicity will soon be available. The international discussions about toxicity have been lively in conjunction with the coming legislation from the European Union and it appears that toxicity is definitely on the agenda.

Electrical appliances such as switch cabinets, cable splices, etc., are often weak points where a fire may start due to an electrical fault. However, the knowledge of the fire behaviour of these products is limited and even if they are a part of an installation in a building in a similar way as a cable they are not considered an equal fire risk as a cable. Therefore it is believed that further development in this area will take place.

International requirements and test methods are well established regarding the behaviour of the products when they burn. However, more research is needed about the function of the cable when exposed to fire. Cables in certain systems must continue to function even if they are exposed to fire. Competent test methods and fire models need to be developed in this area. There is little activity of fire modelling in relation to cable fires. Modelling efforts are expected to grow. There is a need, for example, to predict the time that a cable would still function when exposed to a fire.

8.6   References

1. FIPEC Final Report on the Europeans Commission, SMT Programme SMT4-CT96-2059, 410pp, ISBN 0 9532312 5 9, London 2000.
2. Axelsson, J., Van Hees, P., Blomqvist, P., *Cable fires in difficultly accessible areas*, SP-REPORT 2002: 12.
3. Andersson, P., Persson, B., *Performance of cables subjected to elevated temperatures*, SP-REPORT 2001: 36.
4. Fardell, P.J., Colwell, R., Hoare, D. and Chitty, R., *Study into cable insulation fires in hidden voids*, published by BRE as reference CI 38/19/131 cc985, 1999.
5. Wyant, J.F., Nowlen, S.P., *Cable Insulation Resistance Measurements Made During Cable Fire Tests*, Sandia National Laboratories, NUREG/CR-6776, SAND2002-0447P.
6. Hirschler, M.M., 'Comparison of large-and small-scale heat release tests with electrical cables', *Fire and Materials*, Vol. 18, No. 2, 1994, pp. 61–76.
7. Hirschler, M.M., 'Can the Cone Calorimeter be used to predict full scale heat and smoke release cable tray results from a full scale test protocol?', in *Proc. Interflam*, 2001, Sept. 17–19, Edinburgh, Scotland, Interscience Communications, London, UK.
8. Khan, M.M., Bill, R.G., Alpert, R.L., 'Screening of plenum cables using a small-scale fire test protocol', *FM Global*, 1151 Boston-Providence Turnpike Norwood, MA 02062, USA.

9

Flammability tests for electrical appliances

M S I M O N S O N, SP Fire Technology, Sweden

9.1 Introduction

Efforts were first made in the 1950s to regulate fire growth through the building codes. These efforts were limited to controlling the flame spread and smoke developed from the burning of the interior finish (walls and ceiling) of structures. They also included provisions for the fire resistance of walls and floors (and roofs) in some of these same structures. In the United States, the regulations included the 'flame spread' (ASTM E84[1]) and the fire resistance (ASTM E119[2]) fire tests; in Germany the 'Brandschacht' (DIN 4102[3]) fire test, in France the 'epiradiateur' (NF P 92-501[4]), and in the UK the 'spread of flame' (BS 476, Part 7[5]) and the fire resistance (BS 476, Part 4[6]) fire tests with numerous similar examples available in other countries worldwide. These regulations were introduced in response to the high number of fires and associated loss of life and property. The regulations resulted in a clear decrease in these fire losses before typically reaching a plateau.

Subsequently, active fire protection was included in the building codes. This included the use of smoke detectors, alarms, and sprinklers. Again, these regulations resulted in decreases in the fire losses until a new plateau was reached. To further improve or reduce the fire losses, regulators are now turning their attention to building contents. It has always been known that fires often start in the building contents and not in the building structure itself. Therefore, building contents have become the focus for building code changes. In the USA and UK, this has taken the form of actual regulation and/or study for regulation of the fire performance of, for example, upholstered furniture, mattresses, and bed clothing (e.g. BS 5852[7] and Cal TB 603[8]).

While upholstered furniture, bedding and mattress fires represent some of the most pressing causes of domestic fires and fire deaths, electrical fires answer for some 20% of all fires in homes and are a major cause of domestic fire losses.[9–12] Electrical equipment is no longer the sole province of the electrical engineer, whose concern has been limited to preventing such equipment from being an ignition source (i.e. from a fire that begins in the internal electrical components).

Such building contents, however, now have the attention of the building and fire officials for their role in fire growth in structures.

The concern with this item of building content, as it is with other items of building content, is that of it being the 'first item ignited' by an external ignition source. In the case of electrical and electronic appliances (E&E equipment) these present a particular risk over that of, e.g., upholstered furniture, due to the possibility of internal ignition through an electrical fault. This has long been the only concern for regulators with standards typically addressing internal ignition as the only area they could control. Indeed, traditionally external ignition was seen to be outside the scope of industry and regulator responsibility. In recent years this attitude has begun to change. The Low Voltage Directive (73/23/EEC) has been recently expanded to include provisions to allow regulators to place contraints on the material in electrical products so that they could resist an external ignition source.[13] This is largely in recognition of the fact that while many electrical fires occur in common household appliances they are often initiated through an external ignition.[9]

The normal course of events in a room fire (fire scenario) are: ignition source, first item ignited, second item ignited or interior finish ignited, further ignition of other items and/or interior finish, complete room involvement or flashover. It is therefore critical in controlling fire growth to 'break the chain' at the first item ignited.

The external ignition source has now been recognised as a key part in this chain. In this context the ignition source that needs to be considered is the small open flame in the form of either a match flame, or a lighter flame, or a candle flame.[14] A candle flame is the most appropriate ignition source for standardisation based on its size and duration of flame, and ubiquitous domestic use and fire statistics. This issue is particularly pertinent to IT equipment and consumer electronics (e.g. TV sets), which is the domain of this chapter. Certainly, the increasing quantities of ignition sources, in particular candles, and the increasing quantities of IT equipment and consumer electronics, in residential occupancies, can only suggest increasing 'collisions' between the two, and therefore increasing fire ignitions. In the instance of large electrical appliances such as refrigerators and washing machines the external ignition source is still outside the regulators' domain and internal ignition is the main source of interest. Presently, there is no indication that external ignition of these products is common.

This chapter will attempt to provide an understanding of the need for standard testing in order to verify that regulations provide a suitable framework for design of products with appropriate performance. To this end it will provide a summary of both previously published and some new heat release data for the most common appliances in our homes. Finally, the ever-present tug-of-war between fire safety and environmental considerations will be addressed with a view to new approaches that will hopefully provide a much needed basis for environmentally sound, or 'green', fire safety in future.

9.2   Fire statistics

A large body of fire statistics is available worldwide from a variety of sources.[15] These statistics contain information of varying detail and accuracy depending on the source. Statistics from fire brigades tend to focus on larger fires, as the fire brigade is generally not called in to deal with a very small fire. In contrast, statistics from insurance companies generally contain both small and large fires. In the case of TVs, for example, where different safety standards govern enclosure material internationally, differences in TV fire statistics between Europe and the US give an indication of the effect of the use of high fire safety requirements in reducing the frequency of fires.

The available statistics are generally defined based on a variety of ignition sources. For the sake of clarity these are discussed in the next section. One important point in common for all electrical appliances is that the focus is traditionally on the elimination (or minimisation) of the risk of internal ignition due to the presence of electricity. In order to understand the fire development in a domestic fire it is important to understand relevant ignition sources. Therefore these are catalogued in terms of internal or external impingement as discussed next.

9.2.1   Ignition sources

Internal ignition sources

A recent and very thorough study, carried out by Sambrook Research International and commissioned by the UK Department of Trade and Industry (DTI),[17] identified the following causes of TV set fires, based on the historical record:

- solder joints ageing causing arcing
- mains switch, worn contacts
- electromechanical stress in 'heavy' components
- overheating due to circuit component imbalances
- capacitor failure (one design)
- line output transformer
- poor design of circuit layout (early TVs)
- cathode ray tube (CRT)
- mains lead
- standby function, especially in old sets.

Many of these ignition sources are no doubt applicable even in the case of other electrical and electronic household appliances. Safety standards relating to these appliances therefore tend to focus on the elimination of this risk. In the case of consumer electronics (CE) and IT equipment this has been regulated through IEC 60950 and IEC 60065. Presently, the IEC is developing a single performance based standard to address the fire performance of all these products. The fact that

Table 9.1 Examples of TV set recalls, 1989–1997

Country	Manufacturer	Recall year	Period of manufacture	Number of sets
Denmark	N/A	1992/3	N/A	40,000
France	Philips	1993	1983–1987	40,000
Germany	N/A	1989	N/A	200,000
Netherlands	Philips	1993	1983–1987	300,000
Sweden	Philips	1993	1983–1987	75,000
UK	Sony	1989	1985–86	N/A
UK	A	1993	1983–1986	21 models
UK	B	N/A	1986–1988	1 model
UK	C	1993	N/A	7 models
UK	D	N/A	>1992	2 models
UK	F	1993	>1992	2 models
UK	W	1993	1983–86	1 model
UK	Dixons/Matsui	1997	1993	'1000s'

the previous prescriptive standards (which have required the use of specific materials, specific minimum distances or 'fire enclosures') have failed to remove the risk of internal ignition is illustrated in Table 9.1 listing major recalls of TV sets between 1989 and 1997. This table is indicative rather than comprehensive as no systematic record of TV set recalls is kept in any country. This example from the UK demonstrates that recalls are not uncommon.

External ignition sources

In the case of CE and IT equipment there is evidence to indicate that many ignitions occur from external ignition sources.[9] While these have not traditionally been the domain of fire safety standards they have now become a major focus in the standardisation work being undertaken by IEC TC108 (Safety of electronic equipment within the field of audio/video, information technology and communication technology) who are in the process of producing a single performance-based safety standard for CE and IT equipment. This standard recognises the importance of external ignition in many CE and IT products and has expanded the scope of the proposed standard to include external ignition from small open flames. In this context they refer specifically to accidental ignition from candle flames although the logic could be expanded to include other small open flames such as matches and lighters in the hands of juvenile fire setters.

Consumer misuse

Manufacturers and fire brigades inform consumers about the safe use of various appliances where there is a perceived hazard. They are warned against using the top of such appliances as a shelf, for supporting vases, candles, or a cloth that could reduce ventilation. Fire brigades indicate the following causes of fire due to consumer misuse:

- lack of ventilation, especially when the TV sets are 'boxed' into furniture
- lack of maintenance, to remove accumulated dust (dampness can lead to electrical failure in case of dust accumulation)
- extensive use of the standby function on TVs, especially by families with children.

Electrical fires

Household appliances constitute an important origin of fires. While the problem of TV fires is well known and documented,[15–18] fires from other electrical appliances in the home are by no means negligible.[9] A Finnish study in 1999[19] addressing electrical fires found that while TVs account for some 12% of all household fires, washing machines or dishwashers account for 10% (i.e., marginally less), and refrigerator/freezer fires make up 3% of the fires. While extensive data is available from TV fires,[18–20] only little is available on specific pieces of IT equipment,[21,22] and common white goods.[19]

Available fire performance information will be presented in the next section for the above-mentioned products. This will provide the backdrop for a discussion of modern standard performance requirements on these electrical items and how these requirements address the hazard the products present in domestic and office applications.

9.3   Fire performance

9.3.1   CE and IT equipment

Fires in TV sets have received a great deal of interest in recent years due to the large number of such fires that can be found in the fire statistics. An important study by the Sambrook Institute first brought this issue to public awareness.[16] Additional data was provided through a detailed investigation into electrical fires in Sweden published in 1997.[9] A compilation of this data and presentation of the first fire data was made in 1999[15] with additional studies corroborating this initial work.[18–20] Some data is presented in this chapter for the fire characteristics of common household TV sets, both traditional cathode ray tube (CRT) and more modern liquid crystal display (LCD) TV. Table 9.2 summarises the main fire performance parameters for the large-scale experiments on the burning standard European 27-inch CRT TV and LCD TV. In both cases the experiments were conducted directly under the furniture calorimeter[23] and ignition was performed using a small open flame, approximately the size of a candle flame.

Both TVs were ignited easily with a small ignition source (candle or tea-light) with the plastic beginning to burn within approximately 30 seconds from

Table 9.2 Fire performance parameters for the large-scale experiments

Parameter	CRT TV	LCD TV
Time for average (min)	5–50	2–23
Ignition source power (kW)	1	1
Time to ignition (s)*	30	30
Peak HRR (kW)	240	60
Average HRR	60	30

* Time after application of ignition source

application of the ignition source. The HRR curves shown in Figs 9.1 and 9.2, however, do not show a major increase in the HRR for these TVs until approximately two minutes after application of the ignition source. This is because the initial fire development is slow with the major increase coming approximately 1.5 minutes after initial ignition of the plastic housing. This implies that an occupant would have approximately 1.5 minutes to react and extinguish a small fire after which the fire development is very rapid.

It should be noted that while the voluntary standard dictating fire performance of TV sets in the US (UL1510) requires that they are essentially resistant to external ignition from a small open flame this is not the case for TV sets throughout most of the rest of the world. The TV fire performance data presented in Figs 9.1 and 9.2 is for European TV sets purchased in Sweden where no such requirement is found. In this case the TV must adhere to the requirements set down in IEC 60065, which has no provision in the 2004 edition for resistance to external ignition. Typically US TV sets would not be

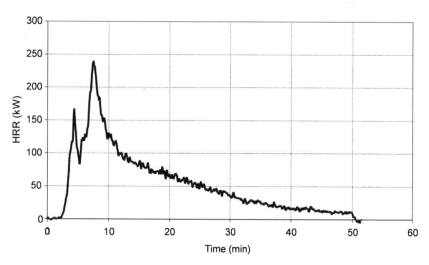

9.1 Mass loss and HRR for 28-inch CRT TV. Ignition with a candle-sized flame occurs at time = 0.

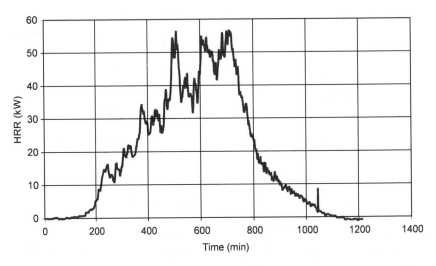

9.2 HRR for LCD TV. Ignition with a candle-sized flame occurs at time = 0.

easily ignited using the small ignition sources shown here (i.e., candle or tea-light).

Little information is available in the literature concerning the fire performance of specific pieces of IT equipment.[21] It has been speculated that this is due to the fact that the present fire performance requirements for IT equipment provides a sound basis for the production of equipment with a high level of fire safety. While this may be true of TV sets in the US, where UL 1510 requires V0 performance of TV housings, it is not true for IT equipment where the US standard (UL 1950) and the international standard (IEC 60950) have been harmonised and neither require resistance to external ignition. Rather, this is no doubt due to a voluntary adoption of higher fire performance materials than required by these minimum standards. Figures 9.3 to 9.5 show the fire performance (HRR and mass loss) for a variety of pieces of IT equipment when the external housing is rated as UL 94 HB (i.e., corresponds to the minimum requirement). The small ignition source used (candle-sized flame) to ignite the products in Figs 9.1–9.5 (coupled to the rapid increase in HRR) shows that these pieces of equipment are easily ignited and present a varied level of risk in a home or office application depending on the peak HRR.

9.3.2　White goods

There is clear anecdotal and statistical evidence to suggest that certain categories of white goods do cause domestic fires.[24] A Finnish study published in 1999 indicated that some 10% of household fires are caused by washing machines or dishwashers while refrigerator/freezer fires account for 3%.[19] Similar evidence

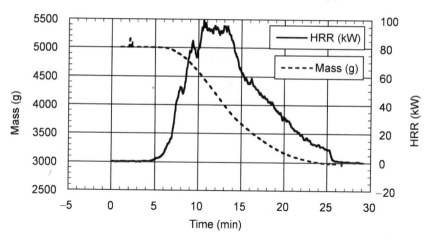

9.3 HRR and mass loss for inkjet printer. Ignition with a candle-sized flame occurs at time = 0.

is available in Sweden indicating that white goods collectively (excluding stoves and ovens) are the second most common cause of domestic fires, surpassed only by TV fires.[9] The fire performance of washing machines is pictured in Fig. 9.6 while that for dishwashers is shown in Fig. 9.7 and for refrigerator/freezers is shown in Fig. 9.8. In all cases tests were performed on off-the-shelf, domestic white goods. The location of the burner is cited in the figure text.

While the results presented in Figs 9.6–9.8 exhibit some spread of fire performance based on the specific ignition characteristics of the equipment, some important conclusions can be drawn concerning the hazard represented by

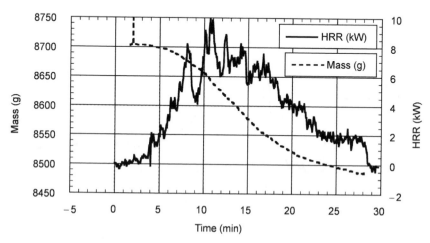

9.4 HRR and mass loss for CPU. Ignition with a candle-sized flame occurs at time = 0.

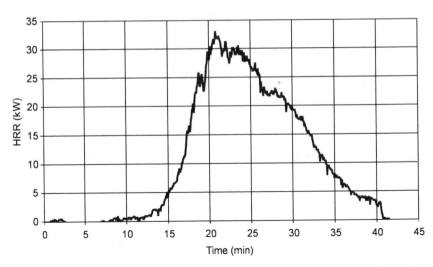

9.5 HRR for keyboard. Ignition with a candle-sized flame occurs at time = 0.

fires started in these common electrical appliances based on their peak heat release. This data is summarised for each piece of equipment in Table 9.3.

The results presented in Table 9.3 clearly show that while washing machines, dishwashers (both free burning and installed in a cupboard) do not provide sufficient heat at their peak to cause flashover in a standard room (ISO 9705) they are sufficiently high to cause secondary ignition of other combustible

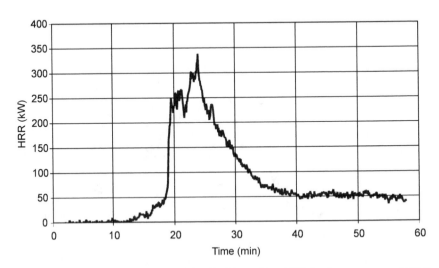

9.6 HRR for washing machine. Ignition with a 1 kW line burner occurs at time = 0. The burner was placed below the plastic washing basin through a hole of dimensions 29 mm × 110 mm cut into the back wall. Reproduced with the permission of Tuula Hakkarainen (VTT) from ref. 19.

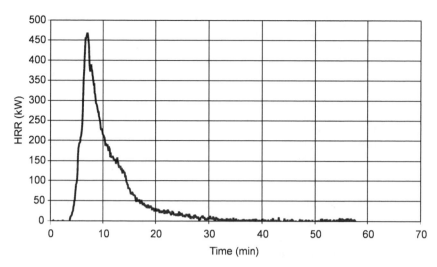

9.7 HRR for dishwasher. Ignition with a 1 kW line burner occurs at time = 0. The burner was placed inside the motor space below plastic items located in the motor space through a hole of dimensions 40 mm × 15 mm, cut into the rear wall. Reproduced with the permission of Tuula Hakkarainen (VTT) from ref. 19.

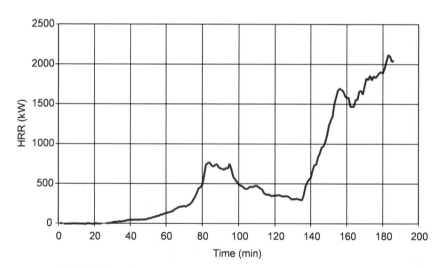

9.8 HRR for refrigerator/freezer. Ignition with a 1 kW line burner occurs at time = 0. The burner was placed in the middle of the motor space at a distance of 10–20 mm below the separating durface between the motor space and the freezer. The distance of the burner from the rear wall was 100–400 mm. Test terminated above 2 MW for safety reasons. Reproduced with the permission of Tuula Hakkarainen (VTT) from ref. 19.

Table 9.3 Collation of mass loss (ΔM) and peak HRR for white goods. Reproduced with the permission of Tuula Hakkarainen (VTT) from reference 19

Apparatus	ΔM (kg)	max. HRR (kW)
Washing machines	10.9	333
Dishwashers (free burning)	7.3	411
Dishwashers (in a cupboard)	18.9	548
Refrigerator-freezers (free burning)	16.2	1970
Refrigerator-freezers (in a cupboard)	24.5	1526

material in the home thereby leading to a potentially large fire. In the case of refrigerator/freezer units, however, they produce sufficient heat in themselves that they could easily cause flashover in a standard room and therefore represent a significant risk should they become ignited in a home.

9.4   Standardisation/regulation

Internationally speaking the major source of standardisation within electrical and electronic technologies is the International Electrotechnical Commission (IEC). Within the European Union, the European Committee for Electrotechnical Standardisation (CENELEC), is the leading standardisation organisation of electrical and electronic technologies. CENELEC's mission is to prepare voluntary electrotechnical standards that help develop the Single European Market/European Economic Area for electrical and electronic goods and services removing barriers to trade, creating new markets and cutting compliance costs.

A Resolution of 7 May 1985 of the European Council formally endorsed the principle of reference to European standards within the relevant European regulatory work (Directives), thereby paving the way to a new approach in the philosophy of regulations and standards in Europe. In the light of this new approach, CENELEC is developing and achieving a coherent set of voluntary electrotechnical standards as a basis for the creation of the Single European Market/European Economic Area without internal frontiers for goods and services. There is naturally a large amount of overlap between the work of CENELEC and the IEC. In many cases standards will be developed in a CENELEC or IEC committee and sent out for parallel voting in both organisations. As members within both organisations come from the national committees there is significant mirror representation on the various committees.

While CENELEC is the leading European standardisation organisation in this field, the IEC is the leading global organisation that prepares and publishes international standards for all electrical, electronic and related technologies. These serve as a basis for national standardisation. As the IEC is most active in the standardisation of the products dealt with in this chapter the main focus of the following presentation will be on the IEC work, although the CENELEC mirror committee is identified in each case.

In the case of TV sets and IT equipment the IEC has traditionally dealt with these products separately within committees IEC TC74 (Office equipment, i.e. IT) and IEC TC92 (consumer electronics, i.e. TVs). In 2001 these two committees were combined in recognition of the fact that the boundaries of use for the products traditionally covered by these two committees were no longer clearly defined, a single joint committee was started (IEC TC108) to deal with the safety of these products. This committee continues to work with the original standards developed by TC74 (IEC60950) and TC92 (IEC 60065) and have simultaneously been charged to develop a new 'performance based' standard which will address the question of safety in a more holistic manner in contrast to the prescriptive approach adopted in the previous standards. IEC 60950 and IEC 60065 have focused on the provision of fire safety through minimisation of the risk of internal ignition through the judicious use of fire enclosures or prescription of ignition resistant material where appropriate. The CENELEC mirror committee is CENELEC TC108.

In no case, has resistance to external ignition been considered prior to the instigation of TC108. With the beginning of TC108 the committee has made a strategic decision to include consideration of external ignition from a small open flame and provisions to resist this in recognition of the fact that many of the products covered within the scope of this committee have been shown to be susceptible to external ignition. It has been suggested that external ignition was previously outside the scope of all IEC documents as this was not within the realm of what a producer of a product could reasonably be expected to be able to foresee. Recent suggested changes to the Low Voltage Directive,[13] however, expand the responsibility of the producer to include foreseeable risks such as ignition by a small open flame (e.g. candle flame) in the home.

In the case of washing machines, dryers and dishwashers, IEC TC59 deals with these products (and others). The committee deals with the performance of household electrical appliances and focuses on minimisation of the risk for internal ignition through the use of fire enclosures where an ignition source exists or the prescription of ignition resistant material where appropriate. In no case has resistance to external ignition been considered. The CENELEC mirror committee dealing with these three products is CENELEC TC59. Although there is not complete overlap between these two committees the differences deal with other products.

Refrigerator/freezer units are covered by IEC TC61. IEC TC61 was created from IEC TC59 in recognition of the fact that TC59 could not cover all household appliances. In this case the focus of safety has been on the motor or compressor unit and not on the material choices within the units themselves. Again, to date there is no provision for resistance to external ignition. In light of the available experimental results characterising the fire performance of these products it can be concluded that they represent a major fuel load in the home and further consideration of their fire performance and limiting the ignition propensity and fire spread characteristics would be wise. As for the other committees, the CENELEC mirror committee is CENELEC TC61. Although there is not complete overlap between these two committees the differences deal with other products.

9.5   Environmental considerations

Environmental requirements have been included in product procurement for some time. Fire performance requirements have, however, typically not been included in such systems. In some cases the environmental focus, however, has potentially had an effect on the fire performance or, vice versa, attention to the fire performance has potentially had an effect on the environmental impact of the product. Traditionally, safety standards have focused on areas that lend themselves to engineering, such as fire safety, electrical safety, etc. The eco-labels have focused more on minimising the environmental effect (real or perceived) of a particular product. In many cases this has been achieved in a simplistic manner through the exclusion of specific species rather than through a holistic approach. Little or no regard has been given in the eco-standards to possible effects on the safety of the product due to the ecological considerations.

Recently it has become apparent that this separate approach (safety through standardisation and environmental protection through eco-labels) is insufficient to address issues posed by complex products moving in international markets. In response to this a new system has been proposed which aims to address both environmental and fire safety issues simultaneously, i.e., to ensure maximum environmental and fire performance. It has been developed to promote environmentally sound fire safety.

Fire code enforcement and environmental officials from the United States, with assistance from Swedish environmental and safety authorities, are now proposing a programme to encourage manufacturers to develop, use and sell products capable of simultaneously meeting high standards for human health, fire safety and environmental quality; Green Flame[TM]. The challenge accepted by the partnership is to move away from chemicals posing unacceptable harm to the environment while maintaining a high level of fire protection through safe alternative chemicals and/or techniques. The Green Flame[TM] programme shall establish and promote voluntary market incentives to encourage industry to develop fire safety technologies, which do not compromise human health and environmental quality. The goal of the system is to create clear incentives for manufacturers to design, produce and sell products capable of simultaneously meeting high standards for fire safety, health and environmental quality.

9.6   References

1. ASTM E 84, Surface Burning Characteristics of Building Materials. Available from webstore.ansi.org/ansidocstore. Equivalent to NFPA 255 and UL 723.
2. ASTM E119, Fire Tests of Building Construction and Materials. Available from webstore.ansi.org/ansidocstore.Equivalent to NFPA 251 and UL 263.
3. DIN 4102, Fire Performance of building materials and components. Available from DIN (Deutches Institut für Normung, D-10772 Berlin).
4. NF P 92-501, Radiation test used for rigid materials, or for material on rigid substrates (finishes) of all thicknesses, and for flexible materials thicker than 5 mm (Epiradiateur Test). Available from AFNOR (Association Francaise de Normalisation, Tour Europe, Cedex 7, F-92080 Paris la Défense).
5. BS 476-7, Fire tests on building materials and structures Method for classification of the surface spread of flame of products. Available from BSI (British Standards Institute, UK).
6. BS 476-4, Fire tests on building materials and structures Non-combustibility test for materials. Available from BSI (British Standards Institute, UK).
7. BS 5852, Assessment of the ignitability of upholstered seating by smouldering and flaming ignition sources. Available from BSI (British Standards Institute, UK).
8. Cal TB 603, Requirements and test procedure for resistance of a residential mattress/box spring set to a large open-flame. Available from California Bureau of Home Furnishings and Thermal Insulation, 3485 Orange Grove Ave, North Highlands, Ca 95660-5595).
9. Enqvist, I. (Ed.), *Electrical Fires – Statistics and Reality*. Final report from the Vällingby project, Electrical Safety Commission ('Elsäkerhetsverket'), 1997. Available in Swedish only.
10. Simonson, M., Andersson, P., Rosell, L., Emanuelsson, V. and Stripple, H., Fire-LCA Model: Cables Case Study SP Report 2001:2 available at http://www.sp.se/fire/br_reports.HTM.
11. Nurmi, V.-P., *Risk Management of Electrical Fires*. TUKES publication Series 3/2001, Ph.D. Thesis, ISBN 952-5095-46-0. Available in Finnish only.
12. Andersson, D., Axelsson, J., Bergendahl, C.-G., Simonson, M., *FIRESEL Fire safety in electronic and electrical products and systems*, IVF Research Publication 01812, ISSN 0349-0653 (2000).
13. LVD Update 5 (October 2003), 'Chapter II.2 Protection against fire hazards'. Wording available from internet: http://europa.eu.int/comm/enterprise/electr_equipment/lv/direct/lvdupdate5.pdf.
14. Holborn, P.G., Nolan, P.F., Golt, J., 'An analysis of fatal unintentional dwelling fires investigated by London Fire Brigade between 1996 and 2000', *Fire Safety Journal* 38, pp 1–42 (2003).
15. De Poortere, M., Schonbach, C. and Simonson, M., 'The Fire Safety of TV Set Enclosure Materials, A Survey of European Statistics', *Fire and Materials* 24, pp 53–60 (2000).
16. TV Fires (Europe), Department of Trade and Industry (UK), Sambrook Research International, 14 March 1996.
17. Television Fires, DEMKO (Danish Electrical Equipment Control Office), 1995.
18. Causes of fires involving television sets in dwellings, Department of Trade and Industry (UK), Consumer Affairs Directorate, URN 01/745 (2001).
19. Hietaniemi, J., Mangs, J. and Hakkarainen, T., 'Burning of Electrical Household

Appliances, An Experimental Study', *VTT Research Notes* 2084 (2001).

20. Simonson, M., Blomqvist, P., Boldizar, A., Möller, K., Rosell, L., Tullin, C., Stripple, H. and Sundqvist, J.O., 'Fire-LCA Model: TV Case Study', SP Report 2000:13 (2001).

21. Simonson, M., Andersson, P. and Bliss, D., 'Fire performance of Selected IT-Equipment', *Fire Technology* 40(1), pp 27–38 (2004).

22. Hall, J., 'Fires Involving Appliance Housings – Is there a Clear and Present Danger?', *Fire Technology* 38, pp 179–198 (2002).

23. NT FIRE 032, Upholstered furniture: Burning behaviour – full scale test. Available from Nordisk Innovation Centre (www.nordicinnovation.net).

24. Stålbrand, K., 'Common household appliances cause thousands of fires', *Aktuell Säkerhet* 1, pp 24–28 (1997). Available in Swedish only.

Part II

Flammability tests for transportation and mining

Flammability tests for regulation of building
and construction materials

K S U M A T H I P A L A , American Forest & Paper Association, USA
and R H W H I T E , United States Department of Agriculture, USA

10.1   Introduction

The regulation of building materials and products for flammability is critical to ensure the safety of occupants in buildings and other structures. The involvement of exposed building materials and products in fires resulting in the loss of human life often spurs an increase in regulation and new test methods to address the problem. Flammability tests range from those in which the sample is ground to a powder prior to testing to the full-scale room corner test. Variations in test methods include the specific measurement of flammability parameters being considered, the intensity and characteristics of the fire exposure, the relative scale of the test specimen, and many other factors. As a result of specific details such as specimen orientation and fire exposure intensity, some existing regulatory test methods have been shown not to classify a type of building material or product in a manner consistent with full-scale tests that simulated actual conditions. Such failures have resulted in the development of alternative test methods for specific applications or products. As a result there is a wide range of tests used to classify building materials and products for flammability or reaction to fire.

Unfortunately, the results of flame-spread tests are often characteristics of the test procedure (Clark, 1981). In a 1970s comparison of international test methods for flammability, the ranking of materials by the different regulatory methods was little better than a random number generator (Emmons, 1974; Karlsson *et al.*, 2002). To achieve harmonization within the European Union, it became necessary to develop a new test – the Single Burning Item test. The scope of this chapter is limited to those flammability test standards that are relevant to the regulation of building materials and products. As such, it is only a subset of the many fire tests that have been used or are being used to evaluate building materials (Tewarson *et al.*, 2004; Eickner, 1977). This chapter focuses on the compliance criteria used in the regulatory classifications of building materials for flammability. More details on many of the applicable test methods can be found in other chapters of the book.

10.2 United States flammability requirements

The regulations for building materials and products to address the hazards associated with their flammability and related characteristics are found in the International Building Code and other building codes, the National Fire Protection Association (NFPA) 101 Life Safety Code, and other such documents. Acceptance criteria and classifications in this section generally refer to provisions of the International Building Code, which is the dominant model building code in the United States. Two main sets of requirements address the combustibility and the flame spread characteristics of building materials. These two main tests are described in the standards of ASTM International: ASTM E136 for combustibility and ASTM E84 for flame spread. The ASTM E84 test method also addresses the regulation for visible smoke characteristics. To address certain materials or applications, other test methods are also used in the regulation of building materials in North America.

10.2.1 Combustibility

The classification of a material as a non-combustible material is based on ASTM E136 Standard test method for behavior of materials in a vertical tube furnace at 750 °C. In the test, a dry 38 mm × 38 mm × 51 mm specimen is heated in a small vertical tube furnace. A material is accepted as non-combustible, if the material passes the criteria of ASTM E136. ASTM E136 requires that three of the four specimens must pass the criteria. If the weight loss of the test specimen is 50% or less, the criteria are (i) the specimen surface and interior temperatures during the test do not increase more than 30 °C above the temperature measured on the surface of the specimen prior to the test and (ii) no flaming from the specimen after the first 30 s. If the weight loss is greater than 50%, the criteria are no temperature rise above the stabilized temperature measured prior to the test and no flaming from the specimen at any time during the test. The commentary in ASTM E136 standard discusses the development of the test method and the rationale for the criteria.

As noted by Carpenter and Janssens (2005), one important limitation of ASTM E136 is that the test method does not provide a quantitative measure of heat generation or combustibility, but only a pass/fail result. The ASTM E136 is a severe test in that composite materials with even a small amount of combustible component will often fail to satisfy the criteria. Thus, materials such as mineral wood insulation with combustible binder, cinder concrete, cement and wood chips, and wood-fibered gypsum plaster are all classified as combustible (Institute for Research in Construction (IRC), 1995). The core of gypsum wallboard will satisfy the criteria for non-combustible. Paper-faced gypsum wallboard does not pass the non-combustible flaming criteria (Canadian Wood Council, 1991). As a result, prescriptive provisions to the codes are needed to permit the use of paper-faced gypsum in non-combustible construction. Fire-retardant treatment of a combustible material is not a viable option to satisfy the requirements of ASTM E136 and be classified non-combustible. Tewarson *et al.* (2004) considered E136 to be capable of providing quantitative data for the performance-based fire codes for the assessment of fire hazards associated with the use of products and protection needs.

To accommodate materials that are of low combustibility, alternative test methods have been developed. One such method is the test for potential heat of building materials, NFPA 259. The potential heat is the difference in the gross heat of combustion (oxygen bomb calorimeter) of the original material and the material after a two-hour exposure in a furnace at 750 °C. In the NFPA 101 Life Safety Code, the test is used as a criterion for limited-combustible materials. The criterion is a potential heat less than 8,140 kJ/kg. The use of the cone calorimeter as a test for combustibility has also been investigated (Carpenter and Janssens, 2005). As noted by Carpenter and Janssens (2005), one of the biggest obstacles to the use of the cone calorimeter for combustibility is the implementation of a classification system that does not disrupt the status quo.

10.2.2   Interior finish flammability

The main test for surface burning characteristics of building materials in North America is the 25-foot (7.6 m) Steiner tunnel. The standards are ASTM E84, NFPA 255, and UL 723. A 1.35 m long test flame at one end provides the ignition source and the fire exposure. The test provides measurements of the surface flame propagation and a measure of the optical smoke density. Based on the observations of the flame propagation, a comparative dimensionless flame spread index (FSI) is calculated from the areas under the flame spread distance-time plot. For the purpose of the calculations, the flame front is assumed to never recede. The comparative smoke developed index (SDI) is calculated from the area under the smoke density-time plot. The areas for reinforced cement board (SDI = 0) and red oak flooring (SDI = 100) are used in the calculation of the smoke developed index. Historically, both the flame spread index and the smoke development index were based on red oak flooring being 100 and asbestos board being zero. The times for the flames to reach the end of a red oak flooring specimen is still used to calibrate the equipment but the results for red oak flooring are no longer part of the calculation of the flame spread index. With the specimen located on the top of the rectangular test furnace, the question of mounting methods has been a controversial aspect of the test method. A non-mandatory appendix of the ASTM E84 standard provides a guide to mounting methods. Tewarson *et al.* (2004) consider the major limitation of E84 (and likewise for E648) to be that it is very difficult to assess the fire behaviors of products for heat exposures and environmental conditions and shape, size, and arrangements of the products other than those used in the test. For the purpose of research and development of new products, the correlation of other test methods to the ASTM E84 results is an area of continuous interests (Eickner, 1977; Stevens, 1998; White and Dietenberger, 2004).

Based on the ASTM E84 FSI, three classes of interior finish are specified. Class A (or I) requires a FSI of 25 or less. For Class B (or II), the FSI must be 26 to 75. Class C (or III) has a FSI range of 76 to 200. Materials not meeting the upper limit of Class III are considered unclassified and not permitted where the flammability of the material is regulated. The requirement for the SDI is 450 or less for all three classes. The primary purpose of the fire-retardant treatment of combustible materials is to reduce the flame spread index classification. Values for flame spread index can be found in Galbreath (1964), American Forest & Paper Association (2002), and the publications of listing services such as Underwriters' Laboratory, Underwriters' Laboratory Canada, and Intertek. Non-fire-retardant paints and other thin coatings can have a negative or positive effect on the flame spread index. The effect is usually not sufficient to change the classification unless the classification of the uncoated material is low or marginal. For example, the classification of brick, concrete, aluminum, and gypsum plaster goes from FSI of zero unfinished to 25 with the application of a 1.3 mm alkyd or latex paint or one layer of cellulosic wallpaper (IRC, 1995). The FSI classifications of 150 for lumber and various types of plywood and 25 for gypsum wallboard do not change with the application of a 1.3 mm alkyd or latex paint or one layer of cellulosic wallpaper (IRC, 1995).

As noted by Belles (2003), materials that may not be adequately evaluated in the tunnel test include foam plastics, textile wall coverings, and vinyl film with foam backing. The large-scale room-corner tests of NFPA 265 and NFPA 286 provide an alternative to the tunnel test. These tests are similar to the standard ASTM and ISO room-corner tests that use a 2.4 m × 3.6 m × 2.4 m room with a single door opening for ventilation. A single propane burner is placed in a corner of the room.

For non-textile wall and ceiling finishes, the room-corner test of NFPA 286 is an alternative method to achieve Class A classification. In NFPA 286, the burner protocol is 40 kW for five minutes followed by 160 kW for ten minutes. The test materials fully cover the three walls and/or the ceiling depending on the intended application of the material being tested. The acceptance criteria are (i) no flame spread to ceiling during 40 kW exposure, (ii) no flame spread to the outer extremity of sample wall or ceiling and no flashover during the 160 kW exposure, and (iii) total smoke released must not exceed 1,000 m^2. The criteria for flashover are heat release rate of 1 MW, heat flux to floor of 20 kW/m^2, average upper air layer temperature of 600 °C, 'flames out door' and ignition of paper targets on floor.

Alternative acceptance criteria for Class A textile wall coverings are based on NFPA 265. In NFPA 265, the burner protocol is 40 kW for five minutes followed by 150 kW for ten minutes. Method A is a corner test in that the test material only covers portions of the two walls adjacent to the corner burner. Method B is like NFPA 286 in that the test material fully covers the three walls. No test material is on the wall with the door in either method. Observations needed for the acceptance criteria of NFPA 265 include flame spread to the outer extremity of the test material, observation of burning droplets, observation of flashover, and peak heat release rate. Criteria for flashover include 25 kW/m^2 to floor, average upper air temperature of 650 °C, 'flames out door', and ignition of paper targets on floor.

Besides the slight difference in the burner protocol, a second difference between NFPA 286 and NFPA 265 is that the burner is directly placed against the wall in NFPA 286 while it is 51 mm (2 in.) from the wall in NFPA 265. The NFPA 265 is considered unsuitable for the testing of ceiling materials since the flame from the burner alone will not touch the ceiling. The involvement of the ceiling in the NFPA test is somewhat due to the higher energy release rate of the NFPA 286 burner, but primarily due to the NFPA 286 burner being in direct contact with the walls, thereby reducing the area over which the flames can entrain air and increase the overall flame height (Janssens and Douglas, 2004). For decorative materials or other materials in a hanging configuration, one of the two tests described in NFPA 701 is used to determine the acceptance of the materials as flame-resistant materials. Selection of the specific test depends on the type of fabric or film being used.

The critical radiant flux apparatus is used to regulate carpets and other floor covering materials. Floor coverings such as wood, vinyl, linoleum, and other resilient non-fiber floor coverings are exempt from such requirements. The standard designations are ASTM E648 and NFPA 253. Because of the 30° incline of the radiant panel, the radiant heat flux profile on the test specimen ranges from 11 kW/m^2 to 1 kW/m^2 along the length of the specimen. The critical radiant flux is the flux that corresponds to the maximum distance of the flame front. As noted by Blackmore and Delichatsios (2002), the flooring radiant panel characterizes opposed flow horizontal flame spread and the pyrolysis is induced by both the external heat flux and convection/conduction from the flame at the front. The classes are Class 1 for a critical flux of 4.5 kW/m^2 or greater and Class 2 for a critical flux of 2.2 kW/m^2 or greater. Units of W/cm^2 are used in the U.S. codes and standards. The ASTM D2859 'pill test' is used as the least restrictive requirement. In the 'pill test' a small methenamine tablet is ignited in the center of a 230 mm square sample of the carpet. The specimen passes the test if the charred portion of the carpet does not extend beyond a 178 mm circle around the center of the sample. Requirements for carpets are also specified by the U.S. Federal Consumer Product Safety Commission (CPSC). The flooring radiant panel test is also part of the European regulations for floor coverings.

The duration of the standard ASTM E84 test is ten minutes. As a performance requirement for fire-retardant treated wood, the test is extended to 30 minutes. Fire-retardant-treated wood is required to have a FSI of 25 or less and show no evidence of significant progressive combustion when the test is continued for the additional 20 minutes. In addition, the flame front must not progress more than 3.2 m beyond the centerline of the burners at any time during the test. For fire-retardant-treated wood intended to be exposed to weather, damp, or wet locations; the listed FSI must not increase when tested after being exposed to the rain test of ASTM D2898.

Due to difficulties in the fire testing of foam plastic insulation that melt, specific requirements beyond those based on ASTM E84 test are specified for all foam plastic insulation. For foam plastic insulation that is part of a classified roof-covering assembly, testing according to FM 4450 or UL 1256 is an option. For exterior walls with foam plastic insulation, test methods include NFPA 268 and NFPA 285. In the NFPA 268 ignitability test, a 1.2 by 2.4 m test specimen is exposed to a 12.5 kW/m^2 heat flux and a piloted ignition source for a 20 minute test period. The NFPA 285 test is an intermediate-scale, multi-storey test apparatus with a minimum 4.6 m height. Large-scale tests such as FM 4880, UL 1040, NFPA 286, or UL 1715 are options to satisfy the provisions for foam plastic via special approval. FM 4880 is a 25 foot or 7.6 m high corner test. For light-transmitting plastics, tests include ASTM D1929 for self-ignition temperature, ASTM D2843 for smoke-development index, and ASTM D635 for combustibility classification (Class CC1 and CC2).

Exposed insulation is also tested using ASTM E84 for flame spread and smoke development. For loose-fill materials that cannot be mounted in the standard ASTM E84 tunnel without a screen or other artificial supports, the Canadian tunnel test for floors (CAN/ULC S102.2) is used to evaluate for flame spread. The Canadian flooring tunnel test addresses the question of testing of loose-fill insulation by placing the test specimen on the floor of the ASTM E84 test furnace and turning the burners down towards the test specimen. Another test for exposed attic floor insulation is the critical radiant flux test described in ASTM E970. It is similar to ASTM E648. The requirement for exposed attic insulation is a critical heat flux of 0.12 W/cm^2 or greater. Cellulose loose-fill insulation requirements are specified by the US Federal Consumer Products Safety Commission requirements that include the test of ASTM E970.

The standard test methods for roof coverings are in ASTM E108 (UL 790). Four separate fire tests for a relative comparison of roof coverings subjected to exterior fire exposure are part of ASTM E108: intermittent flame exposure test, spread of flame test, burning brand test, and flying brand test. The details of the tests are a function of the three classes (A, B, and C). Class A is the most restrictive. While the roof covering tests are primarily considered fire-resistance tests in terms of preventing penetration of the fire to the interior of the structure, there are requirements for surface flammability. The spread of flame test consists of a luminous gas flame burner at the edge of a 1 m wide inclined roof deck and an air current of 5.3 m/s. The length of the deck is 4 m for Class C, 2.7 m for Class B and 2.4 m for Class A. The gas flame and air current is applied in the Class A and B tests until any flaming of the test specimen recedes from its point of maximum flame spread, but no longer than ten minutes. If the test is run to obtain the Class C classification, the gas flame and air current is applied for four minutes. In the spread of flame test, the flaming cannot spread beyond 1.8 m for Class A, 2.4 m for Class B and 4.0 m for Class C. No significant lateral spread of flame from the area directly exposed to the test flame is permitted.

10.3   Canadian flammability requirements

The Canadian Commission on Building and Fire Codes is responsible for the National Building Code of Canada. Administrative and technical support for this model code is provided by the Institute for Research and Construction of the National Research Council of Canada. A new objective-based code is planned for 2005. The fire tests used in the Canadian codes for building materials requirements are similar to those in the United States. The test standard for determination of non-combustibility is CAN4-S114-M80. The tunnel test is CAN/ULC-S102-M88 in the Canadian standards. In addition, there is the tunnel test in which the test specimen is placed on the floor of the tunnel (CAN/ULC-S102.2-M88). It is intended for materials that (i) are intended for horizontal applications in which only the top surface is exposed, (ii) would require supports not representative of those used in actual applications to test the specimen on top of the tunnel (e.g. loose-fill insulation), or (iii) are thermoplastic. Because of melting or dripping, thermoplastics will have lower flame spread index in the ASTM E84 tunnel test with the material on the ceiling of the tunnel than in the Canadian flooring tunnel test (Canadian Wood Council, 1991). The Canadian flooring tunnel test addresses the question of testing of loose-fill insulation by placing the test specimen on the floor of the ASTM E 84 test furnace. Except for a few specific applications, the upper limits for the flame-spread rating are 25, 75, or 150. When regulated, the limits on the smoke developed classification are for values of 50, 100, 300, and 500. The standard for roof covering is CAN/ULC-S107-M87. More recently, the cone calorimeter was proposed as a standard test method for degree of combustibility (ULC-S135-04) (Richardson and Brooks, 1991; Richardson, 1994; Carpenter and Janssens, 2005).

10.4   European flammability requirements

The national fire regulators of the European Union countries developed a new flammability classification system in 2001. The new system is expected to replace the large number of different tests and classification systems used across the European Union. Two classification systems were developed; one for floor coverings (Table 10.1) and the other for all other construction products (Table 10.2). Suffix 'FL' is used to denote the floor covering classifications. This classification system is published in the *Official Journal of the European*

Table 10.1 European Union classification system for floor coverings

Class	Test method	Classification criteria
$A1_{FL}$	ISO 1182 and ISO 1716	$T \leq 300\,°C$ and $m \leq 50\%$ and $t_f = 0$ and, $PCS \leq 2\,MJ/kg$ and $PCS \leq 1.4\,MJ/m^2$
$A2_{FL}$	ISO 1182 or ISO 1716 and ISO 9239-1	$T \leq 500\,°C$ and $m \leq 50\%$ and $t_f = 20$ s or, $PCS \leq 3\,MJ/kg$ and $PCS \leq 4\,MJ/m^2$ and, Critical flux $\geq 8\,kW/m^2$
B_{FL}	ISO 9239-1 and ISO 11925-2	Critical flux $\geq 8\,kW/m^2$ and $Fs \leq 150\,mm$ within 20 s following 15 s exposure.
C_{FL}	ISO 9239-1 and ISO 11925-2	Critical flux $\geq 4.5\,kW/m^2$ and $Fs \leq 150\,mm$ within 20 s following 15 s exposure.
D_{FL}	ISO 9239-1 and ISO 11925-2	Critical flux $\geq 3\,kW/m^2$ $Fs \leq 150\,mm$ within 20 s following 15 s exposure.
E_{FL}	ISO 11925-2	$Fs \leq 150\,mm$ within 20 s following 15 s exposure.
F_{FL}	No requirement	

Communities (OJ L50, 23.2.200, P.14) and published as EN 13501-1. The reference scenario used as the basis for the classification system is a fire starting in a small room and growing to reach flashover subject to a contribution from the room lining material. The reference test protocol is standardized as ISO 9705.

Table 10.2 European Union classification system for building products other than floor coverings

Class	Test method	Classification criteria
A1	ISO 1182 and ISO 1716	$T \leq 300\,°C$ and $m \leq 50\%$ and $t_f = 0$ and, $PCS \leq 2\,MJ/kg$ and $PCS \leq 1.4\,MJ/m^2$
A2	ISO 1182 or ISO 1716 and EN 13823	$T \leq 500\,°C$ and $m \leq 50\%$ and $t_f = 20$ s or, $PCS \leq 2\,MJ/kg$ and $PCS \leq 1.4\,MJ/m^2$ and, $FIGRA \leq 120\,W/s$ and $THR_{600s} \leq 7.5\,MJ$
B	EN 13823 and ISO 11925-2	$FIGRA \leq 120\,W/s$ and $THR_{600s} \leq 7.5\,MJ$ and $Fs \leq 150\,mm$ within 60 s following 30 s exposure.
C	EN 13823 and ISO 11925-2	$FIGRA \leq 250\,W/s$ $Fs \leq 150\,mm$ within 60 s following 30 s exposure.
D	EN 13823 and ISO 11925-2	$FIGRA \leq 750\,W/s$ and $THR_{600s} \leq 15\,MJ$ and $Fs \leq 150\,mm$ within 60 s following 30 s exposure.
E	ISO 11925-2	$Fs \leq 150\,mm$ within 60 s following 30 s exposure.
F	No requirement	

The following series of tests are used for classification.

ISO EN 1182 Non Combustibility Test: identifies products that will not contribute significantly to a fire. Used for classification in Class A1, A2, $A1_{FL}$, $A2_{FL}$.

ISO EN 1716 Calorific Value: determines potential heat release by a product under complete combustion. Used for classification in Class A1, A2, $A1_{FL}$, $A2_{FL}$.

EN 13823 Single Burning Item Test: evaluates the potential contribution to fire growth with a single burning item in a room corner. Used for classification in Class A2, B, C and D.

ISO EN 11925-2 Ignitability: evaluates the ignitability of a product when subject to a small flame. Used for classification in Class B, C, D, E, B_{FL}, C_{FL}, D_{FL} and E_{FL}.

ISO EN 9239-1 Burning behavior of floor coverings exposed to a radiant heat source: evaluates the critical radiant flux at flame extinguishment over a horizontal surface. Used for classification in Class $A2_{FL}$, B_{FL}, C_{FL}, D_{FL} and E_{FL}

10.5   Japanese flammability requirements

Japan uses the Cone Calorimeter test results to classify the interior finish materials. The test method referenced is ISO 5660-1 at a radiant exposure of $50\,kW/m^2$. The three class classification system is summarized in Table 10.3.

Table 10.3 Japanese classification system for interior finish materials

Class	Test method	Classification criteria
Non-combustible materials	ISO 5660-1 with $50\,kW/m^2$ exposure for 20 mins	THR $\leq 8\,MJ/m^2$ and PHR $\leq 200\,kW/m^2$
Quasi non-combustible materials	ISO 5660-1 with $50\,kW/m^2$ exposure for 10 mins	THR $\leq 8\,MJ/m^2$ and PHR $\leq 200\,kW/m^2$
Fire-retardant materials	ISO 5660-1 with $50\,kW/m^2$ exposure for 5 mins	THR $\leq 8\,MJ/m^2$ and PHR $\leq 200\,kW/m^2$

Exception: Peak heat release rate is allowed to exceed $200\,kW/m^2$ provided the time interval where the heat release rate stayed above $200\,kW/m^2$ is less than ten seconds.

10.6   Chinese flammability requirements

China has established a non-combustible test and a three-class classification system for combustible wall and ceiling interior finishes. They have also established a two-class classification system for combustible floor coverings. The following test methods are used for the purpose of classification.

GB 5464-99: non-combustibility test method of building materials. This test is similar in principle to ISO 1182 and ASTM E136. Used for classification in Class A.

GB 8625-88 Combustibility test of burning materials. This test is based on the German fire test apparatus *Brandschacht* (DIN 4102-15) and is used for classification in Class B1 and B2 for wall and ceiling linings.

GB 8626-88: Flammability of building materials. This test is based on the *Kline burner* and is used for classification in Class B3 for wall and ceiling linings.

GB 11785-89: Determination of thermal transfer of flooring material-radiant heat method. The test method is similar in principle to ASTM E648 and ISO 9239-1. Floor coverings with a critical radiant flux $\geq$ 4.5 kW/m^2 are classified B1 while those between 4.5 and 2.2 kW/m^2 are classified B2.

10.7   Future trends

Various drivers will continue to push innovations in fire testing and the adoption of improved test methodologies for flammability of building materials. The first driver is the increased level of engineering involved in the fire protection profession. The increased use of fire models to demonstrate that acceptable levels of fire safety will increase the demand for fire tests that provide suitable test results to use either as input to the models or provide validation data for the models. Tests that provide only a comparative measure of fire performance under specified and limited test conditions will become less acceptable. In the future, Belles (2003) noted that we may use fire models to link the actual hazards created by the use of a product in a specific situation to regulatory limits.

In synergy with increased fire protection engineering is the international movement toward performance-based or objective-based building codes. Many of the traditional tests for flammability that have been acceptable in a prescriptive code environment will hinder the potential benefits of the performance-based codes. For harmonization and reliable assessment of fire resistance of materials, Tewarson *et al.* (2004) noted it is necessary to use quantitative measurements, rather than visual observation. Even if the goals of performance-based codes are not achieved in the near term, Beyler (2001) believes that there is the need to rationalize our prescriptive fire safety requirements by using these methods for assessing fire safety performance.

Grayson and Hirschler (1995) noted (i) the recent trend to replace older tests that only rank materials with new standard fire tests that can provide input into mathematical fire models and fire hazard assessment, and (ii) the use of real-scale tests for validation of the small-scale tests and the models. However, as noted by Karlsson *et al.* (2002), the validity of the models must be demonstrated for a wider range of materials and the models must be made more user friendly and widely available to engineers before there can be widespread performance-based design of interior finish materials. As is happening within the European Union, world globalization of markets will drive the harmonization of regulatory test standards for products. Tewarson *et al.* (2004) expects further harmonization of test methods as many regulatory agencies are considering augmenting or replacing the prescriptive-based fire codes (currently in use) by performance-based codes. FORUM for International Cooperation on Fire Research (Croce, 2001) supports this movement away from *ad hoc* approval tests and towards scientifically based tools (accurate data, tests, and models) as a basis for equitable performance levels.

The increased use of heat release rate measurements to quantify the flammability of building materials is likely to continue. Innovation in building materials will continue to be a driver for improvements in the testing of building products for flammability. One potential impact of a new material or product, such as foam plastic in recent history, is to illustrate the inadequateness of an existing test method to prevent high-hazard materials from the market. It is expected that the introduction of new and improved low-hazard materials will increase the pressure to develop and implement methodologies for degrees of combustibility that will not require prescriptive lists of exceptions to the current noncombustibility requirements.

10.8 Sources of further information and advice.

10.8.1 Standard organizations

ASTM International, 100 Barr Harbor Drive, West Conshohocken, PA 19428 2959, USA, www.astm.org

D 2843 Test for density of smoke from the burning or decomposition of plastics.
D 2859 Test method for ignition characteristics of finished textile floor covering materials.
D 2898 Test method for accelerated weathering of fire-retardant-treated wood for fire testing.
E 84 Test methods for surface burning characteristics of building materials.
E 108 Test methods for fire tests of roof coverings.
E 136 Test method for behavior of materials in a vertical tube furnace at 750 °C.
E 648 Test method for critical radiant flux of floor-covering systems using a radiant heat energy source.
E 970 Test method for critical radiant flux of exposed attic floor insulation using a radiant heat energy source.

Consumer Product Safety Commission, 4330 East West Highway, Bethesda, MD 20814-4408, USA, www.cpsc.gov

16 CFR Part 1209 Interim safety standard for cellulose insulation.
16 CFR Part 1630 Standard for the surface flammability of carpets and rugs (FF 1-70).

Factory Mutual Approvals, 1151 Boston-Providence Turnpike, Norwood, MA 02062, USA www.fmglobal.com/approvals

4450 Approval standard for class I insulated steel deck roofs.
4880 American national standard for evaluating insulated wall or wall and roof/ceiling assemblies, plastic interior finish materials, plastic exterior building panels, wall/ceiling coating systems, interior and exterior finish systems.

National Fire Protection Association, 1 Batterymarch Park, Quincy, MA 02269-9101 USA, www.nfpa.org

101 Life safety code
253 Test for critical radiant flux of floor covering systems using a radiant heat energy source.
259 Test method for potential heat of building materials.
265 Standard method of fire tests for evaluating room fire growth contribution of textile wall coverings.
268 Standard test method for determining ignitability of exterior wall assemblies using a radiant heat energy source.
285 Standard method of test for the evaluation of flammability characteristics of exterior non-load-bearing wall assemblies containing combustible components.
286 Standard method of fire test for evaluating contribution of wall and ceiling interior finish to room fire growth.
701 Standard methods of fire tests for flame-propagation of textiles and films.

Underwriters Laboratories, 333 Pfingsten Road, Northbrook, IL 60062-2096, USA, www.ul.com

723 Test for surface burning characteristics of building materials.
790 Standard test method for fire tests of roof coverings.
1040 Fire test of insulated wall construction.
1256 Fire test of roof deck construction.
1715 Fire test of interior finish material.
1975 Fire tests of foamed plastics used for decorative purposes.

Underwriters Laboratories of Canada, 7 Crouse Road, Scarborough, Ontario, M1R 3A9 Canada. www.ulc.ca

CAN/ULC-S102-M88 Test for surface burning characteristics of building materials and assemblies.
CAN/ULC-S102.2-M88 Test for surface burning characteristics of flooring, floor covering, and miscellaneous materials and assemblies.
CAN/ULC-S107-M87 Fire tests of roof coverings.

CAN/ULC-S134 Standard method of fire test of exterior wall assemblies.

CAN/ULC-S114-M80 Test for determination of non-combustibility in building materials.

CAN/ULCS135-04 Standard test method for the determination of combustibility parameters of building materials using an oxygen consumption calorimeter (cone calorimeter).

Canadian Codes Centre, Institute for Research in Construction, National Research Council Canada, Building M-23A, 1200 Montreal Road, Ottawa, Ontario, K1A 0R6 Canada (Institute for Research and Construction of the National Research Council of Canada, *http://irc.nrc-cnrc.gc.ca/irccontents.html* and Canadian Commission on Building and Fire Codes *http://www.nationalcodes.ca/*)

National Building Code of Canada 1995.

International Organization for Standardization http://www.iso.ch/

ISO 1182: 2002: Reaction to fire tests for building products – Non-combustibility test.

ISO 1716: 2002: Reaction to fire tests for building products – Determination of the heat of combustion.

ISO 5660-1:2002: Reaction to fire tests – Heat release, smoke production and mass loss rate – Part 1: Heat release rate (cone calorimeter method).

ISO 9239-1:2002: Reaction to fire tests for floorings – Part 1: Determination of the burning behaviour using a radiant heat source.

ISO 11925-2:2002: Reaction to fire tests – Ignitability of building products subjected to direct impingement of flame – Part 2: Single-flame source test.

European Committee for Standardization: http://www.cenorm.be/

EN 13501-1: Fire classification of construction products and building elements. Classification using test data from reaction to fire tests.

EN 13823:2002: Reaction to fire tests for building products. Building products excluding floorings exposed to the thermal attack by a single burning item.

10.9   References

American Forest & Paper Association (2002), *Flame spread performance of wood products,* Design for Code Acceptance (DCA) No. 1, Washington DC, AF&PA. (Available at *www.awc.org*).

Belles D W (2003), 'Interior finish', in *Fire Protection Handbook*, Section 12, Chapter 3, Quincy, MA, National Fire Protection Association, 12-43–12-60.

Beyler C L (2001), 'Fire safety challenges in the 21st century', *Journal of Fire Protection Engineering*, 11(1), 4–15.

Blackmore J M, Delichatsios M A, (2002), 'Flammability tests for assessing carpet performance', *Journal of Fire Protection Engineering*, 12, 45–59.

Canadian Wood Council (1991), *Wood and Fire Safety*, Ottawa, Canadian Wood Council, 266 p.

Carpenter K, Janssens M (2005), 'Using heat release rate to assess combustibility of building products in cone calorimeter', *Fire Technology*, 41, 79–92.

Clark F R S (1981), 'Fire spread tests – a critique', *Fire Technology*, 17(2), 131–138.

Croce P A (2001), 'The FORUM for International Cooperation on Fire Research: A position paper on evaluation of products and services for global acceptance', *Fire Safety Journal*, 36(7), 715–717.

Eickner H W (1977), 'Surface flammability measurements for building materials and related products', in Kolthoff I M, Elving P J and Stross F H, *Treatise on Analytical Chemistry*, Part 3, Volume 4, John Wiley & Sons (available at http://www.fpl.fs.fed.us/documnts/pdf1977/eickn77a.pdf).

Emmons H W (1974), 'Fire and fire protection', *Scientific American*, 231(1), 21–27.

Galbreath M (1964), *Flame spread performance of common building materials*, Technical Paper No. 170, NRCC 7820, Ottawa, Division of Building Research, National Research Council Canada.

Grayson S J, Hirschler M M (1995), 'Comparison of ASTM fire standards with international fire standards for building and contents', in Grand A F *Fire standards in the international marketplace*, ASTM STP 1163, Philadelphia, PA, American Society for Testing and Materials, 41–60.

IRC (Institute for Research in Construction) (1995), *National building code of Canada 1995*, Ottawa, National Research Council of Canada, Institute for Research and Construction, 571 p.

Janssens M, Douglas B (2004), 'Wood and wood products', Chapter 7, in Harper C A, *Handbook of Building Materials for Fire Protection*, New York, McGraw-Hill, 7.1–7.58.

Karlsson B, North G, Gojkovic D (2002), 'Using results from performance-based test methods for material flammability in fire safety engineering design', *Journal of Fire Protection Engineering*, 12, 93–108.

Richardson L R (1994), 'Determining degrees of combustibility of building materials-National Building Code of Canada', *Fire and Materials*, 18, 99–106.

Richardson L R, Brooks M E (1991), 'Combustibility of building materials', *Fire and Materials*, 15, 131–136.

Stevens M G (1998), 'Cone calorimeter as a screening test for ASTM E-84 tunnel test', in *Proc. 5th International conference in Fire and Materials*, London, Interscience Communications, 147–151.

Tewarson A, Chin W, Shuford R (2004), 'Materials specifications, standards, and testing', Chapter 2, in Harper C A, *Handbook of Building Materials for Fire Protection*, New York, McGraw-Hill, 2.1–2.54.

White R H, Dietenberger M A (2004), 'Cone calorimeter evaluation of wood products', in *Proc. 15th Annual BCC Conference on Flame Retardancy*, Norwalk, CT, Business Communications Co., Inc., 331–342.

Fire testing in road and railway tunnels

H INGASON, SP Fire Technology, Sweden

11.1 Introduction

Knowledge of smoke spread and fire development in tunnels is generally obtained from large-scale testing and laboratory testing (e.g. scale models). The aim is usually to investigate some specific problems such as the influence of different ventilation systems on smoke and temperature distribution along the tunnel, the fire development in different type of vehicles or the heat exposure to the construction. Large-scale testing is generally expensive, time consuming and logistically complicated to perform and therefore the number of large-scale tests in tunnels is limited. The information obtained from these tests is often incomplete due to the low number of tests and lack of instrumentation. Large-scale testing is, however, necessary in order to obtain acceptable verification in realistic scale and the information obtained from many of these large-scale tunnel fire tests provides the basis for the technical standards and guidelines used for tunnel design today.

In this chapter an overview and analysis of large-scale tests performed in road and railway tunnels is given. A general analysis of the results have been carried out and wherever possible, presentation of following parameters is given:

- measured peak heat release rates (HRR)
- measured peak gas temperatures
- flame lengths.

These parameters are important for engineers and scientists working with tunnel fire safety.

11.2 Overview of large-scale tunnel experiments

The diversity of the large-scale tests found in the open literature is in the scales, the type of fire source and size (HRR), instrumentation, documentation, tunnel geometry and ventilation conditions. In order to show this diversity, a summation of all large-scale tunnel fire tests carried out worldwide since the beginning of

Table 11.1 Summation of large-scale fire tests carried out since the middle of the 1960s

Test programme, country, year	No. of tests	Fire source	Cross-section (m²)	Height (m)	Length (m)	Measurements	Range of peak HRR (MW)	Comments
Ofenegg, Switzerland, 1965	11	Gasoline pool (6.6, 47.5, 95 m²)	23	6	190	T, CO, O₂, v, visibility	11–80	Single track rail tunnel, dead end, sprinkler
Glasgow, 1970	5	Kerosine pool (1.44, 2.88, 5.76 m²)	39.5	5.2	620	T, OD	2–8	Disused railway tunnel
Zwenberg, Austria, 1974–1975	30	Gasoline pool (6.8, 13.6 m²), wood and rubber	20	3.9	390	T, CO, CO₂, NOₓ, CH, O₂, v, OD	8–21	Disused railway tunnel
P.W.R.I. Japan, 1980	16	Gasoline pool (4, 6 m²), passenger car, bus	57.3	~6.8	700	T, CO, CO₂, v, OD, radiation	Pool: 9–14* Cars and buses unknown	Special test tunnel, sprinkler
P.W.R.I. Japan, 1980	8	Gasoline pool (4 m²), bus	58	~6.8	3277	T, CO, CO₂, O₂, v, OD, radiation	Pool: 9 Bus unknown	In use road tunnel, sprinkler
TUB–VTT, Finland, 1985	2	Wood cribs (simulate subway coach and collision of two cars)	24–31	5	140	HRR, T, m, CO, CO₂, O₂, v, OD	1.8–8	Disused cavern system

EUREKA 499, Norway, 1990–1992	21	Wood cribs, heptane pool, cars metro car, railcars, HGV trailer and mockup	25–35	4.8–5.5	2300	HRR, T, CO, m, CO_2, O_2, SO_2, $CxHy$, NO, visibility, soot, m, v	2–120	Disused transportation tunnel
Memorial, USA, 1993–1995	98	Fuel oil (4.5–45 m^2)	36 and 60	4.4 and 7.9	853	HRR, T, CO, CO_2, v, visibility	10–100	Disused road tunnel, sprinkler
Shimizu No. 3, Japan, 2001	10	Gasoline pool (1, 4, 9 m^2), cars, bus	115	8.5	1120	T, v, OD, radiation	2–30*	New road tunnel, sprinkler tests
2nd Benelux tunnel, The Netherlands, 2002	14	n-heptane + toluene, car, van, HGV mock up	50	5.1	872	HRR, T, m, radiation, v, OD, visibility	3–26	New road tunnel, sprinklers
Runehamar tunnel, Norway, 2003	4	Cellulose, plastic, furniture	32 - 47	4.7–5.1	1600	HRR, T, PT, CO, CO_2, O_2, HCN, H_2O, isocyanates, OD, radiation	70–203	Disused road tunnel

* the bus was determined to be equal to 20 MW convective and 30 MW total.

HRR = Heat Release Rate, m = mass loss rate, T = temperature, PT = PlateThermometer, CO = Carbon monoxide, CO_2 = Carbon dioxide, CH = Hydrocarbon, HCN = cyanide, H_2O = water vapour, v = velocity, OD = Optical density, visibility = cameras for smoke registration.

1960 is given in Table 11.1. A dozen large-scale fire test programmes have been carried out to date. Except for the Memorial test series, which consisted of 98 tests, all the other test series were less than or equal to about 30 tests. The focus has mainly been on the heat and smoke spread and how different ventilation systems influence these parameters. Nearly half of the test series included sprinkler testing, which is surprisingly high compared to how few sprinkler systems are installed in tunnels today. In Japan over 80 tunnels have sprinkler systems whereas in other parts of the world the total is less than ten.

The quality of large-scale tests carried out in the 1960s to 1980s varies considerably and in all these tests there is a lack of the key fire hazard parameter, the heat release rate (HRR). The boundary conditions were usually not the most favourable for validation of advanced computer models. These tests were performed simply to fill a wide gap of non-exiting knowledge about the influence of ventilation systems on tunnel fires rather than to fulfil the need of academic studies and validation of computer models.

The first series of large-scale tunnel fire test series were performed in the 1960s and the 1970s in Europe. They were mainly directed at solving the fire problems of road tunnels in Europe and they had a major influence on the technical standards at that time and even today. Grant *et al.*[1] considered these tests as 'tantalizing snapshots' primarily due to the inadequate HRR data. The documentation of fuel mass loss rates, combustion efficiency, ventilation flow rates and wind and pressure conditions was not sufficient to completely validate the functional relationships derived theoretically or in laboratory scales at that time.

Among these renowned large-scale tunnel fire test series are the Ofenegg (1965, 24 m², 190 m)*[2] series, the Glasgow series (1970, 40 m², 620 m)[3] and the Zwenberg series (1974–1975, 20 m², 370 m).[3-5] Both the Ofenegg[6] and the Zwenberg[5] test series have been reported with admirable detail on the test data and the test set-up. A less known large-scale test series was carried out in Japan in the late 1970s and at the beginning of the 1980s[7] (P.W.R.I. – Public Works Research Institute). The documentation in English is rather limited. The tests were carried out in a full-scale test tunnel (1980, 57.3 m², 700 m) built by P.W.R.I. and in a full-size road tunnel, the Kakei Tunnel (1980, 58 m², 3277 m). This was the first time full-size vehicles were used in a large-scale test series in tunnels (cars, buses) but unfortunately no HRR measurements were carried out. Sprinkler systems were tested both in the Ofenegg test series and the P.W.R.I. test series but the results from these tests led in two different directions. The consequence of the Ofenegg tests is that there are very few sprinkler systems installed in tunnels in Europe whereas in Japan sprinklers have become a key part of the technical fire safety systems in Japan. In Europe and other parts of the world the ventilation system still constitutes the basic technical fire safety feature of tunnels. In general, we can say that the tests carried out in the 1960s

* Test year, cross-section, tunnel length.

and 1970s had, and still have, a major influence on the standards and guidelines used for fire safety in tunnels.

A new era began in tunnel fire testing with the use of Oxygen Consumption Calorimetry.[8,9] In the early 1980s Oxygen Consumption Calorimetry became widely used in fire laboratory worldwide. Tewarson[10] introduced another gas analysis technique, Carbon Dioxide Generation. This method was not as widely used in fire laboratories as Oxygen Consumption Calorimetry but both these techniques found their way into tunnel fire testing. A German (Technische Universität Braunschweig (TUB)) and Finnish (VTT) cooperation[11,12] (1985, 24–31 m^2, 140 m) led to the performance of two large-scale tests in a tunnel using wood cribs as fuel to simulate fire in a subway car (80 GJ), and in two passenger cars (11.7 GJ) colliding in a tunnel. The original idea was to utilise the oxygen consumption technique, but due to uncertainties in the oxygen and flow measurements it was never completed.[13] The cooperation between TUB-VTT developed and widened later into the EUREKA project EU499 (1990–1992, 25–35 m^2, 2300 m)[12] in the early 1990s. Oxygen consumption calorimetry was used for the first time in the EUREKA project and made it possible to measure the HRR from large vehicles with relatively good accuracy, although not nearly as good as in fire laboratories. Experience from large-scale testing in tunnels shows that the accuracy of HRR measurements is highly dependent on the measuring location and the number of probes used.[14–16] The accuracy has been estimated to be in the order of ±15–25% in large-scale testing whereas in fire laboratories it is in the order of ±7–11%.[17]

When the EUREKA tests were performed at the beginning of 1990s, they became a milestone concerning new valuable information for tunnel engineers, especially the great variety in HRR data for vehicle types such as cars, train coaches, subway coaches and articulated lorries with furniture.[14,16,18] The tests have resulted in significant improvements of information regarding HRR levels for single vehicles in tunnels. The EUREKA tests contain the most comprehensive fire testing of rail and metro vehicles ever performed. Nearly all the large-scale fire tests series conducted prior to the EUREKA tests focused on road tunnel problems. The EUREKA tests have filled the gap between road and railway tunnels although there is still a great need for more testing of vehicles travelling through rail and metro tunnels. In the EUREKA tests there was very little consideration given to the risk of fire spread between vehicles, mainly because prior to, and at the time of, the performance of the tests there had not been that many serious large fires involving multiple vehicles as turned out to be the case in the late 1990s and beginning of the century. The great majority of road tunnel fires consist of fires in one or two vehicles whereas large catastrophic fires involve multiple vehicles. This fact has not yet been considered in today's standards and guidelines for tunnels.

Another milestone in large-scale tunnel fire testing was obtained in the Memorial tunnel test series (1993–1995, 36–60 m^2, 853 m)[19] carried out between

1993–1995 although the fire source consisted of low sulfur No. 2 fuel oil pans (diesel-related fuel) and not real vehicles. There was a need for a well-defined fire source in order to compare the performance of the different ventilation systems. In order to establish the influence of vehicles on the ventilation flow, silhouettes representing vehicles were placed at different locations inside the tunnel. There is no doubt that the Memorial tests demonstrated very well the performance and control of different types of smoke and heat ventilation systems. The tests also provide a very important source for validation of computer models (CFD – Computational Fluid Dynamics). The Memorial test data were made available on a CD-ROM and are the best-documented fire test results ever made available.

A comprehensive instrumentation was located in both the up- and down-stream directions of the fire. The test results were used as the basis for the design of the ventilation system in the Boston Central Artery Tunnel project and they have already had a great impact on the design of smoke control systems worldwide. The usefulness of longitudinal ventilation and exhaust ventilation was clearly shown as well as the positive performance of foam sprinkler systems. A consistent confirmation for the correlation between HRR and 'critical velocity' was established for the first time on a large scale, especially the independence over 3 m/s. To date, these fire experiments are the most comprehensive and most expensive large-scale tests ever performed. It is a little surprising that the Memorial test series was not designed with the aid of small-scale tests in order to minimise the number of large-scale tests and thereby costs. This would have made this large-scale test series much more effective and the focus would have been more on the promising solutions. There is no doubt that the Eureka tests and the Memorial tests are the most well-known and best-regarded large-scale fire test series to date. They have already been established as the 'large-scale fire tests' and provide a new basis for standards and knowledge in tunnel fire safety.

Since 2000 there have been to date three mediocre fire test series performed in large-scale tunnels. Large-scale tests were performed in the No. 3 Shimizu Tunnel (2001, 115 m², 1120 m) on the New Tomei Expressway using gasoline pan fires, cars and a bus.[20] These tests included natural and longitudinal ventilation as well as water sprinklers. The main focus was on heat and smoke spread in a large cross-section tunnel (three lanes). In the Second Benelux tunnel in the Netherlands (2002, 50 m², 872 m, large-scale tests with cars and HGV mock-ups using wood pallets were performed in 2002.[21] Tests with natural ventilation and longitudinal ventilation and water sprinkler systems were performed. These tests provide very important results on the effects of longitudinal ventilation on HRRs in HGVs and on car fires. A large-scale test series was carried out in the Runehamar tunnel (2003, 47 m², 1600 m).[15,22] Four tests using a mock-up of HGV fire loads were carried out. These tests provide important information on fire development in different types of ordinary hazard goods and show that this type of freight can create fires which are similar in size to a gasoline tanker fire. The tests showed clearly that the gas temperature levels from ordinary hazardous goods could easily be in the order of a tank fire. The results from the Runehamar tests have already had implications on the furnace testing of tunnel elements in Scandinavia and will most likely have dramatic effects on the future NFPA 502 standards.[23]

There are numerous tests found in the literature that have been carried out in 'intermediate-sized' tunnels. The cross-sections vary between 5 to 13 m^2, which can be compared to the cross-sections of the large-scale tests series presented which varied between 25 to 115 m^2. Apte and Green[24] presented a detailed study of pool fires in a tunnel (NA, 13 m^2, 130 m) using longitudinal ventilation in a typical mine roadway. These experiments were used for developing a computational fluid dynamics (CFD) approach to modelling tunnel fires. They also show the effects of longitudinal ventilation on burning rate of pool fires.

An extensive series of experiments were carried at the Health and Safety Laboratory (HSL) in Buxton, England (1992–1993, 5.4 m^2, 366 m).[25] Both obstructed- and open-tunnel situations were considered. The former included one-third scale models of a part of a HGV shuttle train from the Channel Tunnel and in the latter kerosene pools. In the second phase of the test programme, even wood cribs were used. The HRRs were measured with the oxygen-consumption technique and mass loss rates combined with a value of combustion efficiency. The objective was to provide data for CFD simulation of the interaction of longitudinal flow and a back-layering smoke flow. The results suggested that the value of the critical velocity tended to some near-constant value with increasing HRR and thus did not conform to the simple theory developed by Thomas.[26] This discovery was very important for the design of longitudinal ventilation systems, especially when this finding was verified in the Memorial tunnel test series.

Ingason et al. (1995, 9 m^2, 100 m)[27] presented results from tests carried out in intermediate-sized tunnel tests. Tests were carried out using wood cribs, pool fires and a passenger car. The aim with these tests was to establish a correlation between optical smoke density and gas concentrations[28] for use in CFD simulations. The CFD codes at that time were not able to predict with any accuracy the optical smoke density but they could predict the concentrations of gas species. The experiments showed a good correspondence between the measured optical density (visibility) and the measured gas concentrations at different locations in the tunnel and, accordingly, that this was a suitable way to predict the smoke optical density or visibility.

There are many tests performed in large-scale tunnels, where the main purpose is to test the ventilation systems of a specific tunnel before it is put into operation. The fire source can consist of pan fires, wood crib fires or car fires. Examples of such well-documented tests can be found in refs 29 and 30. Within the framework of legal enquiry initiated after the catastrophic fire in the Mont Blanc tunnel in 1999, a series of large-scale tests were conducted in the same tunnel (2000, 50 m², 11600).[31] The objective was to investigate the consequence of the fire during the first half-hour. These tests cannot be regarded as a part of a general large-scale test programmes and will therefore not be included in the more detailed presentation given of each of the tests. Finally, worth mentioning, although not large-scale or intermediate-scale tests, are the small-scale tests performed by TNO in an 8 m long tunnel, 2 m high and 2 m wide model.[32] In these tests very high gas temperatures were measured and the Rijkswaterstaat Tunnel Curve (the RWS Curve) in the Netherlands is based on these tests.

11.3   Description of large-scale tests

In the following, more detailed information is given for each of the tests listed in Table 11.1. Wherever possible information on maximum HRRs ($\dot{Q}_{max}$), ambient (T_0) and maximum ceiling temperatures (T_{max}) and maximum horizontal flame lengths (L_f) along the ceiling is given. The maximum horizontal flame lengths along the ceiling are based on ceiling temperature measurements, assumed flame tip at 600 °C, as proposed by Rew and Deaves.[33] A definition of L_f is given in Fig. 11.1 for the case with natural ventilation and longitudinal ventilation. In order to get flames to deflect along the ceiling, and thereby comply with the definition of L_f, the vertical flame height in the open needs to be larger than the tunnel height, H. For calculations of the vertical flame height in the open, $L_{f,free}$ see, e.g., ref. 34:

$$L_{f,free} = -1.02 \cdot D + 0.23 \cdot \dot{Q}^{2/5} \qquad 11.1$$

where D is the diameter in metres of the fire source and Q is the HRR in kW.

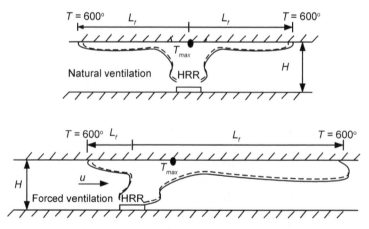

11.1 A definition of the maximum ceiling temperature, T_{max} and the horizontal flame length, L_f, in tunnel fires with natural and forced ventilation. H is the ceiling height, u is the longitudinal ventilation and HRR is the heat release rate.

In many of the large-scale test series presented here this has not been the case and the results are then given with the sign (–).

11.3.1   Ofenegg 1965

The first large-scale tunnel fire test series to obtain scientific and engineering information was carried out in the Ofenegg tunnel in Switzerland 1965.[2] These tests were carried out in order to study the ventilation capacities (natural, longitudinal,* semi-transverse†) in the case of a fire, especially in case of a gasoline tank fire. The tests were expected to give information on the hazardous level for tunnel users, possibilities to rescue people and the impact on tunnel construction and installations. Also the influence of a water sprinkler system was investigated. This type of information was urgently needed in Switzerland due to the large road tunnel projects carried out in the 1960s. The tunnel used for these experiments was a railway tunnel (23 m², 3.8 m wide and 6 m high) with a dead end located 190 m from the portal and the ceiling was 6 m high with a rounded top. A total of 11 tests were performed using gasoline pool fires on a concrete trough with the edge placed 131.5 m from the open entrance. The other end was bricked up. The sizes of the pools used were 6.6 m², 47.5 m² and 95 m², respectively, with the smallest representing the contents of the fuel tanks of two cars and the largest a substantial spill from a gasoline tanker. The width of the trough (fuel pan) was 3.8 m and the length of the trough varied; 1.7 m, 12.5 m and 25 m, respectively.

The experiments showed that large quantities of smoke were generated in all tests. The smoke front travelled along the tunnel at speeds of up to 11 m/s and visibility deteriorated in most cases 10–20 s after the start of the fires. Generally, the greater the fuel quantity, the worse the conditions.[35] It was found that heat evolution was a decisive factor and with a semi-transverse ventilation system supplying up to 15 m³/s the burning rate was virtually unchanged. With a longitudinal ventilation system giving an air velocity along the tunnel of about 1.7 m/s, averaged over the cross-section of the tunnel, the burning rate of a 47.5 m² fire was about twice that for the 47.5 m² fire with no ventilation.

An estimation of the HRR has been made here and the results are presented in Table 11.2. The estimation, which is based on the fuel rates given in ref. 2 and an assumed combustion efficiency of 0.8 in the tunnel, show that the average HRR was 2.1 MW/m² for the 6.6 m² fuel, 0.95 MW/m² for the 47.5 m² fuel and 0.35 MW/m² for the largest one (95 m²). In the open the HRRs are in the order of 2.4 MW/m² (0.055 kg/m²s and $\Delta H_c = 43.7$ MJ/kg).[36] It is obvious that the

---

* Longitudinal ventilation consists of fans blowing in outside air through the rear end duct system with an air quantity of 39 m³/s, i.e., a longitudinal velocity of 1.7 m/s.
† Semi-transverse systems have air inlets at low levels but either no extraction or extraction at only a few points, so that the air and vehicle exhaust gases flow along the tunnel at a velocity which increases along the tunnel length. The fresh air supply is equal to 0.25 m³/s, m.

Table 11.2 Relevant data from the Ofenegg tunnel tests in 1965

Test no.	A_f	Type of ventilation	Sprinkler	Air supply ventilation	T_0	Velocity at portals at max conditions (2 min) (outflow/inflow)	$\dot{m}_f''$ lower-upper	HRR lower-upper*	Average HRR	T_{max}	L_f towards portal	L_f towards end
	(m²)			(m³/s)	(°C)		(g/m²s)	(MW)	(MW)	(°C)	(m)	(m)
1	6.6	Natural	no	0	16	2.2/1.5	62–74	14–17	16	710	18	0
2		Semi-transv.	no	15	17.5	2.2/2.3	46–62	11–14	12	830	23	0
2a		Longitud.	no	39	11	4.2/1.1	46–62	12–16	14	450	NA	NA
3		Natural	yes	0	16	1.9/2.7	NA	NA		950	21	0
5	47.5	Natural	no	0	10	4.8/2.3	21–26	35–43	39	1200	66	11
6		Semi-transv.	no	15	10	NA	19–21	32–35	33	1180	100	11
7a		Longitud.	no	39	11.3	5.8/0.5 (out)	32–43	60–80	70	1325	74	7
7		Natural	yes	0	11.3	–	NA	NA		995	58	11
9	95	Natural	no	0	4.6	4.6/3	10–11	33–37	35	1020	79	23
10		Semi-transv.	no	6	9	5/2	9–10	30–33	32	850	82	23
11		Natural	yes	0	11.2	4.1/2.8	NA	NA	NA	800	NA	NA

* $HRR = \eta \dot{m}_f'' \cdot A_f \cdot \Delta H_c$ where η is the combustion efficiency, $\dot{m}_f''$ is the burning rate per square metre, A_f is the fuel area and ΔH_c is the heat of combustion. We assume $\eta = 0.8$ in tests with natural and semi-transverse ventilation and $\eta = 0.9$ in the tests with longitudinal ventilation. The heat of combustion ΔH_c is assumed to be equal to 43.7 MJ/kg and the fuel density is assumed to be 740 kg/m³ (ref. 36). NA = Not available

burning rate per square metre in these tests is highly influenced by the ventilation and the test set-up. The poor accessibility of the oxygen to the fuel bed in the troughs (pans) used was nearly of equal width (3.8 m) to the tunnel (4.2 m). In a wider tunnel the results may have been quite different. In the case when longitudinal ventilation was used the burning rate increased dramatically, especially for the large fire (test no. 7a, 47.5 m^2), since the oxygen was more effectively mixed with the fuel. The burning rate became slightly less per square metre when the fire was small (6.6 m^2) and the maximum ceiling temperature reached a level of 1325 °C. The average HRR was estimated to be 70 MW. With natural ventilation or semi-transverse ventilation the temperatures were slightly lower or about 1200 °C and the average HRR was between 33 MW to 39 MW. In general, we see that the maximum ceiling temperature varies between 450 °C–1325 °C for average HRRs between 12 MW and 70 MW. Clearly, the temperatures are not only dependent on the level of the HRR but also on the ventilation conditions.

In Table 11.2, an estimation of the flame length, L_f, from the centre of the trough is given as well. The flame length is given both towards the portal where most of the air flow was directed and towards the end of the tunnel. It is calculated from the centre of the pool fire and it is based on linear interpolation of the peak gas temperatures measured in the 0.5 m below the ceiling and represent the 600 °C temperature front.[33] Here the size of the pool in combination with the ventilation conditions plays an important role whether the temperatures become high or low.

In the tests with the 6.6 m^2 and 47.5 m^2 pool fires the temperature in the ceiling increased rapidly, and reached a maximum after about two minutes from ignition. Shortly after reaching the maximum the temperature dropped rapidly and after about ten minutes from ignition the temperature was in all cases without sprinklers less than 200 °C. In the case with the largest pool fire (95 m^2) and no sprinkler the temperature was relatively constant at its high temperatures for about 8–10 minutes. The oxygen measurements indicated that all the oxygen was consumed. This indicates that the 95 m^2 pool fire was ventilation-controlled. That the fire was ventilation-controlled could explain the large difference in HRR data per square metre and temperature data compared to the smaller pool fires.

These tests were very valuable for the design of the tunnel ventilation systems at that time. Much effort was put into analysing data in order to relate it to the conditions of evacuation. These tests had also a major impact on the view of using sprinkler systems in Europe. It was not found feasible to use sprinkler systems in tunnels due to some adverse effects of the sprinkler system. The sprinklers were able to extinguish the fire but the visibility was reduced in the vicinity of the fire and after the fire was extinguished gasoline fuel vapour continued to evaporate. In the last test the critical concentration (20 minutes) was obtained and due to hot particles in the fire zone the vapour cloud ignited. The conflagration created resulted in velocity rates of 30 m/s.

11.3.2   Glasgow tests 1970

In the UK, the Building Research Establishment (BRE) (the former Fire Research Station (FRS)) carried out in collaboration with the Glasgow fire brigade five experimental fires in a disused railway tunnel in Glasgow.[35] The purpose of the tests was originally not tunnel related. The tests were actually carried out to investigate smoke spread in an enclosed shopping mall, and a disused railway tunnel was used because it was a reasonable approximation to certain features of such a building.[35] The disused railway tunnel was 620 m long, 7.6 m wide and 5.2 m high. Fires of one, two or four trays of kerosene were burnt. The trays were square with side length of 1.2 m, or area of 1.44 m^2 with a fuel load of 45-litre kerosene. The estimated HRR in each tray was 2 MW,[35] or 1.39 MW/m^2.

The experimental instrumentation was scattered inside the tunnel. The smoke layer height and the time of arrival of the smoke front were measured at 20 different locations with human observers using breathing apparatus. According to Heselden[35] there were some temperature and smoke obscuration measurements done, but no details are given. Observations from the tests show that the smoke layer was actually quite flat during the tests. Heselden[35] describes thoroughly the smoke conditions within the tunnel after ignition;

> In all the tests the bulk of the smoke formed a coherent layer, which was initially 1–2 m thick depending on the size of the fire, and which gradually deepened as the test progressed, reaching 3–4 m deep for the largest fire ten minutes after ignition. The velocity of advance of the layer was in the region of 1–1.5 m/s, discounting the initial half minute when the burning rate was building up to an equilibrium value. In two tests the smoke nose was followed to the end of the tunnel, a distance of 414 m from the fire. The smoke layer was then quite well defined even though it would have been only some 5 °C above the air beneath. It was found that a layer or plug of smoke reaching to ground level often formed at the tunnel entrance probably due to the mixing and cooling produced by a cross wind; this plug tended to be drawn back into the tunnel with air current induced by the fire. The air below the main smoke layer was not perfectly clear. Although the bulk of the smoke formed a layer, some optically thinner smoke tended to build up in the clear layer below even before the ceiling smoke layer had reached the end of the tunnel. This may have been due to some mixing of smoke downwards at the smoke nose, which was more turbulent than the layer following it, or to mixing at obstructions (which were very few), or to wisps of smoke cooled by contact with the wall, clinging to the wall and moving downwards where they swept up by and mixed into the main air flow to the fire.

The Glasgow tests have not been widely referred to in the tunnel literature, most likely due to the scattered data obtained from these tests and the fact that the tests were not originally performed to improve tunnel fire safety. More detailed information about these tests can be found in ref. 37.

The Fire Research Station (FRS) has been involved in other large-scale tunnel testing in collaboration with local fire brigades. Heselden[35] reports briefly on tests carried out in Hampshire in the UK without giving any further references. These tests were carried out in connection with proposals for the Channel tunnel, which was taken into traffic in 1994. The FRS in collaboration with the Hampshire Fire Brigade and British Railways carried out an experimental fire in a disused railway tunnel near West Meon, Hampshire. The tunnel was 480 m long, 8 m wide and 6 m high and the cars to be burnt were placed 45 m from one of the tunnel portals. During the burning of one car a smoke layer up to 3 m thick formed under the roof but observers were able to remain near the fire without any ill effects except headaches afterwards. The flow of the smoky hot gas was controlled by the wind of about 2 m/s that was blowing through the tunnel.

11.3.3   Zwenberg 1975

A decade after the Ofenegg tunnel tests, a new test series was carried out in the Zwenberg tunnel in Austria 1975.[4] The reason for these tests was similar as for the Ofenegg tests. Large road tunnel projects were planned in the early 1970s in Austria. The aim was to investigate the effects of different types of ventilation (longitudinal, semi-transverse and transverse ventilation*) on the distribution of smoke (visibility), heat and toxic gases and the effects of heat on the ceiling construction and the exhaust fans. The Ofenegg tests were concentrated on studying the conditions during fire with more or less unchanged ventilation patterns, whereas the main objective of the Zwenberg tests was to investigate how changing the ventilation pattern could influence conditions inside the tunnel. For the operation of tunnel ventilation the following two major questions had to be answered:[3]

1. What quantities of fresh air shall be supplied in order to provide the best conditions in case of tunnel fire?
2. What influence has forced longitudinal ventilation on the traffic space?

Beyond that the scope of the research project was to study the effects of a tunnel fire on human beings. In order to do that, temperatures, content of toxic gases and oxygen in the tunnel, the visibility and fire duration were measured. The aim was also to find ways to improve the situation in the tunnel by using different tunnel ventilation. The focus was also on the effects of the fire on the tunnel structure and technical equipment within the tunnel.

The tests were carried out in an abandoned railway tunnel owned by the Austrian Railways. The tunnel was 390 m long with a cross-section of $20\,m^2$

---

* Transverse ventilation system has both extraction and supply of air. Fully transverse ventilation has equal amounts of exhaust and supply air.

(traffic space) and a ventilation duct of 4 m². The tunnel gradient was 2.5% from the south to the north portal. The tunnel height up to the ventilation duct was 3.8 m and the tunnel width was 4.4 m. A fully transverse ventilation system was installed in the test tunnel, designed for a supply of 30 m³/s of fresh air and for the same quantity of exhaust air. An injection fan installed near the southern portal was designed to provide a longitudinal flow up to 7 m/s in the traffic space. Every 6 m alternately a fresh air opening and a polluted air opening were installed.

The fire source was located 108 m from the south portal. It consisted of 12 individual concrete trays in two rows with a total volume of 900 litres of liquid (gasoline, diesel) corresponding to a surface area of 20 m² where the internal dimensions of each tray was 1 m wide and 1.7 m long. Only four trays (alongside each other) were used in the standards test (6.8 m²) and six in the large tests (13.6 m²). A total of 46 measuring points for temperature were mounted, 11 for air and gas velocities, 19 for gas sampling (O_2, CO_2, CO, CH and NO_x) and 7 for visibility observations. A total of 30 tests, see Table 11.3, were performed using gasoline pools of 3.4 m², 6.8 m² and 13.6 m², respectively. The majority of the tests, 23 'standard fire' tests, were run using four trays with a fuel area of 6.8 m² and 200 litres of fuel. This fire size was found to be sufficient to obtain useful data and avoid damage to the installation. In the tests with the 'standard fire' the following parameters were varied:

- location of the fresh air supply (from below or above)
- quantity of polluted air to be exhausted
- quantity of air supply
- forced longitudinal ventilation in traffic space
- conditions in the traffic space (open or obstructed).

The selected combination of different test parameters can be obtained from the second column in Table 11.3 (identification code of test 210) according to the following system:

$$U - 1 - 1/3 - 2 - A$$

U   Location of fresh air supply
- U = from below
- O = from above
- X = no supply

1   Quantity of exhausted air
- 1 = nominal quantity 30 m³/s
- 1/3 = 10 m³/s

1/3   Supplied quantity of fresh air
- 1 = nominal quantity 30 m³/s
- 1/3 = one third of 30 = 10 m³/s

2   Longitudinal flow in the traffic space (2 m/s)

A   Conditions in the traffic space
- F free cross-section
- A test models in the traffic space

The ventilation arrangement, the pool size, the length of the tunnel and that no sprinkler was used are the main parameters that differentiate these tests from the Ofenegg tests. The average burning rate per square metre varied between 32–64 g/m^{2}s with an average value of 43 g/m^{2}s, whereas in the Ofenegg tests it varied between 9–74 g/m^{2}s. In the open a corresponding value for large pool fires is 55 g/m^{2}s.[36] The burning rates in the Zwenberg and the Ofenegg tests are not based on any weighted results, it was calculated as the total fuel consumption divided by an estimated burning time. This will lead to conservative values since the burning rate varies with time, especially in the beginning of the test and during the decay period. In between these periods it should be relatively constant. As shown earlier, the variation in the burning rates per square metre in these tests is much less than in the Ofenegg tests. The main reason is probably that the fire size was not nearly as large as in the Ofenegg tests and also that the tunnel was open at both ends and the total width of the pool (two trays beside each other ~2.5 m) was much less than the width of the tunnel (4.4 m).

Feizlmayr reports that two classes of danger areas were used when analysing the results of the Zwenberg tests; class 1 areas with fatal effects and class 2 areas of potential danger. This type of classification was used in the Ofenegg tests as well. The criteria for class 2 used were the following; 80 °C temperature, 4.3% CO_2 and 1000 ppm (0.1%) of CO at heat level. The results of the Zwenberg tests showed that the extension of the danger area and smoke area (visibility) could be influenced to a great extent by the system of ventilation. The fully transverse ventilation (FTV), when a properly designed air flow supply (throttle), was found to offer the best conditions for getting the fire under control. With semi-transverse ventilation (STV) with only fresh air supply the system gave only modest improvements of the conditions within the tunnel. It was recommended to throttle the fresh air supply in order to improve the conditions. New STV installations should be designed so that in case of fire a quick change over from fresh air supply to air extraction could be achieved. In tunnels with bi-directional traffic it was found that the FTV or STV (if properly designed) would be more effective in case of fire than the longitudinal ventilation system due to the possibility of smoke extraction. Based on the Zwenberg tests it was strongly recommended that longitudinal ventilation should be shut down in case of fire with the exception that meteorological conditions require other measure to prevent the longitudinal flow. In tunnels with uni-directional traffic it was found that longitudinal ventilation systems could protect people on the upstream side of the fire, assuming that the vehicles were not trapped on the downstream side of the fire. The recommendations given after the Zwenberg tests have been a guide for the design of ventilation systems worldwide.

Table 11.3 Relevant data from the Zwenberg tunnel fire tests in 1975

Test no.	Identification code of test*	Test conditions*	Fuel (litre, area, fuel type)	T_0 (°C)	$\dot{m}''_f$ (g/m² s)	Average HRR** (MW)	T_{max} (°C)	L_f towards north portal (m)	L_f towards south portal (m)
101	U-1-1-7-F	TOF	100, 3.4 m², gasoline	NA	64	8	NA	NA	NA
102	U-1-1-2,5-F	TOF	200, 6.8 m², gasoline	NA	51	12	NA	NA	NA
103	U-1-1-0-F	FTV		12	44	10	904	19	6
104	U-1-1/3-0-F			10	52	12	1240	14	60
105	X-1-0-0-F	EO		12	54	13	1320	11	12
106	0-1-1/3-0-F			8	49	12	1222	15	12
107	0-1-1-0-F	FTV		10	35	8	1080	17	10
203	U-1-1-0-A	FTV		8	41	10	856	21	6
204	U-1-1/3-0-A			8	41	10	1118	16	11
205	X-1-0-0-A	EO		10	51	12	1254	17	14
206	0-1-1/3-0-A			8	49	12	1318	20	20
207	0-1-1-0-A	FTV		10	33	8	1134	19	10
208	U-0-1-0-A	STV		12	35	8	822	23	7
209	U-1-1-2-A	FTV		14	48	13	663	15	0
210	U-1-1/3-2-A			12	45	12	563	5	0

211	U-1-1-2-F	FTV		12	44	12	670	16	0
212	X-0-0-2-A	PLV		14	44	12	623	12	0
213	X-0-0-4-A	PLV		12	45	12	312	10	0
214	X-0-0-0-A	EO		16	40	9	1000	23	0
215	0-1-1-2-F	FTV		12	44	12	612	10	0
216	0-0-1-0-A	STV		13	37	9	893	26	5
217	0-0-1/3-0-A	STV		11	32	8	1165	26	10
218	0-1-1/3-2-A			10	40	11	623	12	0
219	X-1-0-2-A	EEO		6	40	11	675	16	0
221	X-1-0-2-A	EO		8	28	7	723	4	0
220	X-1-0-0-A	EO	200, 6.8, diesel	8	41	10	643	13	0
301	X-1-0-0-A	EO	400, 13.6, gasoline	6	42	20	1332	59	12
302	0-1-1/3-0-A	EO		6	35	17	1320	46	31
303	0-0-1/3-0-A	EO		8	44	21	1330	60	21
2000	U-0-1-0-A	STV	Wood, rubber	NA	NA	NA	NA	NA	NA

* TF Test of facility (preliminary tests)   FTV Fully transverse ventilation
EO Extraction only   STV Semi transverse ventilation
PLV Pure longitudinal ventilation   EEV Enlarged extraction opening

** $HRR = \eta \dot{m}''_f \cdot A_f \cdot \Delta H_c$. We assume $\eta = 0.8$ in tests with natural and semi-transverse ventilation and $\eta = 0.9$ in the tests with longitudinal ventilation. The heat of combustion ΔH_c is assumed to be equal to 43.7 MJ/kg and the fuel density is assumed to be 740 kg/m^3 (ref. 36). NA = Not available

11.3.4 P.W.R.I. 1980

The Public Works Research Institute (P.W.R.I.) in Japan performed two series of large-scale tests.[7] The first test series was carried out in P.W.R.I.'s own full-scale test tunnel facility and the second test series was carried out on the Chugoku Highway in the Kakeitou Tunnel. The full-scale tunnel at the P.W.R.I. site has a total length of 700 m, a cross-sectional area of 57.3 m^2 (H $\approx$ 6.8 m) and is equipped with ventilating and sprinkler facilities. The Kakeitou tunnel has a total length of 3277 m, a cross-sectional area of 58 m^2 (H $\approx$ 6.7 m), and is equipped with ventilating and sprinkler facilities. The majority of the experiments were conducted in the full-scale tunnel at P.W.R.I. but also in the Kakeitou tunnel. The main purpose of using the long tunnel was to determine the environment for people evacuating from tunnels.

The fire source consisted of gasoline pool (gasoline) fires, passenger cars and large-sized buses. Gasoline pool fires of 4 m^2 and 6 m^2 were used to generate a HRR equal to the fire for large-sized vehicles, large-sized buses and passenger cars. The pool fires were applied in order to accomplish steady and repeatable fires, which may not be the case in tests using real motor vehicles. Several real motor vehicles were used although for confirmation of the results using pool fires. Four to six sets of gasoline fire pools (trays) were arranged for fires, each having four 0.25 m^2 (1 m^2) fire trays in one set. Further, 18 litres of gasoline were uniformly placed in each fire tray in order to maintain almost the same burning rate for about ten minutes after ignition. In the tests with passenger cars, doors of the driver's seat were left half-opened, while other doors and windows were closed. Approximately 10–20 litres of gasoline were put in the fuel tank of the passenger cars. For large-sized buses, the entrance door, exit doors and the window next to the driver's seat were fully opened, and 50 litres of light oil were put in the fuel tank. With respect to passenger cars and buses, pieces of cloth soaked in advance in a small amount of gasoline were placed on the rear seats and ignited. A comprehensive instrumentation was used in these test series. The gas temperatures (84 points in the Kakei tunnel), concentrations of smoke (78 points in the Kakei tunnel), gas velocities (5 points), concentrations of O^2, CO gases (1 and 3 points, respectively), radiation (1 point) and burning velocity were measured. No HRR measurements were carried out in these tests.

The ventilation system was able to create a longitudinal flow up to 5 m/s. The water sprinkler facilities were set so that comparisons could be made between the presence and absence of sprinkling under the same fire sources and the same longitudinal flow. Duration of sprinkling was set at about 20 minutes. The area sprinkled was that area directly above the fire source. In some tests the sprinkler system was used leeward from the fire source in order to check the cooling effect of sprinkling on hot air currents. The amount of water discharge was set at about 6 litres/min/m on road surface. In order to review the possibility of fire spread to following vehicles congested during the fire, an experimental scenario was carried out where cars were arranged longitudinally and transversely.

The influence of the temperature due to the fire was found to be limited only to the nearby areas of the fire. Table 11.4 presents a summary of all peak HRRs and ceiling temperatures. The data show clearly the effects of the longitudinal flow on the peak temperature in the ceiling. Higher velocity tends to lower the ceiling temperature due to dispersion of the hot air. It was not possible to extract any information about the flame lengths from the information available. An estimation of the free flame height for the pool fires used in this test series indicates that the flames were not impinging on the ceiling. The ceiling temperatures given in Table 11.4 confirm these calculations.

It is pointed out in the report[7] that it is extremely important to determine the behaviour of smoke and to control smoke when considering the evacuation

Table 11.4 The test programme for the P.W.R.I. test series in Japan 1980[7]

Test no.	Test tunnel	Fire source	u	Sprinkler discharge time from ignition	$\dot{Q}_{max}$	T_{max} (+5 m from centre) (no sprinkler)
		(m², litre fuel)	(m/s)	(min)	(MW)*	(°C)
1	P.W.R.I. 700 m	4 m², 288 l	0.65	–	9.6	252
2	"	4, 288	5	–	9.6	41
3	"	4, 288	0.65	3	9.6	NAs
4	"	4, 288	5	3	9.6	NAs
5	"	6, 432	2	–	14.4	429
6	"	6, 432	2	0	14.4	NAs
7	"	Passenger car	1	–	NA	62
8	"	Passenger car	3	–	NA	NA
9	"	Passenger car	5	–	NA	NA
10	"	Passenger car	1	2.4	NA	NAs
11	"	Passenger car	3	2.4	NA	NAs
12	"	Passenger car	5	2.4	NA	NAs
13	"	Large-sized bus	5	–	NA	166
14	"	Large-sized bus	0.65	1.4	NA	NAs
15	"	Large-sized bus	2	10.5	NA	NAs
16	"	Large-sized bus	5	1.37	NA	NAs
17	Kakei 3277 m	4, 288	0	–	9.6	511
18	"	4, 288	2	–	9.6	199
19	"	4, 288	5	–	9.6	69
20	"	4, 288	0	3	9.6	NAs
21	"	4, 288	2	3.16	9.6	NAs
22	"	4, 288	5	3	9.6	NAs
23	"	Large-sized bus	0	–	NA	186
24	"	Large-sized bus	0	2.5	NA	–

* Due to the good ventilation conditions we assume free burning conditions, i.e., 2.4 MW/m² for gasoline (0.055 kg/m²s × 43.7 MJ/kg (ref. 36))
NA = Not available, NAs = Not available temperature due to the sprinkler

possibilities during a tunnel fire. It was concluded that in the case of a $4\,m^2$ gasoline fire or a large-sized bus fire, the conditions for evacuation could be maintained near the road surface for about ten minutes and over a distance of 300 m to 400 m, if the longitudinal velocity was lower than 2 m/s. However, if the wind velocity increased, the smoke spread over to the entire section was such that any type of evacuation would become difficult.

It was also found that the wind velocity in order to prevent backlayering was 2.5 m/s and that increasing the wind velocity would influence the fire so that the amount of heat and smoke would increase. It was found that the water sprinkler facilities of the present scale were not able to extinguish gasoline fires and roofed motor vehicles but they were able to lower the nearby temperature and to prevent fire spread to nearby motor vehicles. It was also shown that water sprinkling may cause smoke to descend and deteriorate the evacuation environment near the road surface. Therefore precautions should be taken concerning the method of operation of sprinkling facilities.

11.3.5   TUB-VTT tests 1986

As a part of German–Finnish cooperation on tunnel fires, the Technische Universität Braunschweig (TUB) in Germany and the Technical Research Centre of Finland (VTT) performed two large-scale tunnel fire tests in 1985 in Lappeenranta in the south-eastern part of Finland. This cooperation developed and widened later into EUREKA project EU499. Two pilot tests were carried out in a tunnel in a limestone quarry 45 m below ground. The tunnel was 140 m long, 6 m wide and 5 m high ($30\,m^2$), and had natural calcite rock surfaces which were unprotected and without reinforcements.

The first experiment was designed to simulate a fire in a subway car stalled in a tunnel. The second experiment simulated the case when one car in a queue of cars in a tunnel catches fire. Forced ventilation of fresh air at the rate of $7\,m^3/s$ was used. This generated a longitudinal flow of 0.2 m/s to 0.4 m/s over the cross-section prior to ignition. At the maximum HRR an inflow of 0.3 m/s was measured in the lower part of the cross-section and outflow of about 6 m/s in the upper part of the cross-section at the same location, i.e., 19 m inside the exit portal. The fire load was made of wood cribs (moisture 17%) nailed together in a way that allowed an air space of 50% of the total volume. Temperatures of air, rock surface of the walls and the ceiling and temperatures of the steel and concrete columns placed on the floor were recorded at several locations. Also, concentrations of O_2, CO_2 and CO and air flow velocities were measured close to the exit of the tunnel. Fuel burning rate was determined by measuring the mass loss of wood on a weighing platform. The original idea was to utilise the oxygen consumption technique, but due to large uncertainties in O_2 and flow measurements it was never completed.[13] Smoke level and visibility were observed visually close to the exit.

In the first test (F1-1) the fire load of 7600 kg was distributed over an area of 3.2 m × 48 m (spread as a layer on light concrete blocks 0.47 m above ground). After ignition at the upstream end of the fire load the wood cribs burned without flashover with a constant velocity of 0.66 mm/s for 21.5 hours. In the second test (F1-2) the fire load consisted of eight separate piles (clusters) of wood cribs, 1.6 m × 1.6 m in area and 0.8 m in height, each with a mass of 500 kg. The free space between the piles was 1.6 m and the lower end of the wood cribs was 0.5 m over the tunnel floor.

Two adjacent piles were ignited simultaneously at the upstream end of the fire load. The fire growth rate was quite steep and reached a peak HRR of 8 MW after about 15 minutes into the test and then started to decay. The two wood piles burned out since fire never spread to the adjacent wood piles. Therefore, a new ignition was done at the other end (downstream side) of the wood crib cluster. The fire growth rate was slower this time and reached a HRR of about 3 MW after about 20 minutes. The HRR was relatively constant at 3 MW (except one short peak at 4 MW) for about 45 minutes. The main difference between the first ignition and the second ignition was that the fire spread downwind after the first ignition and upwind in the second ignition. The highest gas temperature in the ceiling after the first ignition was obtained after about 20 minutes into the test (F1-2); 679 °C and in the second ignition it was 405 °C obtained after about 26 minutes from the second ignition.

During the both experiments (F1-1 and F1-2) 10–20 cm thick layers of rock scaled off the walls and the ceiling in regions close to the fire. One of the main conclusions from these experiments was that the theoretical calculations based on existing room fire codes did not reliably predict occurrence of flashover.

11.3.6   EUREKA EU499 tests 1990–1992

The EUREKA EU499 test programme was performed in an abandoned tunnel named Repparfjord Tunnel in northern Norway. The tunnel was 2.3 km long with a gradient less than 1%, running north to south from the main portal to a vertical shaft of 90 m height (9 m²). The cross-section of the tunnel was horseshoe shaped to rectangular with a flattened roof. The tunnel is approximately 5.3 to 7.0 m wide with a maximum height in the centre between 4.8 m and 5.5 m.

The test programme included 21 large-scale tests, which were carried out in 1990, 1991 and 1992. The majority of the tests were performed in 1992 as can be observed in Table 11.5. The main objectives of the EUREKA EU499 test programme were to investigate the fire behaviour of different type of fuels including real vehicles, to see the possibilities of escape and rescue, and to see the damage to tunnel structure caused by fires, etc. The fire behaviour of trains and HGVs revealed by these tests has had major effects on many design studies of large tunnel projects today.

The main results of the EUREKA EU499 project relate to the unique data of measured HRR for real vehicles where oxygen consumption calorimetry was applied for the first time in large-scale tunnel tests. It also contained well-defined fire sources such as wood cribs and heptane pool fires, which are very valuable for scientific analysis of the results. The wood crib tests showed a tendency to increased fire growth rate with increased ventilation rate whereas it was not as apparent for the peak HRR. Results showed that generally the temperature of vehicles with a body structure that can melt, e.g., the aluminium subway coach and the school bus (GFRP), could reach ceiling temperatures from 800 up to about 1100 °C and HRR of about 30–50 MW (tests 7, 11 and 14). For trains with steel body structures the HRR was less than 19 MW and the fire duration longer and the ceiling temperatures tended to be lower than 800 °C (tests 4, 5, 12 and 13). For the passenger car, the highest temperature was between 210 and 480 °C and the HRR was up to 6 MW (tests 3 and 20). The same tendency towards the influence of the body type on the results is found for the plastic car and the steel body passenger car. The estimated flame lengths are given in Table 11.5 as towards the portal and towards the vertical shaft. It is based on a 600 °C flame tip obtained from maximum temperature graphs as a function of the distance from the centre of the fire given by Ekkehard.[38]

The EUREKA tests show the importance of glazed windows on fire growth in steel-bodied trains. Fire growth rate is apparently governed by the sequence and timing of window cracking. This can be shown by analysis of temperature development inside the train compartments. The type of interior material (former or new design) appears not to be as important for fire growth as expected. The type of body and the quality of the windows appears to be more important than the type of interior materials. For a heavy goods load (furniture), which is not contained by any steel or aluminium body, the corresponding data were about 1000 °C and a HRR of 120–128 MW. The propagation speed of the smoke front was constant along the tunnel, implying that the behaviour of smoke propagation was similar to the movement of gravity currents.

11.3.7 Memorial Tunnel tests 1993–1995

The Memorial Tunnel Fire Ventilation Test Program (MTFVTP) consisted of a series of large-scale fire tests carried out in an abandoned road tunnel. Various tunnel ventilation systems and configurations of such systems were operated to evaluate their respective smoke and temperature management capabilities. The Memorial Tunnel test programme was performed in a two-lane, 853 m long and 8.8 m wide road tunnel built in 1953, taken out of traffic 1987 and was a part of the West Virginia Turnpike. The tunnel has a 3.2% rising gradient from south to north portal. The tunnel was originally designed with a transverse ventilation system, consisting of a supply fan chamber at the south portal and an exhaust fan chamber at the north portal. An overhead air duct, formed by a concrete ceiling

Table 11.5 The test programme for the EUREKA EU499 test series

Test no.	Date of test	Fire load	u (m/s)	E_{tot} (GJ)	$\dot{Q}_{max}$ (MW)	T_0 (°C)	T_{max} (0 m) (°C)	T_{max} (+10 m from the centre) (°C)	L_f towards portal (m)	L_f towards shaft (m)
1	07.12.90	Wood cribs no 1.	0.3	27.5	NA	~5	NA	500	NA	NA
2	24.07.91	Wood cribs no 2.	0.3	27.5	NA	~5	NA	265	NA	NA
3	08.08.91	Private car (steel body)	0.3	6	NA	~5	210	127	NA	NA
4	19.08.91	Metro car F3 (steel)	0.3	33	NA	4.5	480	630	NA	~17
5	29.08.91	Half railcar F5 (steel)	0.3	15.4	NA	1.7	NA	430	NA	NA
6	04.09.91	Half railcar F6 (steel)	0.3	12.1	NA	4	NA	NA	NA	NA
7	23.08.92	School bus (GFRP)	0.3	40.8	29	3	800	690	0	~17
8	28.08.92	Wood cribs no 3.	0.3	17.2	9.5	~8	NA	480	NA	NA
9	30.08.92	Wood cribs no 4.	3–4	17.9	11*	8.2	NA	440	NA	NA
10	31.08.92	Wood cribs no 5.	6–8	18	12*	10.4	NA	290	NA	NA
11	13.09.92	1.5 rail cars F2Al+F7 (Aluminium + steel)	6–8/3–4	57.5	43	3.3	980	950	0	~20
12	25.09.92	Rail car F2St (steel)	0.5	62.5	19	4.7	650	830	0	~20
13	07.10.92	Rail car F1 (steel)	0.5	76.9	13	2.2	450	720	0	~20
14	14.10.92	Metro car F4 (Aluminium)	0.5	41.4	35	1.6	810	1060	~11	~22
15	23.10.92	Mixed load simulating truck load	0.5	63.3	17	~0	NA	400	NA	NA
16	27.10.92	1 m² heptane pool no 1.	0.6–1.0	18.2	3.5†	~0	NA	540	–‡	–
17	28.10.92	1 m² heptane pool no 2.	1.5–2.0	27.3	3.5†	~0	340	400	–	–
18	29.10.92	3 m² heptane pool no 3.	1.5–2.0	21.2	7†	~0	NA	NA	–	–
19	29.10.92	3 m² heptane pool no 4.	2.0–2.5	54.5	7†	~0	NA	NA	–	–
20	04.11.92	Private car (plastic)	0.5	7	6	0	480	250	NA	NA
21	12.11.92	Heavy goods vehicle (HGV) with furniture	6–8/3–4	87.4	128	~0	925	970	~19	38

* fuel mass loss rate times ΔH where $\Delta H = 17$ MJ/kg for wood.
† measured average burning rate 78 g/m²s multiplied with $\Delta H = 44.6$ MJ/kg and 1 m².
‡ – sign indicates that there were no horizontal flame lengths, L_f, registered by the thermocouples.

4.3 m above the roadway, was split into supply and exhaust section by a vertical concrete dividing wall. In some of the tests the horizontal ceiling was removed in order to put in place 24 reversible jet fans in groups of three equally spaced over the tunnel. The cross-section changed from rectangular shape with cross-sectional area of $36.2\,m^2$ to more of a horseshoe shape with a height of 7.8 m and a cross-sectional area of $60.4\,m^2$. These fans had a 56 kW motor, an outlet velocity of 34.2 m/s and a volume flow of $43\,m^3/s$. They were designed to withstand air temperatures of about 300 °C.

The test programme consisted of 98 tests where the types of ventilation, fuel size and sprinklers were changed. The ventilation systems were modified and run with the following system configurations:

- full transverse ventilation (FTV)
- partial transverse ventilation (PTV)
- PTV with single point extraction
- PTV with oversized exhaust ports
- point supply and point exhaust operation
- natural ventilation
- longitudinal ventilation with jet fans.

The tunnel was equipped with instrumentation and recording equipment for data acquisition. Sensors measuring air velocity, temperature, carbon monoxide (CO), carbon dioxide (CO_2) and total hydrocarbon content (THC) were installed at 12 cross-sections along the tunnel. In total there were approximately 1400 measuring points, each point was recorded once every second during the test (the test time ranged from about 20 to 45 minutes). Smoke generation and movement and the resulting effect on visibility was assessed using seven remote-controlled television cameras with associated recording equipment.

Ventilation system effectiveness in managing smoke and temperature movement was tested for the previously calculated fire sizes 10, 20, 50 and 100 MW. The corresponding fuel surface area is $4.5\,m^2$, $9\,m^2$, $22.2\,m^2$ and $44.4\,m^2$, respectively, meaning an average heat release rate of $2.25\,MW/m^2$. The fire source consisted of low-sulfur No. 2 fuel oil in different pools. In addition to varying the fire size, systematic variations were made in airflow quantity, longitudinal air velocity near the fire, and fan response time for each ventilation system. Tests were also conducted to assess the impact of longitudinal air velocities on the effectiveness of a foam suppression system. Various smoke management strategies and combinations of strategies were employed, including extraction, transport, controlling direction of movement, and dilution to achieve the goals of offsetting buoyancy and external atmospheric conditions and to prevent backlayering (critical velocity).

It is not possible here to present all the test data from the Memorial Tunnel tests due to the large of number of tests performed. An extract of data for T_0, T_{max} and L_f is given in Table 11.6. The data is collected after the mechanical

Table 11.6 The Memorial test programme with different types of ventilation system. In the case of mechanical ventilation the peak temperature and flame lengths are obtained after the start of ventilation

Test ID	Type of ventilation	u (m/s)	T_0 (°C)	H (m)	$\dot{Q}_{max}$ (MW)	T_{max} (°C)	L_f toward north portal (m)	L_f toward south portal (m)
101CR	Full transverse		21	4.4	10	574	–	–
103	Full transverse		19	4.4	20	1361	10	10
113A	Full transverse		20	4.4	50	1354	37	0
217A	Partial transverse (PTV)		13	4.4	50	1350	45	6
238A	PTV-two zone		23	4.4	50	1224	21	13
239	PTV-two zone		21	4.4	100	1298	54	15
312A	PTV-single point extraction		13	4.4	50	1301	42	7
318A	Point supply and point extraction		11	4.4	50	1125	22	20
401A	PTV-oversized exhaust ports		21	4.4	50	1082	21	12
605	Longitudinal	2.2	6	7.9	10	180	–	–
607	Longitudinal	2.1	6	7.9	20	366	–	–
624B	Longitudinal	2.3	14	7.9	50	720	–	21
625B	Longitudinal	2.2	15	7.9	100	1067	–	85
501	Natural ventilation		13	7.9	20	492	–	–
502	Natural ventilation		10	7.9	50	923	27	–

– sign indicates that there were no horizontal flame lengths, L_f, registered by the thermocouples.

Table 11.7 The Memorial test programme. The table shows data from tests with mechanical ventilation where the data is taken prior to the start of mechanical ventilation (pre-burn time)

Test ID	T_0	H	$\dot{Q}_{max}$	T_{max}	L_f toward north portal	L_f toward south portal
	(°C)	(m)	(MW)	(°C)	(m)	(m)
101CR	21	4.4	10	281	–	–
103	19	4.4	20	1053	8	7
217A	13	4.4	50	1169	8	9
239	21	4.4	100	1210	41	17
606A	6	7.9	10	152	–	–
618A	11	7.9	20	378	–	–
624B	10	7.9	50	829	10	7
615B	8	7.9	100	957	27	9

– sign indicates that there were no horizontal flame lengths, L_f, registered by the thermocouples.

ventilation system has been started. For the full transverse ventilation, longitudinal ventilation and the natural ventilation tests nominal HRRs of 10, 20, 50 and 100 MW were used. For partial transverse ventilation systems only 50 MW and 100 MW (if available) tests are presented. For comparison, data from tests with mechanical ventilation where the data is taken during the pre-burn time (the period prior to to the start of the mechanical ventilation when there was natural ventilation) is presented in Table 11.7. The main findings from these tests according to the test report[19] are listed below.

• The Memorial Tunnel fire ventilation tests have shown that longitudinal airflow near a fire is equally important as extraction rate for temperature and smoke management. Therefore, specifying a ventilation rate for temperature and smoke management, solely on its extraction capabilities, is insufficient. Further, any criteria established for emergency ventilation should include the impact of tunnel physical characteristics and tunnel ventilation system.

• Longitudinal ventilation using jet fans was shown to be capable of managing smoke and heat resulting from heat releases up to 100 MW. The required longitudinal air velocity to prevent backlayering in the Memorial Tunnel was approximately 3 m/s for a 100 MW fire.

• Jet fans positioned downstream of, and close to, the fire were subjected to temperatures high enough to cause failure. Accordingly, this condition needs to be considered in the system design and selection of emergency operational modes.

• Full transverse ventilation systems can be installed in single-zone or multi-zone configurations and can be operated in a balanced or unbalanced mode. Single-zone, balanced (equal flow rates for supply and exhaust air) full transverse systems indicated very limited smoke and temperature management capability. Multiple-zone full transverse systems have the inherent capability to manage smoke and temperature by creating longitudinal airflow.

- Partial transverse ventilation systems can be installed in single-zone or multi-zone configurations and can be operated in supply or exhaust mode. Single-zone partial transverse systems capable of supplying only air (no possible reversal of fans to exhaust air) were relatively ineffective in smoke or temperature management. Single-zone partial transverse systems which can be operated in the exhaust mode provided a degree of smoke and temperature management.

- Longitudinal airflow is a significant factor in the management of smoke and heat generated in a fire. Ventilation systems which effectively combine extraction and longitudinal airflow can significantly limit the spread of smoke and heat.

- Single point extraction (SPE) is a ventilation system configuration capable of extracting large volumes of smoke from a specific location through large, controlled openings in a ceiling exhaust duct, thus preventing extensive migration of smoke.

- Oversized exhaust ports (OEP) are a modification to transverse type systems which provides smoke extraction capability in the immediate location of a fire. Significant improvement in temperature and smoke conditions were obtained using OEPs relative to the basic transverse ventilation system using conventional size exhaust ports. The OEP enhancement is also applicable to tunnels with bi-directional traffic.

- Natural ventilation resulted in extensive spread of heat and smoke upgrade of the fire. However, the effects of natural buoyancy are dependent on the fire size and the physical characteristics of the tunnel.

- The restricted visibility caused by smoke occurs more quickly than does a temperature high enough to be debilitating. Carbon monoxide (CO) levels near the roadway never exceeded the guidelines established for the test programme.

- The effectiveness of the foam suppression system was not diminished by operation in strong longitudinal airflow.

- Adequate quantities of oxygen to support combustion were available from the tunnel air. The possible increase in fire intensity resulting from the initiation of ventilation did not outweigh the benefits.

11.3.8   Shimizu No. 3 2001

In 2001, ten fire experiments were conducted in the three-lane No. 3 Shimizu tunnel on the New Toumei expressway in Japan.[39] The tunnel was 1119 m long with a slope of 2% down from west to east. The cross-sectional area was 115 m^2 and the width and height were 16.5 m and 8.5 m, respectively. The cross-section was shaped as a semicircle. The reason for performing these tests was to investigate fire behaviour in tunnels with a large cross-section regarding combustion rate, formation of smoke layer, interaction of longitudinal flow on smoke distribution and behaviour of sprinklers on the smoke layer and risk of fire spread. Comparison with the P.W.R.I. tests (two-lane tunnel) was one of the main arguments for performing these tests. Numerous studies have been published from these tests focusing on different subjects concerning convective heat release rate and numerical simulations,[40] smoke descent,[41] plume fires in large tunnel cross-section[42] and bus fire.[43]

The fire source consisted of gasoline pools with an area of $1 m^2$, $4 m^2$ and $9 m^2$. In the $1 m^2$ pool fire, no forced ventilation was used. In the $4 m^2$ pool fire case, tests were carried out both with and without forced ventilation. The forced ventilation consisted of longitudinal ventilation of 2 m/s and 5 m/s from the west to east portal. In the $9 m^2$ case, longitudinal ventilation of 2 m/s was used. When no forced ventilation was used the west portal was blocked. One test with three passenger cars and a longitudinal velocity of 5 m/s was carried out as well as a single large bus with a longitudinal flow of 2 m/s. Jet fans installed in the west portal created the longitudinal flow in the tunnel. Measurements were made at a number of points throughout the tunnel. Temperature (91 points) was measured by type K thermocouples, optical smoke density (57 points) was measured by optical penetration type absorption density meters, heat radiation was measured by a radiation meter located on the floor 30 m west of the fire, and longitudinal air velocity was measured by means of a vane anemometer (measurable range 0.3–15 m/s) located 100 m east of the fire.[40]

A summary of the information obtained from refs 39–43 is given in Table 11.8.

Table 11.8 Test programme and data for the No. 3. Shimizu Tunnel tests in 2001

Test no.	Test ID	Fire source	u	T_0	Sprinkler discharge time from ignition	$\dot{Q}_{max}$	ΔT_{max}
		(m^2)	(m/s)	(°C)	(min)	(MW)*	(°C)
1	1G-0	1	0	NA	NA	2.4	110
2	4G-0	4	0	NA	NA	9.6	577
3	4G-2	4	2	NA	NA	9.6	144
4	4G-5	4	5	NA	NA	9.6	58
5	4G-0	4	0	NA	NA	9.6	NA
6	4G-2	4	2	NA	NA	9.6	NA
7	4G-5	4	5	NA	NA	9.6	NA
8	9G-2	9	2	NA	NA	21.6	300
9		3 passenger cars	5	NA	NA	NA	NA
10		Single large bus	2	NA	NA	30**	283

* Due to the good ventilation conditions we assume free burning conditions, i.e., 2.4 MW/m^2 for gasoline (0.055 $kg/m^2 s$ × 43.7 MJ/kg (ref. 36)).
** This is estimated from the convective HRR of 20 MW derived by Kunikane *et al.*[43] because a sprinkler system was activated when the convective HRR was 16.5 MW. We assume that 67% of the HRR is convective and thereby we can estimate the HRR = 20/0.67 = 30 MW.
NA = Not available.

There was no information available for the ambient temperature, T_0, or the discharge time of the sprinklers – only the temperature differences, ΔT are given. There is not enough information available to obtain any horizontal flame length. Most likely there were no horizontal flames along the ceiling in these tests, which can be shown by using free burning flame height equations, see, e.g., ref. 35. The information obtained from these tests is by no means unique. One exception is the test with the large bus and the fact that these tests were performed in a tunnel with a very large cross-section. Since the fires used were relatively small it is difficult to see any dramatic effects of the size of the cross-section on temperatures or smoke distribution.

11.3.9   Second Benelux Tunnel tests 2002

Fourteen large-scale tests were carried out in the Second Benelux Tunnel in the Netherlands in 2002. The tests were designed to assess the tenability conditions for escaping motorists in the case of tunnel fire and to assess the efficiency of detection systems, ventilation systems and sprinkler systems for numerous types of fire sources. These were pool fires, passenger cars, a van and mock-ups with truck loads. Temperatures, radiation levels and optical densities in the tunnel were measured, as well as smoke velocities and heat release rates.

The tests were carried out in a sink tunnel outside Rotterdam and are given in Table 11.9. The tunnel has a rectangular cross-section with a height of 5.1 m, a width of 9.8 m and a length of about 900 m. The tunnel has a maximum slope of 4.4% and was equipped with longitudinal ventilation. A total of six jet fans were installed at the upstream portal of the tunnel in order to create air velocities up to 6 m/s. The test site was located at 265 m from the downstream portal. The test program included four pool fire tests with ventilation rates between 0 m/s to 6 m/ s. The pool fires consisted of a mixture of n-heptane/toluene. The pool fire source consisted of two and four fuel pans, respectively, where each pan measured 1.8 m long and 1 m wide and the fuel level was 0.5 m above the road surface. The total fuel surface was 3.6 m^2 in tests 1 and 2 and 7.2 m^2 in tests 3 and 4.

The effects of ventilation were tested in tests 5 to 10 using cars and covered truckloads. Passenger cars (tests 5, 6 and 7) and covered truckloads (tests 8, 9 and 10) were tested under different ventilation conditions. Each truckload consisted of 800 kg wooden pallets (total of 36 EURO-pallets, four piles with nine pallets in each pile), with four tyres placed on the top. The fire load was mounted on a mock-up of a truck with a cover of tarpaulin and the rear end open. The total length of the mock-up was 4.5 m, the width was 2.4 m and the height was 2.5 m. Longitudinal ventilation was varied between 0 m/s to 6 m/s. In tests 12 to 14, different sprinkler systems were tested for different ventilation rates. In test 11, a van loaded with 800 kg of wooden pallets (36 pallets) and three tyres on the top was tested. In tests 12 to 14, a covered truckload was tested with the same fire load as in tests 5 to 10, using aluminium covering. In test 14 no covering was used and the fire load was doubled to 1600 kg of wooden pallets.

Table 11.9 The test programme for the 2nd Benelux Tunnel tests

Test no.	Fire source	Type of ventilation	Sprinkler discharge time from ignition (min)	E_{tot} (GJ)	u (m/s)	T_0 (°C)	$\dot{Q}_{max}$ (MW)	T_{max} (°C)	L_f downstream (m)	L_f upstream (m)
1	n-heptane/toluene 3.6 m²	No LTV	No sprinkler	NA	~1.5	~13	4.1	218	–	–
2	n-heptane/toluene 3.6 m²	LTV	No sprinkler	NA	4	~15	3.5	220	–	–
3a	n-heptane/toluene 7.2 m²	No LTV	No sprinkler	NA	1.9	~12	11.5	470	–	–
3b	n-heptane/toluene 7.2 m²	LTV	No sprinkler	NA	5	~12	11.5	250	–	–
4	n-heptane/toluene 7.2 m²	LTV	No sprinkler	NA	6	~11	11.4	210	–	–
5	Passenger car	No LTV	No sprinkler	NA	~1.0	10	NA	230	–	–
6	Passenger car	No LTV	No sprinkler	NA	~1.5	10	4.9	210	–	–
7	Passenger car	LTV	No sprinkler	NA	6	10	4.8	110	–	–
8	Truck load, 36 wood pallets, 4 tyres	No LTV	No sprinkler	~10	~1.5	10	13.2	400	–	–
9	Truck load, 36 wood pallets, 4 tyres	LTV	No sprinkler	~10	5.3	10	19.5	290	–	–
10	Truck load, 36 wood pallets, 4 tyres	LTV	No sprinkler	~10	5	10	16.2	300	–	–
11	Van	No LTV	Sprinkler activated at 14 min	NA	~1.0	10	7.4 (at 14 min)	300 (at 14 min)	–	–
12	Truck load – aluminium cover, 36 pallets, 4 tyres	LTV	Sprinkler activated at 4 min	NA	3	11	6.2 (at 4 min)	270 (at 4 min)	–	–
13	Truck load – aluminium cover, 36 pallets, 4 tyres	LTV	Sprinkler activated at 10 min	NA	3	12	13.4 (at 10 min)	~500 (at 10 min)	–	–
14	Truck load – aluminium cover, 72 pallets, 6 tyres	LTV	Sprinkler activated at 21 min	19	~2.5	10	26 (at 12 min)	~600	10	–

– sign indicates that there were no horizontal flame lengths, L_f, registered by the thermocouples.

In all the tests, except for the fuel pans, the fire sources were mounted on a weighing platform in order to measure the HRR. The HRR for the pans was obtained from the mass loss rate of the supply fuel tank. The centreline temperatures were measured at five different heights at distances of 10 m, 20 m and 50 m upstream from the fire and at 10 m, 20 m, 50 m and 200 m downstream from the fire. The radiation heat flux from the fire was measured with cooled heat flux meters at eye level at distances of 5 m, 10 m and 20 m from the fire centre. Ventilation velocities were measured at three positions upstream of the fire with hot wire anemometers and at three positions downstream the tunnel using bi-directional probes.

The main conclusions from these tests consider the backlayering of smoke was prevented by 3 m/s for all cases. This conclusion agrees well with other investigations presented here. For a small truck fire, deadly conditions due to radiation exposure could be obtained within 10 m from the truck but not at 50 m downwind of the fire. Visibility was reduced within a few minutes at distances 100 m to 200 m downwind of the fire. The escape routes were obscured due to the smoke. An open deluge system reduced the temperature considerably. The risk of fire spread between adjacent vehicles was therefore not deemed to be high. Smoke temperatures downwind did not reach fatal levels and steam production was insignificant. Visibility was however reduced such that escape routes would become difficult to find.

11.3.10 Runehamar 2003

Large-scale tunnel tests were carried out with HGV-trailer cargos in the Runehamar tunnel in Norway. The tunnel is a two-way-asphalted road tunnel that was taken out of use and is 1600 m long, 6 m high and 9 m wide, with a slope varying between 1–3%. The tunnel was a blasted rock-tunnel with a cross-section varying between 47–50 m^2. In total four tests were performed with fire in a HGV-trailer mock-up and the results are given in Table 11.10. The specific commodities used consisted of four different materials, each representing a category of material typically found in the cargo of a HGV-trailer. These commodities were standardised wood pallets, plastic pallets made of polyethylene (PE), a standard-ised test commodity consisting of polystyrene cups (PS) in compartmented cardboard cartons and polyurethane mattresses (PUR). In three tests mixtures of the various cellulosic and plastic materials were used, and in one test a commodity consisting of furniture and fixtures was used. A polyester tarpaulin covered the cargo in each test. The HGV trailer mock-up was 10.45 m long, 2.9 m wide and 4.5 m high with the trailer floor at 1.2 m above the road surface.

In Test 1 the fire load consisted of 10.9 tonnes of wooden and plastic pallets. At a distance of 15 m from the downstream side (rear end of the trailer-mockup) there was a target consisting of one pallet row of the same test commodity as used in the test. In Test 2 the fire load consisted of 6.8 tonnes of wooden pallets

Table 11.10 The test programme for the Runehamar tests

Test no.	Fire source	Target	E_{tot} (GJ)	u (m/s)	T_0 (°C)	$\dot{Q}_{max}$ (MW)	T_{max} (°C)	L_f downstream (m)
1	360 wood pallets measuring 1200 × 800 × 150 mm, 20 wood pallets measuring 1200 × 1000 × 150 mm and 74 PE plastic pallets measuring 1200 × 800 × 150 mm − 122 m² polyester tarpaulin	32 wood pallets and 6 PE pallets	240	2–3	12	203	1365	93
2	216 wood pallets and 240 PUR mattresses measuring 1200 × 800 × 150 mm − 122 m² polyester tarpaulin	20 wood pallets and 20 PUR mattresses	129	2–3	11	158	1282	85
3	Furniture and fixtures (tightly packed plastic and wood cabinet doors, upholstered PUR arm rests, upholstered sofas, stuffed animals, potted plant (plastic), toy house of wood, plastic toys). Ten large rubber tyres (800 kg) − 122 m² polyester tarpaulin	Upholstered sofa and arm rest on pallets	152	2–3	9.5	125	1281	61
4	600 corrugated paper cartons with interiors (600 mm × 400 mm × 500 mm; L × W × H) and 15% of total mass of unexpanded polystyrene (PS) cups (18,000 cups) and 40 wood pallets (1200 × 1000 × 150 mm) − 10 m² polyester tarpaulin	No target	67	2–3	11	70	1305	37

and mattresses (include a target at 15 m). In Test 3 the fire load consisted of 8.5 tonnes of furniture on wooden pallets including the target at 15 m. In this test the fire load had ten tyres (800 kg) positioned around the frame at the locations where they would be on a real HGV trailer. In Test 4 the fire load consisted of 2.8 tonnes of plastic cups in cardboard boxes on wooden pallets (no target used in this test). In each test the amount (mass ratio) of plastic materials was estimated to be about 18–19%.

In each test, two fans positioned near the tunnel portal were used to generate a longitudinal airflow, this was about 3 m/s at the start of each test but reduced to about 2–2.5 m/s once the fires became fully involved. At the location of the fire experiments which was approximately 1 km into the tunnel, a 75 m length of the tunnel was lined with fire protective panels, this reduced the cross-sectional area of the tunnel to 32 m^2 in the vicinity of the fire. The tunnel height at the fire location was 4.7 m. The objectives of the test series were to investigate: (i) fire development in HGV cargo loads, (ii) the influence of longitudinal ventilation on fire HRR and growth rate, (iii) production of toxic gases, (iv) fire spread between vehicles, (v) fire-fighting possibilities and (vi) temperature development at the tunnel ceiling.

Heat release rates in the range of 70–203 MW and gas temperatures in the range of 1250–1350 °C were measured using non-hazardous cargoes. Prior to these tests this high temperature level had been observed only in tests with liquid fires in tunnels. These tests show that ordinary trailer loads can generate the same level of heat release rate and ceiling temperatures as a tanker fire. The fire development in all the tests was very fast, despite a relatively small ignition source. The peak heat release rates were reached between 8 and 18 minutes after ignition. Calculation of time to incapacitation 458 m from the fire was found to be about six minutes from the time of arrival of the smoke gases using wood and plastic pallets and about two minutes using PUR mattresses. A 'pulsing' phenomenon was observed in Tests 1 and 2. These tests also indicate that the fire fighters may experience serious problems when trying to fight this type of fire, even with the use of longitudinal ventilation of 2–3 m/s.

11.4   Summation of measured HRR data

In the following a summation of the maximum HRR data from the large-scale tests is given per square metre fuel surface area (MW/m^2). The reason is that it is convenient to compare the maximum HRR between different fuels and different conditions. It also allows estimation of the peak HRR in tunnel fires. The HRR data will be divided into three different groups based on the fuel type. These are liquid pool fires, solid materials such as wood pallets and wood cribs and road and rail/metro vehicles.

11.4.1   Liquid fires

In Table 11.11 a summary of the HRR per square metre is given for tests that included liquid fuels. The free burning rate of gasoline per square metre fuel area is 55 g/m²s, see Babrauskas.[36] In the Ofenegg tests the influences of the ventilation conditions and the tunnel geometry on the burning rate were clearly demonstrated. The average burning rate was 59 g/m²s for the 6.6 m² fuel pan which is slightly higher than given by Babrauskas. In the tests with the 47.5 m² fuel pan the burning rate for the natural and semi-transverse ventilation was 22 g/m²s, whereas in the test with longitudinal ventilation (test 7a in Table 11.2) the average burning rate was much higher or 38 g/m²s. This value can be compared to 54 g/m²s obtained in test 2a with longitudinal ventilation and fuel size of 6.6 m². In the 95 m² pan the average burning rate was only 10 g/m²s, which corresponds to a reduction of about 82% compared to free-burning values. It is obvious that the burning rate per square metre in these tests is highly influenced by the ventilation and the test set-up. The poor accessibility of oxygen to the fuel bed as the trough (fuel pan) used was nearly equally wide (3.8 m) as the tunnel (4.2 m) is one of the reasons.

In the Zwenberg tests, where the fuel source was not as wide as the tunnel, the results were much more consistent. The average burning rate per square metre fuel area of gasoline for all the tests was 43 g/m²s with a standard deviation of 7.5 g/m²s. In the case of a strong longitudinal flow inside the tunnel (2–4 m/s) the average burning rate was also 43 g/m²s. The reduction in the burning rate is about 22% compared to the free burning rate. In the calculation of the HRR rates a combustion efficiency of 0.8 for natural and transverse ventilation and 0.9 for longitudinal ventilation was applied both in the Ofenegg tests as well as in the Zwenberg tests. This combustion efficiency needs to be verified in future large-scale tests carried out with gasoline fires. No burning rates were measured in the P.W.R.I. tests but in the No. 3 Shimizu the burning rate was measured only in the 1G-0 tests, which was 1 m². The measurements show a peak value of about 60 g/m²s , which would correspond to 2.6 MW/m², This is slightly higher than given by Babrauskas[36] but calculations of convective HRR based with aid of CFD[40] show a linear correlation between all the pan sizes used and the convective HRR. The convective HRR was estimated to be 1.9 MW/m² which would mean that the convective part is about 0.73 of the total HRR (2.6 MW/m²).

The large-scale tests show that in a real tank fire accident, where it is not unrealistic to expect that the gasoline spreads over the entire tunnel width, we can expect HRR per square metre fuel surface area in the range of 0.35 MW/m² to 2.6 MW/m² depending on the ventilation conditions and spread of the fuel over the road surface. In well-ventilated conditions with pan fuel depth that is larger than 70 mm and where the pan width is smaller than the tunnel width we should expect HRR of about 2.4 MW/m² to 2.6 MW/m² for gasoline. The effects of fuel depth on the burning rate have not been considered here but in a real

Table 11.11 Summation of HRR per fuel surface area for liquid fires in tunnels. The tests were performed in fuel pans with a relatively deep fuel depth

Type of fuel	Test series	HRR per square metre MW/m^2	HRR per square meter in the open[36] MW/m^2
Gasoline	Ofenegg, Zwenberg	0.35–2.6	2.4–2.5
Kerosene	Glasgow	1.4	1.7
n-Heptane	Eureka	3.5	4.5
n-60% heptane/ 40% toluene	2nd Benelux	1.1–1.6	–
Low-sulfur No. 2 fuel oil	Memorial	1.7–2.5	–

accident we would expect the burning rates to reduce due to cooling of the road surface.

Other fuel types applied in the large-scale test series presented here are n-heptane in the EUREKA tests and the second Benelux tests (mixed with 40% by mass of toluene), kerosene in the Glasgow tests and low-sulfur No. 2 fuel oil in the Memorial tests. The average burning rate per fuel surface area in the Memorial tests in the case of longitudinal ventilation varied between $1.7\,MW/m^2$ to $2.5\,MW/m^2$ with an average value of $2.1\,MW/m^2$. In Babrauskas[36] the free burning rate for fuel oil (heavy) is 35 g/m^2s and for kerosene the burning rate is 39 g/m^2 s. Corresponding HRR assuming complete combustion would be $1.4\,MW/m^2$ and $1.7\,MW/m^2$, respectively. The corresponding values for heptane according to Babrauskas is 101 g/m^2s, or $4.5\,MW/m^2$. In the EUREKA 499 n-heptane tests, see Table 11.5, the average burning rate was about 78 g/m^2s, which corresponds to average HRRs of $3.5\,MW/m^2$. In the second Benelux tests the average HRR for the 7.2 m^2 60% n-heptane/40% toluene fuel pans were $1.6\,MW/m^2$ and $1.1\,MW/m^2$ for the 3.6 m^2 fuel pan.

The analysis presented here shows that the variation in the results is considerable and it is difficult to assume one value for each type of liquid fuel. Parameters that influence the burning rate for each fuel type are the pan geometry, the fuel depth, the ventilation conditions, the reciprocal tunnel and the fuel pan geometry. Further, in the case when the tunnel cross-section is large and the width of the tunnel is larger than the width of the fuel pan the influence of longitudinal ventilation on the burning rate appears to be small.

11.4.2   Solid materials

In many of the large-scale tests presented here solid materials such as pallets, cartons or wood cribs have been used. It is of interest to compare the HRR per square metre fuel surface area at peak conditions in order to see if these values

Table 11.12 Summary of HRR per fuel surface area for solid materials applied in large-scale tunnel fire tests

Type of fuel	Test series	Estimated fuel surface area (m²)	HRR per square metre fuel surface area at maximum MW/m²
Wood cribs	Eureka (tests 8, 9 and 10)	140	0.07–0.09
Wood pallets	2nd Benelux (tests 8, 9, 10 and 14)	120 (36 pallets) 240 (72 pallets)	0.11–0.16
82% wood pallets and 18% PE pallets	Runehamar (test 1)	1200	0.17
82% wood pallets and 18% PUR mattresses	Runehamar (test 2)	630	0.25
81% wood pallets and cartons and 19% plastic cups	Runehamar (test 4)	160	0.44
HGV – furniture	Runehamar (test 3)	240	0.5
HGV – furniture	Eureka (test 21)	300	0.4

are comparable between the tests series. It is postulated here that this type of information could be used when, e.g., estimating the peak HRR from HGV trailers with tarpaulin as coverage. In Table 11.12 a summary of the HRR per square metre is given for tests that included solid materials. Tests that included solid wood cribs or pallets are found in the TUB-VTT tests, the EUREKA test series and the second Benelux tests. In the Runehamar tests wood pallets (about 82% of the total mass) were integrated with other types of solid materials such as plastics (18% of the total mass), cartons and furniture and fixtures.

In the second Benelux tests with wood pallets the HRR per square metre fuel area varied between 0.11–0.16 MW/m² with an average value of 0.13 MW/m². This value tended to increase with the increased ventilation rate. The fuel itself was not densely packed and thus could be regarded as fuel surface controlled. For wood cribs the opposite conditions would be crib porosity controlled fire or ventilation controlled. In the Eureka tests with simulated truckloads the wood sticks were so densely packed that the fire itself became crib porosity controlled (ventilation controlled) under normal conditions. This means that the peak HRR become lower than if it was fuel surface controlled. In the simulated truck load fire test the HRR per square metre was estimated to be in the order 0.04 MW/m². In the wood crib tests (8, 9, 10 in Table 11.5) in the Eureka test series the HRR varied between 0.07–0.09 MW/m² depending on the longitudinal velocity. It was not possible to establish with any certainty if the wood cribs were fuel surface or crib-porosity controlled.

In the HGV test in the EUREKA test series using furniture the HRR per fuel surface area was estimated to be approximately $0.4\,MW/m^2$. The ventilation velocity dropped from 6 m/s to about 3 m/s with peak HRRs of 120 MW and 128 MW, respectively. The total fuel surface of the furniture commodity was estimated to be about $300\,m^2$. The HRR per square metre fuel area in the Runehamar tests was estimated to be about $0.17\,MW/m^2$ for test 1 with wood and plastic pallets, $0.25\,MW/m^2$ for test 2 with wood pallets and mattresses, $0.5\,MW/m^2$ for test 3 with furniture and fixtures and $0.44\,MW/m^2$ for test 4 with plastic cups in cartons.

In the large-scale tests presented here, the peak HRR for solid materials ranges from $0.07\,MW/m^2$ to about $0.5\,MW/m^2$. An interesting observation is that the furniture tests in the EUREKA and Runehamar test series appear to be of the same order of magnitude. The reason is that both tests were performed under good ventilation conditions and that the fuel surface area was of the same order. The fuel surface area was estimated to be roughly $300\,m^2$ in the Eureka test and about $240\,m^2$ in the Runehamar test. With this type of information it would be easy to estimate the peak HRR for a given type of fuel in a HGV.

11.4.3   Vehicle fires

In the following a summary of the peak HRR for different types of vehicles is given for the large-scale tests presented here. The data is presented in Table 11.13 as HRR per square metre of fuel surface area. It is not possible to present all the vehicle tests in this way. It is possible in those cases where the fire was not ventilation controlled at the maximum conditions. In many of the vehicle fires the enclosure structure of the vehicle (body) is burned off (e.g. bodies made of aluminium, plastic, composite materials etc.) allowing oxygen to entrain to the fire plume or as in some of the cases where the enclosure is kept intact the windows are large enough to preserve a fuel-controlled fire. In some of the tests the opening area of the windows controlled the HRR. In tests 12 and 13 with railway coaches in the EUREKA test series the fire developed very slowly due to the windows. The

Table 11.13 The summary of HRR per square metre fuel surface area of vehicles with fuel-controlled fires

Type of fuel	Test series	Estimated fuel surface area (m^2)	HRR per square metre at maximum MW/m^2
Passenger cars		12–18	0.3–0.4
Passenger car plastic	Test 20 in Eureka	17 (no ceiling)	0.35
Bus	Test 7 in Eureka	80	0.36
Train	Test 11 in Eureka	145	0.30
Subway coach	Test 14 in Eureka	130	0.27

fire became ventilation-controlled and spread along the coach body at the same speed as the windows cracked in the heat. In these tests the information on the fuel surface area is impossible to estimate and will therefore be excluded.

Fully developed fires in passenger cars with a steel body can be regarded as fuel surface-controlled fires due to the large window area in comparison with the fuels surface area and the window height. The opening factor[44] for medium-sized passenger cars is estimated to be in the range of 1.2 to 1.8, which is considerably higher than the limits for fuel-controlled enclosure fires 0.29 with wood cribs.[44] The maximum HRRs for single passenger cars (small and large) vary from 1.5 to 8 MW, but the majority of the tests show peak HRR values less than 5 MW.[45] When two cars are involved the peak HRR varies between 3.5 to 10 MW. There are great differences in the times to reach peak HRR. It varies between 10 and 55 minutes. The fuel's surface area of the interior of a medium-sized passenger car can be estimated to be in the range of $12\,m^2$ to $18\,m^2$. This includes the floor and ceiling area, instrument panel area, door area and the seat area (double sided). This would mean that the HRR per square metre fuel surface area of a passenger car developing a peak value of 5 MW can vary between $0.3\,MW/m^2$ to $0.4\,MW/m^2$. The only test in a tunnel available is test no. 20 in the Eureka test programme. The car was a Renault Espace J11 with a plastic body. This car developed a maximum HRR of 6 MW and the fuel surface area was estimated to be about $17\,m^2$, not including the ceiling.

Other vehicles with fuel-controlled fires were the tests 7, 11, 14 and 20 in the Eureka programme. In these tests the main contribution is from the floor material and the seats. In tests carried out for different clients at SP Fire Technology in Borås Sweden we see that seats develop between 0.2 to $0.5\,MW/m^2$. This includes both bus seats and train seats. In Table 11.13 we observe that the total HRR per square meter fuel surface is in line with these values. It ranges between 0.27 and $0.36\,MW/m^2$. In a train we have numerous different materials in the interior of a carriage. It can be everything from textile, rubber, foam padding, PVC, cork, etc. What is interesting here is that the HRR per square metre fuel surface area in fuel controlled fires of different vehicles is rather narrow, or between 0.3 to $0.4\,MW/m^2$. This is also in line with the HRR per square metre fuel surface area for the solid materials presented in Table 11.12. The HRR per square metre fuel surface area of the individual materials have a greater variation, both lower and much higher, but it appears that the total effect of the mixed material is not so broad.

11.5 Summation of the temperature data

The results from the large-scale tests presented here show the gas temperatures up to 1365 °C. Such high temperatures are usually not obtained in enclosed fires. Usually it is in the range of 900–1100 °C. In enclosure fires the highest temperatures are usually obtained when the ventilation conditions are optimal (opening factor ~10–$12\,\mathrm{m}^{-1/2}$). We should expect similar behaviour in tunnel fires. The large-scale tests presented here show that the highest temperatures were obtained in those tests with high HRRs (> 20 MW) and low ceiling height (~4 m to 5 m) in combination with mediocre mechanical ventilation. For high HRR the flames impinge on the ceiling and the combustion zone, where the highest temperatures are usually obtained, is situated close to the ceiling, even when the longitudinal ventilation deflects the flames. The geometrical shape and size of the fuel, the tunnel cross-section (especially the height), the ventilation rate and the combustion efficiency appear to be the main parameters that determine the temperature level at the ceiling height.

In the Memorial tunnel tests the peak temperatures of about 1350 °C are obtained when a relatively ineffective ventilation system (Table 11.6; test ID 103, 113A and 217A) was activated in the tests with low ceiling height (4.4 m) and HRR equal to or larger than 20 MW. When the ceiling height increased from 4.4 m to 7.9 m the temperature drops notably for the corresponding HRR, or well below 1100 °C. The corresponding gas temperatures for ceiling height of 4.4 m, measured before mechanical ventilation was started, show that the temperatures were in the range of 1053 °C–1210 °C instead of 1350 °C. Also it is noticed that increased longitudinal flow tended to lower the temperature due to the mixing effects of the cold flow. This indicates that there exists optimal ventilation conditions that are needed to obtain these extremely high temperatures.

In the Ofenegg tests, with a ceiling height of 6 m, the highest temperatures were measured in the tests with good ventilation and HRRs between 33 MW to 70 MW (tests 5, 6, 7a) whereas in the case with poor ventilation conditions but HRRs of the same magnitude, 35 MW and 32 MW (tests 9 and 10), respectively, the temperature drops down from 1200 °C–1325 °C to 850 °C–1020 °C. The same tendency can be observed in the Zwenberg tests. The ceiling height was 3.9 m and the HRR ranged from 7 MW to 21 MW. The highest temperatures were obtained in tests 301–303, with transverse ventilation and HRRs in the range of 17–21 MW. The temperature varied between 1320 °C–1332 °C. For similar ventilation conditions but HRRs in the range of 8 MW to 13 MW the temperature varied between 1222 °C and 1320 °C.

In general we can say that when the combustion zone (usually up to about two fifths of the free flame height) extends up to the tunnel ceiling and we attain optimal ventilation in the combustion area the highest ceiling temperatures are obtained. When the longitudinal ventilation is increased the cooling effects predominate and the temperature drops again.

11.6   Summation of flame length data

The flame length in tunnel fires is very important when considering fire spread between vehicles. The large-scale tests presented here show that the type of ventilation and HRR govern the spread of the flames along the ceiling. The extended length of the flames is related to the HRR, the ceiling height and the longitudinal velocity.

Rew and Deaves[46] presented a flame length model for tunnels, which included heat release rate and longitudinal velocity but not the tunnel width or height. Much of their work is based on the investigation of the Channel Tunnel fire in 1996 and test data from the HGV-EUREKA 499 fire test and the Memorial tests. They defined the horizontal flame length, L_f, as the distance of the 600 °C contour from the centre of the HGV or the pool, or from the rear of the HGV. A possible way of representing both the tunnel height and width is by using the hydraulic diameter of the tunnel, D_h. In Fig. 11.2 the results from the Runehamar tests, assuming a 600 °C flame tip, are compared to the data obtained from the Memorial and Runehamar tests with longitudinal ventilation using the tunnel hydraulic diameter. The results were found to be improved by plotting them as a function of the hydraulic diameter.[47]

The results show that the flame length tends to increase nearly linearly with the heat release rate. The discrepancy in the results between the EUREKA tests and the Runehamar tests is not known. In the Runehamar tests the cross-section was smaller ($32\,m^2$) in the first 53 m from the centre of the fire, due to the

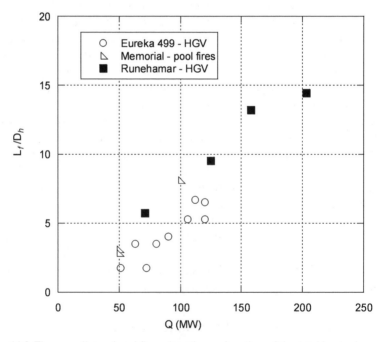

11.2 The non-dimensional flame length as a function of the total heat release rate for the HGV-EUREKA 499, Memorial tests 615B, 624B and 502 and the Runehamar tests T1, T2, T3 and T4. These tests were performed with longitudinal ventilation flow.

protection of the rock. After this point it was $47\,m^2$. The hydraulic diameter used here is the average of the two cross-sections.

11.7   Conclusions

A dozen large-scale fire test programmes have been carried out to date. The focus has mainly been on the heat and smoke spread and how different ventilation systems influence these parameters. Nearly half of the test series included sprinkler testing. The quality of large-scale tests carried out in the 1960s to 1980s varies considerably and in all these tests there is a lack of the key fire hazard parameter; the heat release rate (HRR). In the analysis carried out here new estimated HRRs are given. There is no doubt that the Eureka tests and the Memorial tests are the most well-known and well-reputed large-scale fire test series to date. They have already been established as the 'large-scale fire tests' and provide a new basis for standards and knowledge in tunnel fire safety.[48,49] The use of oxygen consumption calorimetry has increased the quality in the HRR results and made it possible to measure HRRs from vehicles.

The analysis presented here show that the variation in the results is considerable and it is difficult to assume one value for each type of liquid fuel. Parameters that influence the burning rate for each fuel type are the pan geometry, the fuel depth, the ventilation conditions and the reciprocal tunnel and the fuel pan geometry. Further, in the case when the tunnel cross-section is large and the width of the tunnel is larger than the width of the fuel pan the influence of the longitudinal ventilation on the burning rate appears to be small.

The HRR per square metre fuel surface area in fuel-controlled fires of different vehicles is rather narrow, or between 0.3 to $0.4\,MW/m^2$. This is also in line with the HRR per square metre fuel surface area for solid materials. The HRR per square metre fuel surface area of the individual materials have a greater variation, both lower and much higher, but it appears that the total effect of the mixed material is not so broad.

In general we can say that when the combustion zone (usually up to about two fifths of the free flame height) extends up to the tunnel ceiling and we attain optimal ventilation in the combustion area the highest ceiling temperatures are obtained. When the longitudinal ventilation is increased the cooling effects dominate and the temperature drops again. The results show that the flame length tends to increase nearly linearly with the heat release rate.

11.8   References

1. Grant, G. B., Jagger, S. F., and Lea, C. J., 'Fires in tunnels', *Phil. Trans. R. Soc. Lond.,* **356**, 2873–2906, 1998.
2. Haerter, A., 'Fire Tests in the Ofenegg-Tunnel in 1965', International Conference on Fires in Tunnels, Borås, Sweden, 1994.

3. Feizlmayr, A., 'Research in Austria on tunnel fire, Paper J2, BHRA', 2nd Int Symp on Aerodynamics and Ventilation of Vehicle Tunnels, 19–40, Cambridge, UK, 1976.

4. Pucher, K., 'Fire Tests in the Zwenberg Tunnel (Austria)', International Conference on Fires in Tunnels, 187–194, Borås, Sweden, 1994.

5. Lässer-Feizlmayr, I.-I., 'Brandversuche in einem Tunnel', Bundesministerium f. Bauten u. Technik, Strassenforschung, P0244, 1976.

6. 'Schlussbericht der Versuche im Ofenegg Tunnel von 17.5–31.5 1965', Kommission fur Sicherheitsmassnahmen in Strassen Tunneln (two volumes), 1965.

7. 'State of the Road Tunnel Equipment in Japan – Ventilation, Lighting, Safety Equipment', Public Works Research Institute, Technical note, Vol. 61, 1993.

8. Huggett, C., 'Estimation of Rate of Heat Release by Means of Oxygen Consumption Measurements', *Fire and Materials,* **4,** 2, 61–65, 1980.

9. Parker, W. J., 'Calculations of the Heat Release Rate by Oxygen Consumption for Various Applications', *Journal of Fire Sciences,* **2,** September/October, 1984.

10. Tewarson, A., 'Experimental Evaluation of Flammability Parameters of Polymeric Materials'. In *Flame Retardant Polymeric Materials* (M. Lewin, S. M. Atlas, and E. M. Pearce, eds), Plenum Press, 130–142, New York, 1982.

11. Keski-Rahkonen, O., 'Tunnel Fire Tests in Finland', International Conference on Fires in Tunnels, 222–237, Borås, 10–11 October 1994, 1994.

12. *Fires in Transport Tunnels: Report on Full-Scale Tests,* edited by Studiensgesellschaft Stahlanwendung e. V., EUREKA-Project EU499:FIRETUN, Düsseldorf, Germany, 1995.

13. Mikkola, E., email correspondence to the author, 10 September 2004.

14. Ingason, H., 'Heat Release Rate Measurements in Tunnel Fires', International Conference on Fires in Tunnels, 86–103, Borås, Sweden, October 10–11, 1994, 1994.

15. Ingason, H., and Lönnermark, A., 'Large-scale Fire Tests in the Runehamar tunnel – Heat Release Rate (HRR)', International Symposium on Catastrophic Tunnel Fires (CTF), *SP Report 2004:05,* pp. 81–92, Borås, Sweden, 2003.

16. Steinert, C., 'Smoke and Heat Production in Tunnel Fires', International Conference on Fires in Tunnels, 123–137, Borås, Sweden, 10–11 October, 1994.

17. Axelsson, J., Andersson, P., Lönnermark, A., Van Hees, P., and Wetterlund, I., 'Uncertainties in Measuring Heat and Smoke Release Rates in the Room/Corner Test and the SBI', SP Swedish National Testing and Research Institute, *SP Report 2001:04,* Borås, Sweden, 2001.

18. Grant, G. B., and Drysdale, D., 'Estimating Heat Release Rates from Large-scale Tunnel Fires', Fire Safety Science, *Proceedings of the Fifth International Symposium,* 1213–1224, Melbourne, 1995.

19. *Memorial Tunnel Fire Ventilation Test Program – Test Report,* Massachusetts Highway Department and Federal Highway Administration, 1995.

20. Takekuni, K., 'Disaster Prevention of Road Tunnel and Characteristics of the Evacuation Environment during Fires in Large Scale Tunnels in Japan', 4th Joint Workshop COB/JTA 2Joint Meeting *JTA/Cob Open Work Shop 2001* in Netherlands, 35–43, 2001.

21. Lemaire, A., van de Leur, P. H. E., and Kenyon, Y. M., 'Safety Proef: TNO Metingen Beneluxtunnel – Meetrapport', TNO, *TNO-Rapport 2002*-CVB-R05572, 2002.

22. Lönnermark, A., and Ingason, H., 'Large Scale Fire Tests in the Runehamar Tunnel

– Gas Temperature and Radiation', International Symposium on Catastrophic Tunnel Fires (CTF), *SP Report 2004:05*, pp. 93–103, Borås, Sweden, 20–21 November, 2003.

23. James Lake, A. G. B., 'Changes in the 2004 Edition of NFPA 502 Standard for Road Tunnels, Bridges and other Limitied Access Highways', Tunnel Fires, Fifth International Conference, 3–8, London, 2004.

24. Apte, V. B., and Green, A. R., 'Pool Fire Plume Flow in a Large-Scale Wind Tunnel', Fire Safety Science – The Third International Symposium, 425–434, Edinburgh, 1991.

25. Bettis, R. J., Jagger, S. F., Lea, C. J., Jones, I. P., Lennon, S., and Guilbert, P. W., 'The Use of Physical and Mathematical Modelling to Assess the Hazards of Tunnel Fires', 8th International Symposium on Aerodynamics and Ventilation of Vehicle Tunnels, 439–469, Liverpool, 1994.

26. Thomas, P. H., 'Movement of Smoke in Horizontal Corridors against an Air Flow', *Inst. Fire Engrs Q*, 30, 1970.

27. Ingason, H., Nireus, K., and Werling, P., 'Fire Tests in Blasted Rock Tunnel', FOA, *Report FOA-R-97-00581-990-SE*, Sweden, 1997.

28. Ingason, H., and Persson, B., 'Prediction of Optical Density using CFD', *Fire Safety Science – Proceedings of the 6th International Symposium*, 817–828, Poitiers, 1999.

29. Perard, M., and Brousse, B., 'Full size tests before opening two French tunnels', *8th Int Symp on Areodynamics and Ventilation of Vehicle Tunnels*, pp. 383–408, Liverpool, UK, 1994.

30. Casale, E., Brousse, B., Weatherill, A., and Marlier, E., 'Full Scale Fire Tests Performed in the Mont Blanc Tunnel – Evaluation of the Efficiency of the Fully Automatic Ventilation Responses', International Conference on Fires in Tunnels, 313–325, 2–4 December, Basel, Switzerland, 2002.

31. Brousse, B., Voeltzel, A., Botlan, Y. L., and Ruffin, E., 'Mont Blanc tunnel ventilation and fire tests', *Tunnel Management International*, 5, 1, 13–22, 2002.

32. 'Rapport betreffende de beproevning van het gedrag van twee isolatiematerialen ter bescheming van tunnels tegen brand', Instituut TNO voor Bouwmaterialen en Bouwconstructies, *Rapport B-80-33*, 1979.

33. Rew, C., and Deaves, D., 'Fire spread and flame length in ventilated tunnels, a model used in Channel tunnel assessments', *Proceedings of Tunnel Fires and Escape from Tunnels*, pp. 385–406, Lyon 5–7, May 1999.

34. Heskestad, G., 'Fire Plumes, Flame Height, and Air Entraiment'. In *The SFPE Handbook of Fire Protection Engineering* (P. J. DiNenno, ed.), National Fire Protection Association, 2-1–2-17, Quincy, Massachusetts, USA, 2002.

35. Heselden, A., 'Studies of fire and smoke behavior relevant to tunnels, Paper J1, BHRA', 2nd Int Symp on Aerodynamics and Ventilation of Vehicle Tunnels, 1–18, Cambridge, UK, 1976.

36. Babrauskas, V., 'Burning rates'. In *In SFPE Handbook of Fire Protection Engineering* (C. L. B. P. J. DiNenno, R. L. P. Custer, P. D. Walton, J. M. Watts, D. Drysdale, and J. R. Hall, eds), The National Fire Protection Association, USA, p3.1–3.15, 1995.

37. Heselden, A. a. H., P. L., 'Smoke travel in shopping malls. Experiments in cooperation with Glasgow Fire Brigade. Parts 1 and 2', Fire Research Station, *Joint Fire Research Organisation Fire Research Notes* 832 and 854, 1970.

38. Ekkehard, R., 'Propagation and Development of Temperatures from Test with

Railway and Road Vehicles', International Conference on Fires in Tunnels, 51–62, Borås 10–11 of October, 1994.

39. Shimoda, A., 'Evaluation of Evacuation Environment during Fires in Large-Scale Tunnels', 5th Joint Workshop COB/JTA, pp. 117–125, Japan, 2002.

40. Kunikane, Y., Kawabata, N., Takekuni, K., and Shimoda, A., 'Heat Release Rate Induced by Gasoline Pool Fire in a Large-Cross-Section Tunnel', International Conference on Fires in Tunnels, 387–396, Basel, Switzerland, 2002.

41. Kawabata, N., Kunikane, Y., Yamamoto, N., Takekuni, K., and Shimoda, A., 'Numerical Simulation of Smoke Descent in a Tunnel Fire Accident', International Conference on Fires in Tunnels, pp. 357–366, Basel, Switzerland, 2002.

42. Kunikane, Y., Kawabata, N., Okubo, K., and Shimoda, A., 'Behaviour of Fire Plume in a Large Cross Sectional Tunnel', 11th Int. Symp. on AVVT, 78–93, Luzern, Switzerland, 2003.

43. Kunikane, Y., Kawabata, N., Ishikawa, T., Takekuni, K., and Shimoda, A., 'Thermal Fumes and Smoke Induced by Bus Fire Accident in Large Cross Sectional Tunnel', The fifth JSME-KSME Fluids Engineering Conference, Nagoya, Japan, Nov. 17–21, 2002.

44. Drysdale, D., *An Introduction to Fire Dynamics*, John Wiley & Sons, 1992.

45. Ingason, H., and Lönnermark, A., 'Recent Achievements Regarding Measuring of Time-heat and Time-temperature Development in Tunnels', 1st International Symposium on Safe & Reliable Tunnels, Prague, Czech Republic, 4–6 February, 2004.

46. Rew, C., and Deaves, D., 'Fire spread and flame length in ventilated tunnels – a model used in Channel tunnel assessments', *Proceedings of the International Conference on Tunnel Fires and Escape from Tunnels*, 397–406, Lyon, France, 5-7 May, 1999.

47. Ingason, H., and Lönnermark, A., 'Fire development of Large Vehicles in Road Tunnels', *Proceedings of the Fifth International Conference on Tunnel Fires*, 203–212, London, UK, 25–27 October, 2004.

48. Anon., NFPA 502: *Standard for Road Tunnels, Bridges and Other Limited Access Highways* (2004 edition), National Fire Protection Association, Quincy MA, USA, 2004.

49. Anon., *Fire and Smoke Control in Road Tunnels* (PIARC O5.O5B-1999), Association Mondiale de la Route (PIARC), Paris, France, 1999.

12

Flammability tests for aircraft

J M PETERSON, Boeing Commercial Airplanes, USA

12.1 Introduction

This chapter addresses the flammability regulations for transport category airplanes, i.e., commercial airplanes used by airlines for transport of goods and people. Although flammability regulations for other aircraft types such as general aviation, commuters, agricultural, etc., are similar but not as comprehensive as those for transport category airplanes, they are beyond the scope of this chapter.

In the United States, the Federal Aviation Administration (FAA) has the responsibility for establishing and enforcing all regulatory requirements for civil aviation. FAA fire safety regulations on transport category airplanes are quite extensive and implementation and enforcement processes are considerably more intricate and involved than those imposed by other regulatory agencies on land-based and water-based transport vehicles. Passenger cabin and engine compartment components are subject to one or more of over a dozen tests.

Beyond the USA, FAA regulations and FAA regulatory changes are commonly adopted by almost all national aviation authorities. Hence, FAA regulations are essentially used worldwide, and for this reason this chapter is limited to FAA flammability requirements for transport category airplanes. A brief history of the evolution of FAA flammability regulations is provided. The original flammability requirements are described.

Over the years, the FAA has greatly increased the stringency of airplane flammability requirements as the state-of-the-art of available materials advanced, and/or as existing fire threats based on large-scale testing were better understood and steps were taken to mitigate them. The development of regulatory flammability requirements in the 1980s were dynamic, and are described.

This chapter also covers FAA processes for approval of design and production of airplanes. These processes or their non-USA equivalents are also used by almost all national regulatory authorities. These are often more of a challenge to applicants for regulatory approval than the tests themselves.

The required FAA flammability tests for transport category airplanes are divided into two groups: twelve that are applicable inside the pressurized vessel, which includes the passenger cabin and cockpit, and four that are applicable outside the pressurized vessel, which includes the engine and auxiliary power unit areas. These tests apply to various components regarding their usage, and sometimes regarding the materials of which the components are made. It is beyond the scope of this chapter to describe these approval processes and flammability tests in detail, so the discussion is limited to a general description of each and its applicability. References to more detailed information for each process and test are provided.

In addition to regulatory requirements, airplane manufacturers have adopted supplemental unilateral fire safety requirements which the FAA and other regulatory authorities have not included in the regulations, primarily for technical/economic reasons. Both Airbus and Boeing have documented flammability requirements which they impose on their designs. A brief description of these criteria is provided. Finally, the current trends in FAA directions for potential future rulemaking developments are discussed.

12.1.1 Abbreviations

The following abbreviations are used in this chapter:

AIA   Aerospace Industries Association – the industry trade group representing commercial airplane manufacturers in the United States.

CAR   Civil Air Regulations – the regulations governing design of commercial airplanes in the United States prior to 1965. Section 4b pertained to flammability requirements for cabin materials.

CFR   Code of Federal Regulations – codification of the general and permanent rules published by the US Federal Government, representing broad areas subject to USA regulation

EASA   European Aviation Safety Agency – the agency that regulates civil aviation in the European Union

FAA   Federal Aviation Administration – the agency that regulates civil aviation in the United States

IAMFTWG   International Aircraft Materials Fire Test Working Group – an informal international group working with the FAA in developing test procedures for aircraft cabin materials

IASFPWG   International Aircraft Systems Fire Protection Working Group – an informal international group working with the FAA in developing test procedures for aircraft systems

JAA   Joint Aviation Authorities – a group of regulatory authorities from European countries overseeing aviation regulatory activities in Europe. It has been superseded by the EASA.

JAR   Joint Aviation Regulations – the aviation regulations effective in the European Union

NBS   National Bureau of Standards (now NIST – National Institute of Standards and Technology) – the laboratory responsible for establishing technical standards in the United States

NPRM   Notice of Proposed Rule Making – a notice published in the United States Federal Register announcing an agency's intent to establish additional regulatory requirements in its area of responsibility

OSU   The apparatus that monitors heat release rate that was developed at The Ohio State University

SAFER   Special Aviation Fire and Explosion Reduction – the committee established by the FAA in 1978 to recommend fire safety research for the FAA to pursue

USA   United States of America

12.2   FAA approvals for design and production of airplanes

The regulations governing civil aviation in the United States are published in Title 14 (Aeronautics and Space) of the Code of Federal Regulations (CFR), Chapter I – Federal Aviation Administration, Department of Transportation, Parts 1-1399.[1] The various parts contain regulatory requirements for specific aspects of aviation such as airworthiness standards and certification procedures.

FAA regulations encompass the design, manufacture, and operation of all civil aircraft, including balloons, helicopters, and airplanes of all types including commuter airplanes. The Parts that affect transport category airplanes include:

- Part 21, Certification Procedures for Products and Parts[2]
- Part 25, Airworthiness standards: Transport category airplanes[3]
- Part 33, Airworthiness standards: Aircraft engines[4]
- Part 121, Operating requirements: Domestic, flag, and supplemental operations[5]

The flammability tests and requirements applicable to transport category airplanes are delineated in Part 25. Procedures for specifying how they are applied to various airplane models are contained in Part 21 and Part 121.

There are four separate FAA approvals which must be in place before an airplane may be produced by a manufacturer and operated by an air carrier to transport passengers. These are

- Type Certificate (Part 21 and Part 25),
- Production Certificate (Part 21),
- Airworthiness Certificate (Part 21), and
- Operating Certificate (Part 119, to operate under Part 121).

12.2.1   Type Certificate

A Type Certificate is the formal FAA approval of the Type Design for a new model airplane.[6] The Type Design consists of 'the drawings and specifications, and a list of those drawings and specifications, necessary to define the configuration and the design features of the product shown to comply with the applicable requirements'.[7] The requirements that the Type Design must meet are defined by the Type Certification Basis, which consists of the Part 25 requirements then in effect, plus any special conditions which the FAA feels are appropriate to cover novel characteristics which the existing Part 25 requirements may not adequately address. Part 25 is amended from time to time to upgrade requirements, so the Type Certification Basis – including the flammability requirements – is typically different for each airplane model.

Reviewing and approving the Type Design involves extensive testing to generate certification data. For flammability certification, each affected part to be tested must be presented to the FAA with its design and determined by the FAA to conform to its design, must be tested using test equipment that has been accepted by the FAA, and the testing must be witnessed and the test results approved and documented by the FAA. There are mechanisms available in Part 183[8] whereby the FAA may delegate conformity findings and test witnessing to non-FAA persons and/or organizations that have passed rigorous screening to demonstrate intimate knowledge of the regulations and FAA policies, are known to and acquainted with FAA specialists, and have worked with the FAA for several years in a non-delegated capacity.

12.2.2   Production Certificate

A Production Certificate is the formal FAA approval that allows a manufacturer to produce copies of airplanes for which he holds a Type Certificates.[9] Part 21 contains the requirements and procedures to obtain a Production Certificate. The primary requirement for a Production Certificate is that the manufacturer has a quality control system capable of ensuring that each airplane produced conforms to the approved Type Design. Other processes are also required, such as procedures to obtain FAA approval for changes to Type Design like material substitutions or changes in passenger cabin layout for a new airline customer.

12.2.3   Airworthiness Certificate

An Airworthiness Certificate is the formal FAA approval that allows an individual airplane to be delivered to and used by an airline for carrying passengers.[10] Before an airplane can be used by an airline, the manufacturer's quality control system must demonstrate to the FAA that that individual airplane as-built conforms to the approved Type Design. When conformity to Type

Design is verified, the FAA issues an Airworthiness Certificate to that individual airplane.

12.2.4   Operating Certificate

An Operating Certificate is the formal FAA approval that allows an airline to operate airplanes and transport passengers.[11] The Operating Certificate signifies that the airline has suitable operating facilities, maintenance facilities, and processes to ensure safe transport.

12.3   FAA fire test requirements inside the fuselage for transport category airplanes

For transport category airplanes, there are fire test requirements on components both inside and outside the pressurized part of the fuselage. 'Inside the fuselage' includes that part of the airplane that is pressurized during flight, such as the passenger cabin and its surroundings (overceiling area, cheek area, etc.). The flammability requirements inside the fuselage are rather extensive and complicated. The thrust of the requirements is to inhibit fire propagation by installed components if they should become ignited. While avoiding ignition sources is always a design and operating objective, it has not been possible to totally eliminate them, and materials that are resistant to fire propagation are required. Ignition sources in the past have included electrical malfunctions such as arcs or short circuits, or much more severe ignition sources such as an external fire fed by jet fuel from ruptured fuel tanks that intrudes into the cabin and ignites interior cabin furnishings.

The FAA flammability requirements for components inside the pressurized area of transport category airplanes are in Part 25 §25.853, §25.855, §25.856, and §25.869. Section 25.855 addresses cargo compartments, §25.856 addresses thermal/acoustic insulation, §25.869 addresses electric wire insulation, and §25.853 addresses essentially everything else (e.g., cabin liners, textiles, seats, stowage bins, etc.).

First a brief description of the evolution of the requirements will be presented. Then a general description of the various tests will be given, followed by a description of the applicability of the various tests to the affected components.

12.3.1   Evolution of requirements

The evolution of flammability tests and requirements for airplane interior fuselage components is quite involved. A timeline of some of the major developments is shown in Fig. 12.1. The 'dash' numbers after 25 and 121 refer to the various amendment levels which upgraded the regulations in Parts 25 and

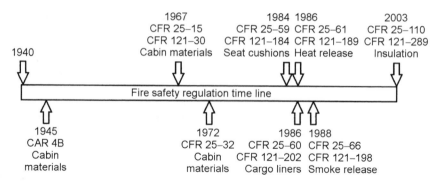

12.1 Timeline of FAA regulatory upgrades for flammability.

121. When Part 25 is amended, the change is applicable only to the Type Certification Bases for new model airplanes. A companion amendment to Part 121 can make the Part 25 change also required for airplane models already type certificated by disallowing airlines to operate newly manufactured airplanes that do not comply with the Part 25 change. If warranted, a Part 121 change can also require existing airplanes to be retrofitted to comply with a Part 25 change.

CAR 4B   Original flammability regulations

Prior to 1965, CAR 4B was the name of the flammability regulations for interior components. In 1965, the regulatory agency and the regulations were reorganized, and the CARs were reissued as Title 14 of the Code of Federal Regulations without a change of requirements. CAR 4B became Part 25.

When the first flammability requirements for airplanes were instituted in 1945 as CAR 4B, all tests for interior fuselage components were Bunsen burner tests.

- Cabin furnishings had to meet a burn rate limit, generally 4 inches/minute (10.2 cm/min), when tested in a horizontal orientation and exposed to a 1-1/2 inch (3.8 cm) long Bunsen burner flame for 15 seconds. The test specimen was 3 inches (7.6 cm) wide and 12 inches (30.5 cm) long.
- Cargo liners in inaccessible cargo compartments had to demonstrate burnthrough resistance when tested in a 45-degree orientation and exposed to a 1-1/2 inch (3.8 cm) long Bunsen burner flame for 30 seconds. The Bunsen burner test specimen was 10 inches by 10 inches (25.4 cm by 25.4 cm) square.
- Waste compartments in galley carts had to be resistant to burnthrough using the same test used for cargo liners, and also as-designed had to contain a fire from a specified fuel load representing waste paper, cups, etc., such as are typically collected by flight attendants from passenger meal service.

Part 25 Amendment 25-15   vertical Bunsen burner test

A Bunsen burner test with the specimen oriented vertically was used first by Boeing beginning in 1962. After lavatory fires in early 707s, Boeing carried out larger scale fire tests and found that a vertical orientation for test specimens was a better discriminator than a horizontal orientation, and added a 12-sec vertical test in addition to the regulatory tests as an internal unilateral requirement. The test specimen was the same size – 3 inches (7.6 cm) wide and 12 inches (30.5 cm) long – as a horizontal test specimen.

In 1965, there was a Boeing 727 accident in Salt Lake City involving a very hard landing that pushed landing gear into the fuselage and ruptured fuel lines going from the wings to the rear engines. Fuel spilled into the fuselage, and the resulting fire resulted in injuries and fatalities. Following that accident, the FAA released a Notice of Proposed Rulemaking (NPRM) to upgrade the flammability regulations,[12] which industry found could not be met with the existing and available state-of-the-art materials. In 1967 the FAA agreed to impose an interim requirement which added the Boeing 12-second vertical Bunsen burner test as Part 25 Amendment 25-15,[13] provided the Aerospace Industries Association (AIA, e.g., Boeing, Douglas, Lockheed) committed to carry out a research program and generate a proposal for what the state-of-the-art could support, and submit its findings to the FAA along with a petition for rulemaking to implement them.

Part 25 Amendment 25-32 – Upgraded Bunsen burner tests

Subsequently, the AIA carried out a two-year Crashworthiness Program which involved both large-scale and laboratory-scale tests, and submitted its report and petition for rulemaking to the FAA. The AIA proposal was adopted with only minor changes as Part 25 Amendment 25-32[14] in 1972. Regulatory additions through this amendment include:

- a 60-second vertical Bunsen burner test for large area, visible cabin components such as sidewalls, ceilings, and partitions
- a 30-second 60-degree Bunsen burner test for electric wire insulation
- an upgraded 12-second vertical Bunsen burner test to include a self-extinguishing time limit for thermoplastics and textiles.

These changes were applicable to Part 25 which affected new Type Designs. There were no changes made to Part 121 that would have affected older designs still in production (e.g., DC-8, DC-9, 707, 727, 737). However, before Part 25 Amendment 25-32 was released, these new requirements were also applied as special conditions to the Type Certification Bases for the first wide-body airplanes (747, DC-10, L-1011).

SAFER Committee

In 1973 and 1974, the FAA proposed three new rulemaking initiatives involving cabin fire safety:

- an NPRM to change Part 121 to require newly manufactured airplanes of older designs (e.g., 707, 727, DC-8, DC-9) to meet the new Part 25 Amendment 25-32 standards
- an NPRM for a Part 25 change to require control of smoke emission by interior components to specified limits using the NBS Smoke Density Chamber[15]
- an ANPRM (Advanced NPRM) requesting input on possible test criteria for a planned Part 25 change to require control of toxic gas combustion products by interior components.

Industry responded that a more unified approach was needed, and the FAA agreed to withdraw the proposals and establish a committee (SAFER – Special Aviation Fire and Explosion Reduction) composed of representatives from industry for guidance in future flammability rulemaking.

SAFER recommended[16] placing a priority on mitigating the effects of post-crash fuel-fed fires, which were considered at the time to be more of a safety issue than in-flight fires. Two SAFER recommendations were pursued immediately by the FAA:

- developing new test standards that would require encapsulating urethane foam seat cushions in a protective fire blocking covering, with acceptance criteria also based on realistic fuel-fed fire tests
- developing new fire test standards involving heat release as the selection criterion for cabin furnishings. The acceptance criteria were to be derived from tests using realistic fire conditions experienced during a fuel-fed fire that might occur following an impact-survivable post-crash accident involving ruptured fuel tanks.

Part 25 Amendments 25-59 Oil burner test for fire blocking of polyurethane foam seat cushions

Following SAFER's recommendations, the FAA conducted large-scale tests using fuel-fed fires to pursue the benefits of encapsulating polyurethane foam seat cushions in a protective fire blocking layer to inhibit their burning.[17] Polyurethane foam is the traditional material used for airplane seat cushions. The FAA released Part 25 Amendment 25-59[18] in 1984 requiring that seat cushions be subjected to an oil burner test as a requirement for new type designs, and Part 121 Amendment 121-184 requiring a retrofit of transport airplanes in operation.

Part 25 Amendment 25-60   Oil burner test for cargo liners

As a result of an in-flight fire involving a Saudi Arabian Airlines L-1011 in 1980, presumed to have started in a cargo compartment, the FAA pursued a more stringent burnthrough test for cargo compartment liners than the existing Bunsen burner test. An oil burner test was developed to simulate the heat from a large cargo fire. Part 25 Amendment 25-60[19] was issued in 1986 using an oil burner test as a requirement for cargo liners.

Part 25 Amendment 25-61   OSU heat release tests for cabin liners

Following SAFER's recommendations, the FAA conducted large-scale fuel-fed fire tests on simulated cabin layouts including sidewalls, stowage bins, ceilings, partitions, and seats.[20] The tests involved exposing a simulated cabin with sidewalls, ceilings, and partitions constructed of various materials to a large fuel-fed pool fire. The cabin atmosphere was monitored, and it was determined that prior to flashover the cabin remained survivable. Therefore, time to flashover was the parameter of importance. The FAA concluded that large-scale time-to-flashover correlated with laboratory-scale heat release rate results. Part 25 Amendment 25-61[21] was issued in 1986 requiring that large-area cabin liners above the cabin floor be tested in the OSU heat release apparatus for new type designs, and Part 121 Amendment 189 was released requiring cabin liners be tested in newly manufactured airplanes of existing designs after 1988.

Part 25 Amendment 25-66 – NBS smoke release tests for cabin liners

The Aerospace Industries Association (AIA) and the ATA (Air Travel Association) petitioned the FAA to change the OSU heat release requirement to a smoke release requirement using the NBS smoke chamber, citing better correlation between large-scale time-to-flashover test results and smoke release tests. Rather than substituting smoke release for heat release however, the FAA added smoke release to heat release. Part 25 Amendment 25-66[22] and Part 121 Amendment 121-198 were issued in 1988 requiring that affected cabin liners be tested in the NBS smoke chamber apparatus as well as in the OSU heat release apparatus for new type designs and for airplanes newly manufactured after 1990.

Part 25 Amendment 25-111   Radiant panel and burnthrough tests for thermal/ acoustic insulation

Part 25 Amendment 25-111[23] was released in 2003, which introduced changes to the requirements for thermal/acoustic insulation inside the fuselage. A primary driver for this change was the SwissAir MD-11 accident in 1998. Thermal/acoustic insulation typically is produced as fiberglass batting encapsulated in a plastic film. The FAA determined that encapsulating film materials that passed the existing regulatory requirement (12-sec vertical Bunsen burner) could propagate fire in realistic configurations. This upgrade included two new requirements:

- a test for resistance to fire propagation applicable to all thermal/acoustic insulation inside the fuselage using a radiant panel apparatus
- a test for resistance to burnthrough applicable to all thermal/acoustic insulation installed next to the airplane skin in the lower half of the fuselage. This new burnthrough resistance test was intended to provide protection for the passenger cabin against intrusion of a post-crash fuel-fed fire.

Part 121 Amendment 121-289 was also released requiring that newly manufactured transport airplanes of existing designs comply with the new fire propagation resistance requirements beginning 2 September 2005, and with the new burnthrough resistance requirement beginning 2 September 2007.

12.3.2   FAA/industry working groups

The FAA has established two FAA/industry working groups to discuss and develop flammability tests for new and projected regulatory requirements.

Materials group

Since 1984, seven new flammability test requirements for interior cabin materials have been added to the regulations. These have all been variations on existing industry standards that were modified by the FAA. All are somewhat complex and many difficulties have been encountered with them regarding interlaboratory consistency of test equipment, test procedures, and test results. To facilitate the introduction and utilization of these tests, in 1988 the FAA established the International Aircraft Materials Fire Test Working Group[24] (IAMFTWG) to bring together affected people and organizations to discuss the difficulties and provide advice to the regulatory authorities on matters involving flammability tests. The group is chaired by the FAA, and provides input for the FAA's use. Participants include regulatory authorities, airplane manufacturers, airlines, material suppliers, test laboratories, and any other interested parties that wish to participate.

Systems group

Another group, the International Aircraft System Fire Protection Working Group (IASFPWG),[25] was established in 1993 to address issues having to do with airplane systems. Originally the group's charter was to work on minimum performance standards and test methodologies for replacements of non-Halon fire suppression agents/systems. The group's focus has since been expanded to include all system fire protection research and development for aircraft. The group is chaired by the FAA, and provides input for the FAA's use. Participants, as with the materials group, include regulatory authorities, airplane manufacturers, airlines, material suppliers, test laboratories, and any other interested parties that wish to participate. The group meets twice a year. The meetings are hosted alternately by the FAA Technical Center in Atlantic City, New Jersey, USA, and by an interested organization at another location.

12.3.3   Flammability tests and acceptance criteria inside the fuselage

This section lists the various fire tests used for components inside the pressurized fuselage of transport airplanes. The next section (Section 12.3.4) describes how these tests apply to the various components inside the fuselage. These tests are described in some detail in Part 25, Appendix F.[26] However, since the descriptions of the tests in Appendix F are in some cases not totally clear, the FAA has released a *Fire Test Handbook*[27] that expands on these descriptions, and contains slightly different FAA-approved versions of the tests. It is beyond the scope of this chapter to go into detail with these requirements and test procedures. The reader is referred to the regulations (Part 25, Appendix F) and/or the FAA *Fire Test Handbook* for additional information. The various tests used are identified in Table 12.1. The handbook contains sketches of the tests and test equipment, which have been modified here and are shown in the indicated figures.

12.3.4   Requirements for components inside the fuselage

The FAA flammability requirements on the various components inside the pressurized section of an airplane are listed below. These requirements are found in CFR 25 §25.853, §25.855, §25.856, and §25.869.

- Interior ceiling and wall panels, other than lighting lenses and windows; partitions, other than transparent panels needed to enhance cabin safety; galley structure, including exposed surfaces of stowed carts and standard containers and the cavity walls that are exposed when a full complement of such carts or containers is not carried; and large cabinets and cabin stowage compartments, other than underseat stowage compartments for stowing small items such as magazines and maps must meet the 60-sec vertical Bunsen burner test, the OSU heat release test, and the NBS smoke release test.
- Passenger blankets and pillows are not included in Type Design, so they are not affected by regulations. However, the FAA has developed a test for blankets.[28]

Table 12.1 Flammability tests for components inside the fuselage

Test type	Acceptance criteria	Appendix F/Handbook
12-sec vertical Bunsen burner (Fig. 12.2)	Burn length $\leq$ 8 in (20.3 cm) Self-extinguishing time $\leq$ 15 sec Drip extinguishing time $\leq$ 5 sec	Part I (b)(4)/Chapter 1
60-sec vertical Bunsen burner (Fig. 12.2)	Burn length cm $\leq$ 6 in (15.2 cm) Self-extinguishing time $\leq$ 15 sec Drip extinguishing time $\leq$ 3 sec	Part I (b)(4)/Chapter 1
45° Bunsen burner (Fig. 12.3)	No flame penetration Self-extinguishing time $\leq$ 15 sec Glow time $\leq$ 10 sec	Part I (b)(6)/Chapter 2
2.5 inch/min. horizontal Bunsen burner (Fig. 12.4)	Burn rate $\leq$ 2.5 inch/minute (6.4 cm/minute)	Part I (b)(5)/Chapter 3
4.0 inch/min. horizontal Bunsen burner (Fig. 12.4)	Burn rate $\leq$ 4.0 inch/minute (10.2 cm/minute)	Part I (b)(5)/Chapter 3
60° Bunsen burner (Fig. 12.5)	Burn length $\leq$ 3 inches (7.6 cm) Self-extinguishing time $\leq$ 30 sec Drip extinguishing time $\leq$ 3 sec	Part I (b)(7)/Chapter 4
OSU heat release (Fig. 12.6)	Max heat release rate $\leq$ 65 kW/m^2 Max total heat released in first two minutes $\leq$ 65 kw-min/m^2	Part IV/Chapter 5
NBS smoke release (Fig. 12.7)	Specific optical density $\leq$ 200	Part V/Chapter 6
Seat cushion oil burner (Fig. 12.8)	Burn length $\leq$ 17 in (43.2 cm) Weight loss $\leq$ 10%	Part II/Chapter 7
Cargo liner oil burner (Fig. 12.9)	No flame penetration Peak temperature 4 inches above horizontal specimens $\leq$ 400 °F	Part III/Chapter 8
Escape slide radiant heat (Fig. 12.10)	Time to fabric failure $\leq$ 90 sec	NA/Chapter 9
Insulation blanket radiant panel (Fig. 12.11)	Flame propagation $\leq$ 2 in (5.2 cm) Self-extinguishing time for each specimen $\leq$ 3 sec	Part VI/Chapter 23
Insulation blanket oil burner (Fig. 12.12)	No flame penetration Heat flux 12 in (30.5 cm) from cold surface $\leq$ 2 BTU/ft^2-sec (22.7 kw/sqm)	Part VII/Chapter 24
Not applicable	Contain standard test fire in actual enclosure	FAR 25 §25.853(h)/ Chapter 10
No test required	N/A	Small parts

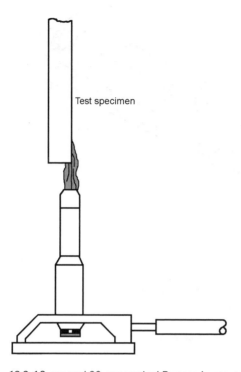

Test specimen

12.2 12-sec and 60-sec vertical Bunsen burner tests.

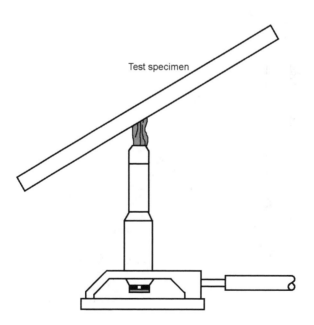

Test specimen

12.3 45-degree Bunsen burner test.

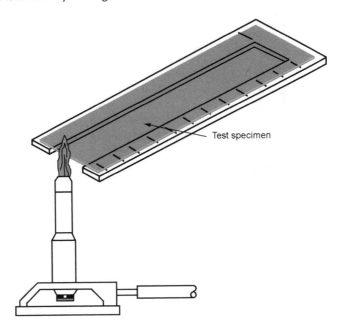

Test specimen

12.4 Horizontal Bunsen burner tests.

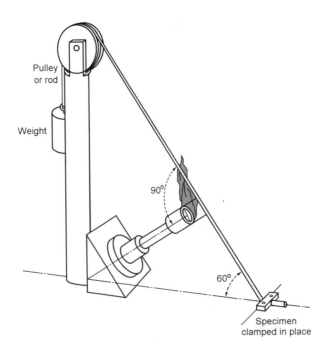

Pulley
or rod

Weight

90°

60°

Specimen
clamped in place

12.5 60-degree Bunsen burner test for electric wire insulation.

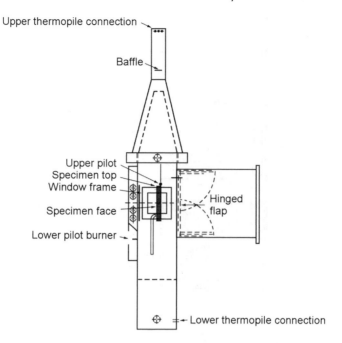

Upper thermopile connection

Baffle

Upper pilot
Specimen top
Window frame

Specimen face

Lower pilot burner

Hinged
flap

Lower thermopile connection

12.6 OSU heat release test.

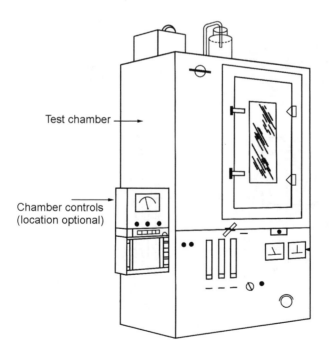

Test chamber

Chamber controls
(location optional)

12.7 NBS smoke release test.

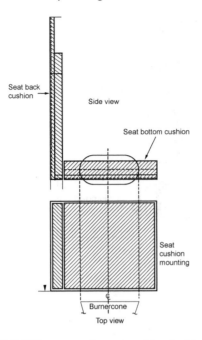

12.8 Oil burner test for seat cushion fire blocking.

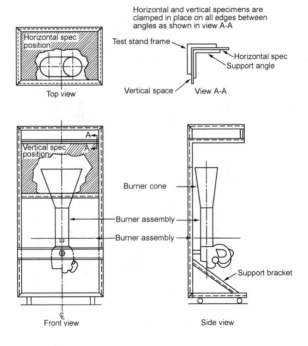

12.9 Oil burner test for cargo compartment liners.

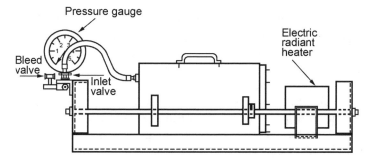

12.10 Radiant heat test for escape slide fabric.

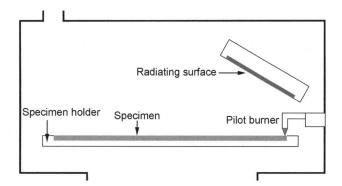

12.11 Radiant panel test for thermal/acoustic insulation.

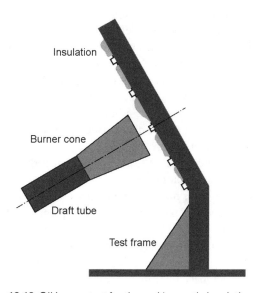

12.12 Oil burner test for thermal/acoustic insulation.

- Floor covering, draperies and upholstery, seat cushions, padding, thermal/acoustic insulation, thermoformed parts not used in applications where heat and smoke release are required, air ducting, and trays and galley furnishings must meet the 12-sec vertical Bunsen burner test.
- Structural flooring must meet the 60-sec vertical Bunsen burner test.
- Receptacles used for disposal of waste material must be fully enclosed, and meet the 45-degree Bunsen burner and fire containment tests.
- Cargo liners forming the walls and ceilings of Class C cargo compartments must meet the cargo liner oil burner test.
- Cargo liners of Class B and Class E compartments, and liners forming the floor of Class C cargo compartments must meet the 12-sec vertical Bunsen burner and 45-degree Bunsen burner tests.
- Escape slides, escape ramps, and life rafts must meet the 12-sec vertical Bunsen burner test. In addition, if the escape slides are manufactured under a Technical Standard Order (TSO), they must also meet the evacuation slide material pressure test.[29] A TSO is an FAA approval whereby a manufacturer of a specific component, such as escape slides, is approved to design and manufacture the component using its own engineering and quality control procedures.
- Clear plastic windows and signs, parts made of elastomers (provided they are not used in applications where heat and smoke release are required), seat belts, and cargo tiedown equipment must meet the 2.5 inch/minute (6.4 cm/minute) horizontal Bunsen burner test.
- Small parts such as knobs, handles, rollers, fasteners, clips, grommets, rub strips, pulleys, and small electrical parts that would not contribute significantly to the propagation of a fire do not need to be tested. (Note that the FAA has never established a firm universal definition of a 'small part'. One set of criteria used by industry defines 'small parts' as parts that are relatively two-dimensional and small enough to fit inside a 3 inch by 3 inch (7.6 cm by 7.6 cm) square, and parts that are relatively three-dimensional and small enough to fit inside a 2 inch by 2 inch by 2 inch (5.1 cm by 5.1 cm by 5.1 cm) cube. Such parts are treated as 'small' if they meet the criteria and there are not many of them close together; otherwise, they are not 'small'.)
- Electric wire insulation must meet the 60-degree Bunsen burner test. There is no small part exclusion for electric wire.
- Miscellaneous parts not covered in the list above must meet the 4 inch/min (10.2 cm/min) horizontal Bunsen burner test.
- All thermal/acoustic insulation for new design airplanes, and for airplanes of existing designs that are newly manufactured after 2 September 2005, must meet the radiant panel test.
- Thermal/acoustic insulation installed in the lower half of the fuselage for new design airplanes, and for airplanes of existing designs that are newly manufactured after 2 September 2007, must meet the insulation oil burner burnthrough test.

12.3.5   Unilateral flammability requirements used by airplane manufacturers

In the design of an airplane, the regulatory requirements have to be met. However, airplane manufacturers in addition establish unilateral requirements and criteria beyond the regulations that address many design issues, including fire safety.

Airbus requirements

In 1979, Airbus established unilateral requirements in ATS1000.001 that went beyond the existing regulations to select materials for use in new airplane models. These requirements were applied to all models beginning with the A300-600. In 1993, ATS 1000.001 was superseded by ABD0031, Fire – Smoke – Toxicity,[30] which contains essentially the same criteria. The criteria address smoke and toxic gas emission of components installed inside the fuselage. Both the smoke emission and toxic gas emission tests are performed in the smoke emission apparatus used by the regulations (NBS smoke chamber) in both the smoldering and the flaming mode.

Boeing requirements

After adopting the 12-second vertical Bunsen burner test unilaterally in 1962, Boeing established additional unilateral requirements beyond the existing regulations during the development of the 747 in the 1960s. Some of these, such as the vertical Bunsen burner test, were reflected in Amendments 25-15 (1967) and 25-32 (1972). The Amendment 25-32 provisions were applied to the 747 by the FAA as special certification conditions. In 1976, Boeing established an updated set of unilateral criteria to serve as guidelines for materials suppliers in developing new materials for use in new airplane models. These 1976 guidelines were superseded in 1983 with a set of design criteria which were applied to new designs such as 737, 747, 757, and 767 derivatives, and to the 777.[31] The criteria address smoke and toxic gas emission of components installed inside the fuselage, and various other design features such as fire stops and fire extinguishers. As in the Airbus procedures, the smoke emission and toxic gas emission tests are performed in the NBS smoke chamber in the flaming mode. The Airbus and Boeing acceptance criteria are similar.

12.4   FAA fire test requirements outside the fuselage

FAA requirements for fire safety outside the pressure vessel address principally those areas where lines carrying flammable fluids, such as engine fuel, are present. These areas are called 'designated fire zones' and they include areas around engines, areas containing auxiliary power units (generators powered by a turbine engine to provide electrical power when the main engines are not operating), and other areas subject to fires fed by flammable fluids. The requirements assume that a leakage of such fluids is a possibility, and mandate measures to protect the structure and operation of the airplane in case of leakage. These include measures to remove leaking flammable fluids, which do not involve flammability tests, and measures to protect structure in case leaking flammable fluids should ignite, which do involve flammability tests. Effective firewalls are required to protect airplane structure from potential fuel-led fires originating in these areas, and to ensure survival of critical items such as fluid-carrying hoses and electrical components. Much more stringent fire exposures are used in these tests than in the tests for interior components, except for the new burnthrough requirement on thermal/acoustic insulation.

12.4.1 Evolution of requirements

Aside from the issuance of non-mandatory advisory material having to do with test procedures, there have been no changes in requirements for components in 'designated fire zones' since the requirements were first established.

12.4.2 Flammability tests and acceptance criteria outside the fuselage

The fire test procedures and acceptance criteria for components in 'designated fire zones' are shown in Table 12.2. The next section (section 12.4.3) describes which tests apply to the various components outside the fuselage. It is beyond the scope of this discussion to go into detail with these requirements and test details. The reader is referred to the regulations (Title 14 CFR 25) and/or the *FAA Fire Test Handbook*[32] for additional information. The various tests used are identified in Table 12.2. The handbook contains sketches of the tests, which have been modified here and are shown in the indicated figures.

12.4.3 Requirements for components outside the fuselage

Fire protection requirements in the non-pressurized areas of the airplane are applicable in places where there is risk of fire from flammable fluids, particularly engine fuel. CFR §25.1181 identifies 'designated fire zones' as:

- the engine power section
- the engine accessory section
- except for reciprocating engines, any complete powerplant compartment in which no isolation is provided between the engine power section and the engine accessory section
- any auxiliary power unit compartment

Table 12.2 Flammability tests for components outside the fuselage

Test type	Acceptance criteria	Handbook
Firewalls, fire resistant (Fig. 12.13)	For 5 minutes, no burnthrough	Chapter 12
Firewalls, fireproof (Fig. 12.13)	For 15 minutes, no burnthrough	Chapter 12
Fire resistance of electric connectors on firewalls (Fig. 12.14)	For 20 minutes test duration, no burnthrough, no loss of current below 1 amp no short-circuit no backside ignition	Chapter 13
Fire resistance of Class A hoses (Fig. 12.15)	For 15 minutes test duration, no leakage of fluid	Chapter 11
Fire resistance of Class B hoses (Fig. 12.15)	For 5 minutes test duration, no leakage of fluid	Chapter 11
Fire resistance of electric wire insulation during test (Fig. 12.16)	Insulation: no excessive flake off minimum resistance $\geq$ 10,000 ohms flame travel within outer bands conductor able to carry > 2 amps	Chapter 14

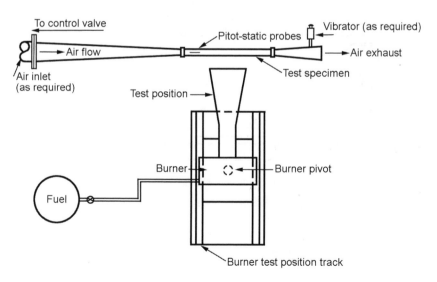

12.13 Test for fireproof and fire-resistant firewalls.

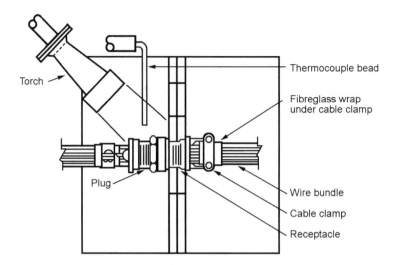

12.14 Test for fire resistance of electrical connectors.

- any fuel-burning heater and other combustion equipment installation described in §25.859
- the compressor and accessory sections of turbine engines
- combustor, turbine, and tailpipe sections of turbine engine installations that contain lines or components carrying flammable fluids or gases.

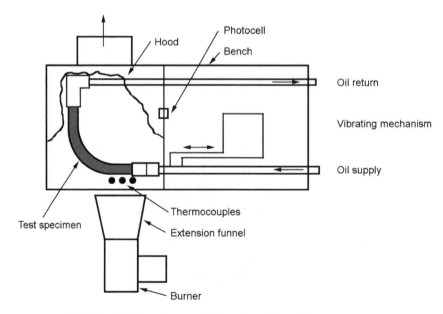

12.15 Test for fire resistance of Class A and Class B hoses.

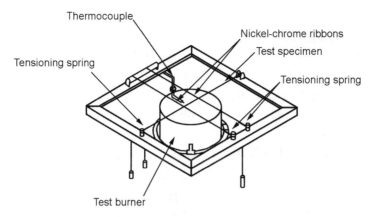

12.16 Test for fire resistance of electric wire insulation.

Each designated fire zone must meet the requirements of §§25.867, and 25.1185 through 25.1203.

The requirements for these 'designated fire zone' areas are specified in the following FAR 25 sections:

- §25.859 – areas containing combustion heaters. The requirements apply mostly to subjective design, and do not necessarily involve testing. Some components that act as a firewall are required to be 'fireproof'. 'Fireproof' is demonstrated by resisting burnthrough for 15 minutes when subjected to a large propane or oil burner.
- §25.863 – areas where flammable fluids might leak. The requirements apply mostly to subjective design to avoid buildup of flammable liquids. Critical components such as hoses and electrical wire and connectors have to be able to 'withstand fire and heat'. Testing, if required, involves the firewall test, the electrical connector test, and the electric wire insulation test for a test duration of 15 minutes.
- §25.865 – flight controls, engine mounts, and other flight structure. Essential flight controls, engine mounts, and other flight structures located in designated fire zones or in adjacent areas which would be subjected to the effects of fire in that zone must be constructed of 'fireproof' material, or shielded so that they are capable of withstanding the effects of fire. Testing involves the firewall test, for a test duration of 15 minutes.
- §25.867 – areas around engine nacelles. Surfaces to the rear of the nacelles, within one nacelle diameter of the nacelle centerline, must be at least fire resistant. This involves the firewall test for a test duration of 5 minutes.
- §25.1103 – air induction duct systems upstream of engine superchargers and auxiliary power units. Depending on location, these are required to be either fire resistant or fireproof. Testing involves the firewall test for a test duration of either 5 minutes or 15 minutes, respectively.

- §25.1181 through §25.1191 – firewalls and flammable fluid drainage for engines and auxiliary power units.
- §25.1195 through §25.1203 – fire extinguishing systems for designated fire zones (e.g., engines and auxiliary power units).

Section 25.1207 requires that compliance with §25.1181 through §25.1203 be demonstrated through large-scale testing.

12.5   Future FAA directions

When airplane fire test regulations were first established in 1945 in the USA, the Bunsen burner test procedures and acceptance criteria were in some cases rather arbitrary, especially for components in areas inside the pressurized vessel. The FAA approach to fire test requirements since the mid-1980s has been to develop lab-scale tests whose results can be correlated with large-scale fire test results, and to establish appropriate acceptance criteria for the lab-scale tests.

The FAA has initiatives under way addressing new or upgraded fire safety requirements on airplanes. Two of the initiatives currently under way at the FAA involving fire safety are given below.

12.5.1   Halon replacement

Fire extinguishing agents are required for deployment in airplanes in the passenger cabins, cockpits, lavatories, most cargo compartments, and engine nacelles. The traditional agents of choice have been Halon 1211 and water for cockpits and passenger cabins, and Halon 1301 for the other areas. Since these Halons are strong ozone depleters, they must eventually be replaced with other agents. The FAA has established acceptance criteria for replacing Halon 1301 in cargo compartments and lavatories, and is currently working to define suitable test and acceptance criteria for other applications and replacement agents.

12.5.2   Center wing tank inerting

The traditional design strategy for fuel tank safety has been to eliminate the possibility of ignition sources entering the fuel tank ullage, which may contain a concentration of fuel vapor between the upper and lower explosive limits. This is particularly true for the center wing tank, which sits directly under the fuselage and over heat-producing machinery such as air-conditioning units. The FAA and industry have worked to develop a means of inerting the center wing tank by effectively separating nitrogen and oxygen from air, and adding the nitrogen to the ullage in the center wing tank to greatly decrease the probability of an explosion. It is anticipated that newly manufactured airplanes will be required to install such an inerting system, and retrofit to the existing fleet has been discussed.

12.5.3   Upgraded flammability requirements for hidden materials

The FAA has established a program to address components installed inside the pressurized vessel in areas not visible or necessarily accessible to passengers and crew. These are referred to as 'hidden materials'. The FAA strategy is to review currently required fire test procedures and acceptance criteria, and determine whether upgrades are needed. The FAA has established a program to develop new tests that take account of more severe ignition source(s) and fire conditions considered possible for 'hidden materials' other than thermal/acoustic insulation, whose requirements were upgraded by Amendment 25-111.

12.5.4   Contamination and aging of hidden materials

'Hidden areas' where 'hidden materials' are located are subject to accumulation of flammable debris of various types, such as fibers, greases, oils, paper, etc. These areas are typically difficult to access and maintain. Both regulatory authorities and industry have concerns about the resulting potentially flammable contamination. Discussions to address the issue have been initiated in the FAA International Aircraft Materials Fire Test Working Group (IAMFTWG).

12.6   References

1. Code of Federal Regulations, Title 14, Aeronautics and Space, Chapter I – Federal Aviation Administration, Department of Transportation (United States) http:// ecfr.gpoaccess.gov/cgi/t/text/text-idx?&c=ecfr&tpl=/ecfrbrowse/Title14/ 14tab_02.tpl

2. Code of Federal Regulations, Title 14, Aeronautics and Space, Chapter I – Federal Aviation Administration, Part 21 – Certification Procedures for Products and Parts, U.S. Department of Transportation. http://ecfr.gpoaccess.gov/cgi/t/text/text-idx?c=ecfr&sid=db4a17a6ffec340d467d74d9c99702cf&rgn=div5&view=text&node=14:1.0.1.3.8&idno=14

3. Code of Federal Regulations, Title 14, Aeronautics and Space, Chapter I – Federal Aviation Administration, Part 25 – Airworthiness Standards: Transport Category Airplanes, U.S. Department of Transportation. http://ecfr.gpoaccess.gov/cgi/t/text/ text-idx?c=ecfr&sid=456b70bbfa92850d903d925204a22b39&rgn=div5&view= text&node=14:1.0.1.3.10&idno=14

4. Code of Federal Regulations, Title 14, Aeronautics and Space, Chapter I – Federal Aviation Administration, Part 33 – Aircraft engines, U.S. Department of Transportation. http://ecfr.gpoaccess.gov/cgi/t/text/text-idx?c=ecfr&sid=456b70bbfa92850 d903d925204a22b39&rgn=div5&view=text&node=14:1.0.1.3.14&idno=14

5. Code of Federal Regulations, Title 14, Aeronautics and Space, Chapter I – Federal Aviation Administration, Part 121 – Operating Requirements, Domestic, Flag, and Supplemental Operations, U.S. Department of Transportation. http:// ecfr.gpoaccess.gov/cgi/t/text/text-idx?c=ecfr&sid=456b70bbfa92850d903 d925204a22b39&rgn=div5&view=text&node=14:2.0.1.4.19&idno=14

6. Code of Federal Regulations, Title 14, Aeronautics and Space, Chapter I – Federal Aviation Administration, Part 21 – Certification Procedures for Products and Parts, Paragraph §21.41 http://ecfr.gpoaccess.gov/cgi/t/text/text-idx?c=ecfr&sid=8da3d21 f7ec1e5f8e94a646972cead85&rgn=div8&view=text&node=14:1.0.1.3.8.2.11.18& idno=14

7. Code of Federal Regulations, Title 14, Aeronautics and Space, Chapter I – Federal Aviation Administration, Part 21 – Certification Procedures for Products and Parts, Paragraph §21.31 http://ecfr.gpoaccess.gov/cgi/t/text/text-idx?c=ecfr&sid= 8da3d21f7ec1e5f8e94a646972cead85&rgn=div8&view=text&node= 14:1.0.1.3.8.2.11.13&idno=14

8. Code of Federal Regulations, Title 14, Aeronautics and Space, Chapter I – Federal Aviation Administration, Part 183 – Representatives of the Administrator. http:// ecfr.gpoaccess.gov/cgi/t/text/text-idx?c=ecfr&sid=456b70bbfa92850d903 d925204a22b39&rgn=div5&view=text&node=14:3.0.1.4.18&idno=14

9. Code of Federal Regulations, Title 14, Aeronautics and Space, Chapter I – Federal Aviation Administration, Part 21 – Certification Procedures for Products and Parts, Paragraphs §21.131-21.165. http://ecfr.gpoaccess.gov/cgi/t/text/text-idx?c= ecfr&sid=a03798975d94fe93ede655413281859a&rgn=div5&view=text&node= 14:1.0.1.3.8&idno=14#14:1.0.1.3.8.7.11.1

10. Code of Federal Regulations, Title 14, Aeronautics and Space, Chapter I – Federal Aviation Administration, Part 21 – Certification Procedures for Products and Parts, Paragraphs §21.171-21.199. http://ecfr.gpoaccess.gov/cgi/t/text/text-idx?c= ecfr&sid=a03798975d94fe93ede655413281859a&rgn=div5&view=text&node= 14:1.0.1.3.8&idno=14#14:1.0.1.3.8.8.11.1

11. Code of Federal Regulations, Title 14, Aeronautics and Space, Chapter I – Federal Aviation Administration, Part 119, Certification: Air Carriers and Commercial Operators http://ecfr.gpoaccess.gov/cgi/t/text/text-idx?c=ecfr&sid=12a39ff9d4c52 a93f317ba9a69256c26&rgn=div5&view=text&node=14:2.0.1.4.18&idno=14

12. *Federal Register*, July 29, 1966 (Volume 31, Number 146), page 10275, Docket No. 7522, Notice No. 66-26 http://www.airweb.faa.gov/Regulatory_and_ Guidance_Library/rgNPRM.nsf/2ed8a85bb3dd48e68525644900598dfb/ 363e3b2fdb1a2ff18625682000666be2!OpenDocument

13. *Federal Register*, September 20, 1967 (Volume 32, Number 182), page 13255, Docket No. 7522, Amendment Nos. 21-16, 25-15 http://www.airweb.faa.gov/ Regulatory_and_Guidance_Library/rgFAR.nsf/HistoryFARAmendment! OpenView&Start=1&Count=200&Expand=3.16#3.16

14. *Federal Register*, February 24, 1972 (Volume 37, Number 37), page 3964, Docket No. 9605, Amendment No. 25-32 http://www.airweb.faa.gov/Regulatory_and_ Guidance_Library/rgFAR.nsf/HistoryFARAmendment!OpenView&Start= 1&Count=200&Expand=3.33#3.33

15. AMINCO – NBS Smoke Density Chamber, Cat. No. 4-500B, Instruction 941-B, May 1983

16. Federal Aviation Administration, 1980. Special Aviation Fire and Explosion Reduction. FAA-ASF-AT-4, Office of Aviation Safety, Atlantic City, NJ, FAA Technical Center

17. DOT/FAA/CT-83/29, Correlation of Laboratory-Scale Fire Test Methods for Seat Blocking Layer Materials with Large-Scale Test Results, Louis J. Brown, Jr., Richard M. Johnson, June 1983

18. *Federal Register*, October 26, 1984 (Volume 49, Number 209), page 43188, Docket No. 25791, Amendment Nos. 25–59 http://www.airweb.faa.gov/Regulatory_ and_Guidance_Library/rgFAR.nsf/HistoryFARAmendment!OpenView&Start= 1&Count=200&Expand=3.60#3.60

19. *Federal Register*, May 16, 1986 (Volume 51, Number 95, page 18236, Docket No. 24185, Amendment 25-60 http://www.airweb.faa.gov/Regulatory_and_ Guidance_Library/rgFAR.nsf/HistoryFARAmendment!OpenView&Start= 1&Count=200&Expand=3.61#3.61

20. DOT/FAA/CT-85/23, Aircraft Interior Panel Test Criteria Derived from Full-Scale Fire Tests, R. G. Hill, T. I. Eklund, C. P. Sarkos, September 1985

21. *Federal Register*, July 21, 1986 (Volume 51, Number 139), page 26206, Docket No. 24594, Amendment 25-61 http://www.airweb.faa.gov/Regulatory_and_ Guidance_Library/rgFAR.nsf/HistoryFARAmendment!OpenView&Start= 1&Count=200&Expand=3.62#3.62

22. *Federal Register*, August 25, 1988 (Volume 53, Number 165), page 32564, Docket No. 24594, Amendment 25-66 http://www.airweb.faa.gov/Regulatory_and_ Guidance_Library/rgFAR.nsf/HistoryFARAmendment!OpenView&Start= 1&Count=200&Expand=3.62#3.67

23. *Federal Register*, July 31, 2003 (Volume 68, Number 147), page 45045, Docket No. FAA-2000-7909, Amendment Nos. 25-111 http://www.airweb.faa.gov/Regulatory_ and_Guidance_Library/rgFAR.nsf/HistoryFARAmendment!OpenView&Start= 1&Count=200&Expand=3.112#3.112

24. http://www.fire.tc.faa.gov/materials.stm

25. http://www.fire.tc.faa.gov/systems.stm

26. Code of Federal Regulations, Title 14, Aeronautics and Space, Chapter I – Federal Aviation Administration, Department of Transportation (United States) http:// ecfr.gpoaccess.gov/cgi/t/text/text-idx?sid=8da3d21f7ec1e5f8e94a646972 cead85&c=ecfr&tpl=/ecfrbrowse/Title14/14tab_02.tpl

27. *Aircraft Materials Fire Test Handbook*, DOT/FAA/CT-99/15, U.S. Department of Transportation, Federal Aviation Administration. http://www.fire.tc.faa.gov/ handbook.stm

28. DOT/FAA/CT-99/15, Chapter 18, Recommended Procedure for the 4-Ply Horizontal Flammability Test for Aircraft Blankets http://www.fire.tc.faa.gov/pdf/handbook/ 00-12_ch18.pdf

29. DOT/FAA/CT-99/15, Chapter 9, Radiant Heat Testing of Evacuation Slides, Ramps, and Rafts http://www.fire.tc.faa.gov/pdf/handbook/00-12_ch9.pdf

30. Fire-Smoke-Toxicity, ABD0031, Airbus Industrie, Engineering Directorate, 31707 Blagnac Cedex, France, 1992

31. Aircraft Fireworthiness Interior Design Criteria, D6-51377, Boeing Commercial Airplanes Group, The Boeing Company, Seattle, Washington, 1983.

32. *Aircraft Materials Fire Test Handbook*, DOT/FAA/CT-99/15, U.S. Department of Transportation, Federal Aviation Administration. http://www.fire.tc.faa.gov/ handbook.stm

Flammability testing in the mining sector

H C VERAKIS, US Department of Labor, USA

13.1 Introduction

13.1.1 Material usage

Many different types of materials are used in the mining industry and for many different purposes. Some of the materials are of natural origin, such as wood and oil. Many of the materials are manufactured and include plastics, rubber, foams, and steel. The materials may be used for the various aspects of mining operations such as the control of mine ventilation, for mine roof and rib support, for providing electrical power and mechanical protection, for lubrication of equipment, and for haulage.

The mining industry consumes large amounts of timber used in roof and rib support and construction. High water bearing fluids are used in the roof supports of longwall mining equipment. Fuels and lubricants including various types of hydraulic fluids are used to power and lubricate mining equipment. Rubber or elastomer materials are used in a variety of forms such as tires, hoses, cables, and belts. Plastic materials may be used as piping or ventilation curtain to direct fresh air to workers in a mine and to remove gas and dust. Plastic tubing, both flexible and rigid, is also used for ventilation purposes. Materials such as cement blocks and steel panels or panels made from metal and plastic are used for the construction of stoppings to seal openings in underground mine passageways or tunnels. The stoppings are constructed to serve as devices for the control of ventilation in a mine passageway. Cementatious or foam materials may be used to coat stoppings to reduce air leakage.

Ventilation curtain material, know as brattice cloth, consists of several different types of construction. Brattice cloth may be made from jute fibers or cotton duck or synthetic fibers such as nylon or polyester woven into a tight mesh and coated with a plastic compound. Underground coal mines are a large consumer of brattice cloth and to a lesser extent, rigid or flexible plastic ventilation tubing. Metal and nonmetal underground mines typically use more rigid and flexible plastic ventilation tubing than plastic ventilation curtain material. The introduction of plastic materials into the mining industry over the past several decades has grown by leaps and bounds. Many plastics used in the mining industry have withstood the test of time and continue to provide economical and safe applications. With the introduction of newer, lighter, and stronger plastic materials and synthetic composites, their use has continued to expand into the mining industry. New materials developed for other industries will also continue to be introduced and adapted for use in the mining industry.

13.1.2   Fire safety concerns

The materials used in the mining industry generally fall into three basic categories with respect to flammability. These categories are flammable, combustible, and noncombustible. These categories are applicable to both solid and liquid materials, although technical distinctions may divide the category of combustible into subclasses. The technical definitions are specified in the National Fire Protection Association's (NFPA) *Fire Protection Handbook* (NFPA, 2003), the Society of Fire Protection Engineers (SFPE) *Handbook of Fire Protection Engineering* (SPFE, 2002) and in various American Society for Materials and Testing (ASTM) standards. In addition, for materials regulated in the U.S. mining industry, the Mine Safety and Health Administration (MSHA) defines these categories in various sections of Title 30, Code of Federal Regulations (CFR 30, 2005). Title 30 may be viewed on MSHA's website at *www.msha.gov*.

At times 'flammable' and 'combustible' have been used interchangeably to describe a material. By technical definition, however, the two terms should not be used interchangeably. The difference between the two terms is the ease of ignition and quickness of burning. A flammable material would more easily ignite and burn faster than a combustible material. Under the same circumstances, a flammable material would present a more serious flammability hazard than a combustible material. The term 'noncombustible' is also used interchangeably with the term 'incombustible.' The preferred term is 'noncombustible.' Also, the term fireproof has been used and is still referred to in some instances with respect to a material that does not burn. The appropriate technical term is 'noncombustible.'

The potential fire hazard of certain materials is always a concern where safety of mining personnel and others may be compromised. The level of fire safety chosen will differ depending on the type of material and the environment in which it is used or applied. In many cases, the use of materials that can present a fire hazard in mining operations may be governed by regulations of the body that has jurisdiction or by guidelines issued by the governing body. The mine operator may also have certain requirements for materials, even though the material or materials may not be regulated by a governing body with regard to fire resistance.

A regulation impacting the flammability of a material, i.e., its fire resistance, depends in large measure on the purpose and use of the material and sometimes its fire history. Also a primary factor that needs to be taken into consideration is the effect on life safety should the material be involved in a fire. As an example, conveyor belting, or electric cable, or hydraulic oil used at U.S. surface mine operations are not required to be fire or flame resistant. A level of fire risk in surface mine operations is accepted where such products are used. However, fires resulting in injuries have occurred with these types of materials. The concern then arises when a fire occurs and jeopardizes the sense of security and level of fire safety. The fire can adversely affect many – the mining operation, personnel, those dependent on supply of product, insurer, and even neighboring areas. With the costs of insurance premiums and liability escalating, the potential fire hazard of materials should be fully addressed to achieve a desired level of safety and to protect personnel. In the mining industry, safety is a value and thus it should be a primary goal to protect miners from the flammability hazard that certain materials may present.

Where materials are used in unconfined or open areas, which occurs in many cases in surface mining operations, the rules governing the flammability of materials are less stringent and may be nonexistent compared to use in a confined area or in underground mining operations. As an example, conveyor belt used in U.S. and many other countries' underground coal mines is required to be flame resistant. Conveyor belt used in U.S. surface mine operations is not required to be flame resistant. Conveyor belt is also not required to be flame resistant for use in underground metal and nonmetal mines. Research and investigative work in the United States and other countries has been devoted to diminishing the flammability hazard of certain materials. Flammability hazards can and have been reduced by increasing the fire-resistant qualities of materials used in day-to-day operations such as conveyor belts, brattice cloth, electric cable, and hydraulic fluid. However, fire safety for the use of materials is not absolute as there are many factors to consider and there is no universal test method currently to determine the absolute safety of various materials. One must recognize that a single flammability test does not reflect the many factors that can occur in a fire environment.

The difficulty with tests used to determine a material's flammability behavior in a fire environment has been addressed in a number of papers (Clark, 1981; McGuire and Campbell, 1980; Emmons, 1967). In Emmon's paper and shown graphically in McGuire and Campbell's paper, the relative rating of 24 materials tested by six different national standard fire rating tests are shown and the agreement between the tests was poor. Notwithstanding, there have been and continue to be advances made in the development of new tests and in modeling the flammability behavior of materials. An example is the development of the cone calorimeter, especially the work done at the U.S. National Institute of Standards and Technology (Babrauskas, 1996, 2003a,b).

13.2   Regulated materials for flammability

13.2.1 Why the need for certain fire-resistant materials?

The ever present threat of fire is great in mining operations where electrical and mechanical equipment is used, where welding and flame cutting are performed, and where ignition sources are present. An analysis of fires that occurred in the past several years in underground coal mines is shown in Fig. 13.1. Several primary sources involved in the fires shown in Fig. 13.1 were conveyor belts and welding and cutting operations. Even though historically, the number of fatalities and injuries from fires in mining operations in the USA continues to decline, mining personnel need to be vigilant and aware of potential fire hazards from materials used. The potential fire hazard from the materials used can be reduced by a number of means. Ignition sources can be isolated, fire detection and suppression systems can be installed, and fire-fighting measures can be instituted. Another measure is to make products more fire resistant and to require the use of fire-resistant materials in mining applications or operations. If a material is noncombustible, the fire safety concern is nonexistent; if a material is fire resistant, the fire safety concern is lessened. A certain level of fire safety can be gained by the use of fire-resistant materials. The level may be expressed in terms such as 'fire or flame resistant,' 'fire or flame retardant,' 'fire safe,' but the terms do not express a precise meaning.

It is important to understand what gains in fire safety may be obtained by the use of less flammable or less combustible materials. Generally higher costs and changes in the performance of fire-resistant materials are barriers to their use. Also, there is the issue of whether smoke and toxicity is increased from the combustion of flame-resistant materials, thereby creating a potential detrimental effect on fire safety. The 'smoke' issue is complex and has been studied

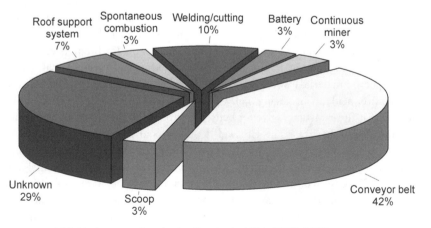

13.1 Underground coal mine fires in the USA, 1997–2002.

extensively. The smoke generated from a flame-resistant material is dependent on quite a few factors such as the formulation of the material, the type of flame retardant(s) used, the source and method of ignition, the amount of material exposed and its orientation in a combustion event, the ventilation in the environment where combustion occurs, and the rate of fire growth. More information and detail is presented in the paper by O'Neill and Sweet (1989) and in the National Bureau of Standards Special Publication 749 (1988) which are both listed in the references. The primary aim to improve fire safety is to increase the resistance to ignition and to reduce the flame propagation of a material or materials. This is the overall goal of using flame or fire-resistant materials. In spite of the issues that have been raised, the cost of one fire may well exceed the cost of using fire-resistant materials.

Various tests are used to determine the flammability characteristics of materials and their response to ignition and combustion under controlled conditions. There are more than 500 tests relating to fire safety published by the NFPA, ASTM, Underwriters Laboratories (UL), and Factory Mutual (FM). Several of the ASTM flammability tests are specified in this chapter with respect to the evaluation of specific materials. Materials used in the U.S. mining industry are regulated for flame resistance by the U.S. Department of Labor, Mine Safety and Health Administration. These regulations, as indicated earlier, are specified in the U.S. Code of Federal Regulations (CFR 30, 2005). A discussion is presented subsequently in this chapter on the certain materials regulated under CFR 30 for flame resistance.

13.2.2   Regulatory measures

Many countries have developed tests and requirements for reducing the flammability hazards of materials used in mining operations. Countries with mining operations such as Australia, Canada, Germany, Great Britain, India, Japan, and the USA have written rules or requirements that address the flammability hazard of materials. The requirements are directed primarily toward specific materials such as conveyor belts, electric cable, brattice cloth and ventilation tubing, and hydraulic fluid that are used in underground coal mines. These specific materials were targeted for flame-resistant requirements because of mine fire incidents involving their use or perceived fire hazards. For example, the need for fire-resistant conveyor belts was brought to the forefront as a result of some disastrous fires. The Creswell Colliery coal mine fire in Derbyshire, England where 80 lives were lost (Creswell, 1952) and Schlagel-Eisen Colliery coal mine fire in Herten, Germany where seven lives were lost (Schlagel and Eisen, 1977) resulted in the development and implementation of stringent requirements for fire-resistant conveyor belts in both Great Britain and Germany. Subsequently, the former U.S. Bureau of Mines followed in the mid-1950s with the development of a flame test method for evaluating the fire resistance of conveyor belts (Pollack, 1956), but a regulation mandating the use of fire-resistant conveyor belts in underground coal mines was not enacted until much later. The Coal Mine Health and Safety Act of 1969 mandated the use of fire-resistant conveyor belts in the underground coal mines in the USA (FCMS&H, 1969). Also, in Canada, a fire performance standard for conveyor belts used in underground mine operations was developed and adopted as a national Canadian standard (CSA, 1987).

Fire safety incidents and hazards with other materials such as electric cables, brattice cloth and ventilation tubing, and hydraulic fluids also resulted in test methods for determining flame resistance and regulations regarding their use in underground mining operations. Although the use of flame-resistant materials has grown, particularly with the introduction of plastics, in many cases regulations were needed to mandate their use. Other alternatives may be available such as fire detection and fire suppression systems, where the flammability of a material is of concern. However, flame-resistant materials provide a first line of defense for fire safety, but cost, performance, and issues on toxic gases produced during combustion are factors affecting their acceptance.

13.3   Definitions and terms

A discussion and explanation of definitions and terms is pertinent to the flammability testing of materials. Earlier in this chapter, several terms were used to describe a material's flammability characteristics. The terms and definitions regarding the flammability of materials that are used in various publications, standards, and books vary to some extent. In the USA, the National Fire Protection Association, the American Society for Materials and Testing, and the Society of Fire Protection Engineers are the primary sources for definitions and terms dealing with the flammability of materials. Also, regulations of the U.S. Department of Labor's Mine Safety and Health Administration and the Occupational Safety and Health Administration, along with other U.S. Government organizations such as the National Institute for Standards and Technology, the Consumer Product Safety Commission and the Federal Aviation Administration specify terms and definitions related to the flammability of products. Another source is the *Encyclopedia of Fire Protection*.

Terms and definitions which include both solids and liquids are listed below. The definition for combustible liquid is divided into several classes based on the flash point of a liquid. Some terms are used interchangeably. The definitions for some terms are general and do not provide a precise level of flammability. The purpose of the definitions is to provide meaning to terms used in discussing the flammability of materials. Not all of the definitions and terms listed are used in this chapter, but a few of the terms not used are included since they are part of the fire and flammability fields of study.

1. Auto-ignition – self-ignition and subsequent combustion of a liquid or solid by a heated surface without the presence of an external ignition source such as a flame or spark.
2. Combustible material – a material that will ignite and burn when subjected to a flame, spark, or other heat source. Typical examples are paper, wood, rubber, and plastics.
3. Combustible liquid – a liquid having a flash point at or above 100 °F.
   Class II liquid – having a flash point at or above 100 °F and below 140 °F.
   Class IIIA liquid – having a flash point at or above 140 °F and below 200 °F.
   Class IIIB liquid – having a flash point at or above 200 °F.
4. Fire – a burning process that liberates heat with the development of flame.
5. Flammable – capable of being easily ignited and of burning rapidly.
6. Flammable liquid – having a flash point below 100 °F and a vapor pressure not exceeding 40 pounds per square inch at 100 °F (Class I liquid).
7. Flame resistant or fire resistance – a measure of resistance to burning or combustion when exposed to an ignition source.
8. Flammability – the characteristic(s) of a material, once ignited, that describes the combustion process such as spread of flame.
9. Flame retardant or fire retardant – a material that has chemical properties or is compounded with certain chemical(s) to reduce the spread of flame or combustion process.
10. Flame spread – the movement of flame over the surface of a material.
11. Flash point – the minimum temperature at which sufficient vapor is released by a liquid or solid to form a flammable vapor-air mixture at atmospheric pressure
12. Ignition – the act of causing a material to begin combustion.
13. Noncombustible – will not ignite or support combustion or release flammable vapors when subject to flame, spark, or heat source. This term has replaced the term 'incombustible.'

Sometimes a definition will incorporate a test method or a criterion which provides a more quantitative level of meaning or a numerical value. An example is the term 'flame spread index' which is defined as a number determined from testing a material by an E-162 radiant panel test, (ASTM E-162-87). As a point of reference, red oak yields a flame spread index of 100. Except for flammable and combustible liquids, the definitions presented do not give precise meaning and may have slight variations in words from definitions published by other organizations and writers. However, they are fairly well understood in the fire community.

13.4 Flammability tests and performance criteria

The materials that must meet prescribed regulations for fire resistance are typically evaluated by a single test method. In many cases, a series of different mandated fire tests are required for only a few materials. Materials used for mine stoppings and non-asbestos packing gland material used in the stuffing boxes of coal mine electrical equipment require a flame test and several other different types of tests. Three flammability tests are required for determining the fire resistance of hydraulic fluids used in coal mining equipment. Most of the tests used to evaluate a material are small-scale laboratory tests. The largest test is a heated furnace test used to determine the fire rating of mine stoppings. The materials that are required to be fire or flame resistant are listed in Table 13.1 along with the specified test. Some materials that are recommended, but not mandated to be flame resistant are also included in Table 13.1. More details on specific materials and some photographs of the tests used and the criteria applied to evaluate their flammability are discussed in the sections that follow.

Table 13.1 Regulated mine materials and flammability test methods

Material	Use	Flammability test
Brattice and vent tubing – 1,3	Control of mine ventilation	30 CFR, Part 7, Subpart B
Conveyor belt – 1 (*4 also conveyor roller covers and lagging)	Material conveying	30 CFR Part 18.65
Cables – 1,3	Electrical equipment and power transmission	30 CFR Part 18.64
Coatings – 1,2	Mine stoppings, ventilation controls and timber support	ASTM E-162-87
Hoses – 1 (includes fire hose, fire suppression hose, hose conduit) *4 hydraulic hose	Fire fighting, fire suppression and mechanical protection of electrical wiring on machines	30 CFR Part 18.65
Hydraulic fluids – *5	Power actuation	30 CFR, Part 35
Noise control – *6	Control of equipment and environmental noise	ASTM E-162-87
Packing material – 1,3	For cable entrances in electrical equipment	30 CFR, Part 18.65
Battery box material – 1,3	Construction of battery boxes and/or insulation	30 CFR Part 18.65
Rib & roof support – *4	Prevent spalling and loose rock	Modified version of 30 CFR, Part 7, Subpart B
Stoppings – 1	Mine ventilation control	ASTM E-119-88

Notes – Flame resistance mandated: 1 = underground coal, 2 = underground metal-nonmetal, 3 = underground gassy metal-nonmetal, *4 = not mandated to be flame resistant, but recommended for underground coal, *5 = fire resistance optional depending on use in underground coal mine equipment, *6 = flame resistance not mandated, but recommended for surface and underground mines.

Test methods developed by the organizations such as the ASTM or NFPA do not specify a criterion or criteria for rating or passing judgment on the flammability of a material. The authority that has jurisdiction or the regulatory body specifies the requirements regarding the flammability of a material depending on its use. In the U.S. mining industry, the federal Department of Labor, MSHA has jurisdiction. As mentioned earlier in this chapter, the regulations governing the flammability requirements for materials used in surface and underground mining are presented in Title 30, Code of Federal Regulations. Some of the tests methods specified in Title 30 are consensus standards such as ASTM E-162 and E-119; others are those developed by the Agency.

Descriptions of flammability tests developed and implemented by MSHA and its earlier predecessors are presented in this section. Also, a synopsis is presented for those consensus standards used to determine the flammability hazard of certain materials.

13.4.1   Brattice and ventilation tubing test

The test used to determine the flame resistance of brattice cloth and ventilation tubing is described in Title 30, U.S. Code of Federal Regulations, Part 7, Subpart B (CFR 30, 2005). The principal parts of the brattice and vent tubing test apparatus are a stainless steel gallery with inside dimensions approximately 58 inches (147 cm) long, 41 inches (104 cm) high, and 30 inches (76 cm) wide. The steel gallery is lined on the top, bottom and both sides with 1/2 inch (1.3 cm) thick Marinite or equivalent insulating material. Two steel J hooks placed in the top of the steel gallery are used to support a steel rod. The steel rod is centrally located along the length of the steel gallery and is used to support the test sample. A tapered stainless steel duct section is connected to the test gallery. The duct section tapers from a cross-sectional area measuring 2 feet 7 inches wide (78.7 cm) by 3 feet 6 inches (106.7 cm) high at the test gallery to a cross-sectional area 1 foot 6 inches (45.7 cm) square over a length of 3 feet (91.4 cm).

A stainless steel fan housing is connected to the tapered duct section. The fan housing consists of a 1 foot 6 inches (45.7 cm) square section 6 inches (15.2 cm) long followed by a 10 inch (25.4 cm) long section which tapers from 1 foot 6 inches (45.7 cm) square to 12 inches (30.5 cm) diameter round and concludes with a 12 inch (30.5 cm) diameter round collar that is 3 inches (7.6 cm) long. A variable speed fan that produces an air velocity of 125 feet per minute (38.1 meters per minute) in the test gallery is secured in the fan housing. A hood with an exhaust system is used over the top front of the test gallery to remove smoke that may escape from the test gallery during a test. The exhaust system remains on during testing, but must not affect the air flow in the test gallery. The ignition source for the test consists of a methane-fueled impinged jet burner. The jet burner measures 12 inches (30.5 cm) long from the threaded ends of the first and last jets and 4 inches wide. There are 12 impinged jets, approximately 1-3/8 inches (3.5 cm) long and spaced alternately along the length of the burner tube. The jets of the burner are canted so that they point toward each other in pairs and the flames from these jets impinge upon each other.

Brattice test method

For the brattice cloth test, six samples are prepared. Each sample is 40 inches (101.6 cm) wide by 48 inches (121.9 cm) long. For each test, the brattice cloth sample is suspended in the test gallery by wrapping the sample around the rod and clamping each end and the center. The brattice cloth is hung so that the distance from the bottom of the sample is 4 inches from the test gallery floor. The methane-fueled impinged jet burner is ignited and adjusted to yield a flame height of 12 inches (30.5 cm) as measured at the outermost tip of the flame. The burner flame is applied to the front lower edge of the brattice cloth and kept in contact with the material for 25 seconds or until 1 foot (30.5 cm) of material, measured horizontally, is consumed, whichever occurs first. The burner flame is moved to maintain contact with 1 foot (30.5 cm) of the sample, if the material shrinks during the application of the burner flame. Sometimes the burner is rotated slightly to prevent melting material from clogging the burner orifices during application of the flame. Three of the brattice cloth samples are tested with no air flow in the test gallery and three samples are tested in the gallery with an average of 125 feet per minute (38.1 meters per minute) of air flowing past the sample.

The duration of burning is the total burning time of the sample during the flame test. The duration of burning includes the burn time of any material that falls on the floor of the test gallery during the igniting period. However, the suspended sample is considered burning only after the burner is removed. If the burning time of a suspended sample and a sample on the floor coincide, the burning time for these coinciding events is counted only once. The length of flame propagation and duration of burning is recorded for each of the six test samples. The average duration of burning is calculated for the three samples tested in still air and for the three samples tested in the air flow of 125 feet per minute (38.1 meters per minute).

Evaluation of test results
For the brattice cloth to qualify as flame resistant, each of the following criteria must be met:

- flame propagation of less than 4 ft (1.2 m) in each of the six tests
- an average duration of burning of less than one minute in both groups of three tests
- a duration of burning not exceeding two minutes in any of the six tests.

Ventilation tubing test method

The ventilation tubing test is conducted using the same test gallery as is used to test brattice cloth. The test procedure for ventilation tubing is similar to the procedure used for brattice cloth. Six samples, 48 inches (1.2 m) in length are prepared with all the flared or thickened ends removed. Any sample with a cross-sectional dimension greater than 24 inches (0.61 m) must be tested in a 24-inch (0.61 m) size. Using the J-hooks in the test gallery, the test sample is suspended in the center of the gallery by running a wire through the 48-inch (1.2 m) length of ventilation tubing sample.

The methane-fueled impinged jet burner is ignited and adjusted to yield a flame height of 12 inches (30.5 cm) as measured at the outermost tip of the flame. The burner is applied to the front lower edge of the ventilation tubing sample so that two-thirds of the burner is under the tubing and the remaining third is exposed to allow the flame to contact the inside of the tubing. The burner is kept in contact with the material for 60 seconds. If melting material might clog the burner orifices, the burner is slightly rotated during the application of the flame. Three ventilation tubing samples are tested with no air flow in the test gallery and three samples are tested in the gallery with an average of 125 feet per minute (38.1 m/min) of air flowing past the sample. The procedures for recording the flame propagation and the duration of burning of the ventilation tubing test samples are the same as used for the brattice cloth test samples. The test criteria for ventilation tubing to qualify as flame resistant are the same criteria as presented earlier in this chapter for brattice cloth. However, this information is repeated below for the reader.

The length of flame propagation and duration of burning is recorded for each of the six ventilation tubing test samples. The duration of burning is the total burning time of the sample during the flame test. The duration of burning includes the burn time of any material that falls on the floor of the test gallery during the igniting period. However, the suspended sample is considered burning only after the burner is removed. If the burning time of a suspended sample and a sample on the floor coincide, the burning time for these coinciding events is counted only once. The average duration of burning is calculated for the three samples tested in still air and for the three samples tested in the air flow of 125 feet per minute (38.1 m/min). The flame test of a rigid ventilation tubing is shown in Fig. 13.2.

Evaluation of test results

For the ventilation tubing to qualify as flame resistant, each of the following criteria must be met:

- flame propagation of less than four feet in each of the six tests
- an average duration of burning of less than one minute in both groups of three tests
- a duration of burning not exceeding two minutes in any of the six tests.

Impinged jet burner

13.2 Flame test of ventilation tubing (CFR 30, Part 7, Subpart B).

13.4.2   Conveyor belt test

The test used to determine the flame resistance of conveyor belts is described in Title 30, Code of Federal Regulations, Part 18, Section 18.65 (CRF 30, 2005). This test was known as the former U.S. Bureau of Mines Schedule 2G flame test. For the purpose of brevity, the acronym SSFT, small-scale flame test will be used at times to designate this test. The principal parts of the SSFT consist of a 21-inch (53.3 cm) cubical test chamber constructed from sheet steel. On the right side of the test chamber is an American Society of Mechanical Engineers flow nozzle of 16 to 8-1/2 inch (40.64 cm to 21.59 cm) reduction. In the front of the test chamber is a steel framed glass viewing door. The door provides the entrance to the test chamber for placing a material sample for test. A photograph of the SSFT is shown in Fig. 13.3.

A Pittsburgh–Universal Bunsen-type burner with an inside diameter tube of 11 mm (0.43 inch) is used to ignite the test sample. The burner is mounted in a burner placement guide so that it may be moved beneath the test sample and retracted from it by a metal rod extending through the front of the test chamber. The burner is fueled by natural gas.

A steel support stand with a steel ring clamp, sample holder, and 20-mesh steel wire gauze, 5 inches square (12.7 cm square) is used to support a sample for test. A variable speed electric fan is used to draw air through the flow nozzle of the test chamber. During a test, the air is drawn across the test sample at a flow rate of 300 feet per minute (91.4 m/min). The electric fan must cause the air flow to go from zero to 300 feet per minute within 8.5 ± 1 second. A mirror is placed in the back of the test cabinet to permit a rear view of the test sample. The test sample is 6 inches long by 1/2-inch wide by the thickness of the material (15.24 cm long by 1.27 cm wide by the thickness of the material).

13.3 Small-scale flame test (CFR 30, Part 18, Section 18.65).

The SSFT is conducted by testing a set of four samples cut from a product. A sample is placed in the holder in the test chamber. The longitudinal axis of the specimen is placed horizontal and its transverse axis is inclined at 45 degrees to the horizontal with the steel wire gauze clamped ¼ inch (0.635 cm) under the test sample. The test sample is located so that ½ inch (1.27 cm) extends beyond the edge of the steel wire gauze. The Bunsen burner is moved away from the test sample and ignited and adjusted to give a blue flame 3 inches (7.6 cm) long. The observation door is closed and the ignited burner is moved under the edge of the test sample. The flame from the burner is applied to the edge of the test sample for one minute. The burner is then retracted and the electric fan is turned on to produce an air flow of 300 feet per minute (91.4 m/min). The duration of flame from the test sample is timed using a stopwatch or timing device. The test sample must remain in the air flow for at least three minutes to determine the presence and duration of afterglow. If a glowing sample exhibits flame within the specified three minutes, then the duration of that flame is added to the duration of any earlier flaming. The test result for each sample is added together in a set of four and averaged. For a material to meet the test criteria, the test result for each set of four test samples must not result in a duration either of flame exceeding an average of one minute or afterglow exceeding an average of three minutes.

13.4.3   Cable and splice kit test

The test used to determine the flame resistance of electric cables, signal cables, and cable splice kits is described in Title 30, Code of Federal Regulations, Part 7, Subpart K (CFR 30, 2005). The principal parts of the apparatus are a test chamber or a rectangular enclosure measuring 17 inches deep by 14 inches high by 39 inches wide (43.2 cm deep by 35.6 cm high by 99.1 cm wide) and open at the top and front. The floor or base of the chamber is lined with a noncombustible material to contain burning matter which may fall from the test specimen during a test. Permanent connections are mounted to the chamber and extend to the sample end location. The connections are used to energize the electric cable and splice specimens. The connections are not used when testing signaling cables. A rack consisting of three metal rods, each measuring approximately 3/16 inch (0.48 cm) in diameter is used to support the specimen during a test. The horizontal portion of the rod which contacts the test specimen shall be approximately 12 inches (30.5 cm) in length. A natural gas type Tirrill burner, with a nominal inside diameter of 3/8 inch (0.95 cm), is used to apply the flame to the test specimen.

For tests of electric cables and splices, a source of either alternating current or direct current is used for heating the power conductors of the test specimen. The current flow through the test specimen is regulated and the open circuit voltage is not to exceed the voltage rating of the test specimen. An instrument is used to monitor the effective value of heating current flow through the power conductors of the specimen. Also, a thermocouple is used to measure conductor temperature while the cable or cable splice kit is being electrically heated to 400 °F (204.4 °C). For the electric cable test, three specimens each three feet (0.91 m) in length are prepared by removing five inches of jacket material and two inches of conductor insulation from both ends of each test specimen.

For splice kits, a splice is prepared in each of three sections of a MSHA-approved flame-resistant cable. The cable used is the type that the splice kit is designed to repair. The finished splice must not exceed 18 inches (45.7 cm) or be less than 6 inches (15.2 cm) in length for test purposes. The spliced cables are three feet in length with the midpoint of the splice located 14 inches (35.6 cm) from one end. Both ends of each of the spliced cables are prepared by removing five inches of jacket material and two inches of conductor insulation. The type, amperage, voltage rating, and construction of the cable must be compatible with the splice kit design.

The test specimen is centered horizontally in the test chamber on the three rods. The three rods are positioned perpendicular to the longitudinal axis of the test specimen and at the same height. This arrangement permits the tip of the inner cone from the flame of the gas burner to touch the jacket of the test specimen. For splices, the third rod is placed between the splice and the temperature monitoring location at a distance 8 inches (20.3 cm) from the midpoint of the splice. The gas burner is adjusted to produce an overall blue flame five inches (12.7 cm) high with a three-inch (7.6 cm) inner cone and without the persistence of yellow coloration. The power conductors of the test specimen are connected to the current source. The connections must be compatible with the size of the cable's power conductors to reduce contact resistance. The power conductors of the test specimen are energized with an effective heating current value of five times the power conductor ampacity rating at an ambient temperature of 104 °F (40 °C).

The electric current is monitored through the power conductors of the test specimen with the current measuring device. The amount of heating current is adjusted to maintain the proper effective heating current value until the power conductors reach a temperature of 400 °F (204.4 °C). For electric cables, the tip of the inner cone from the flame of the gas burner is applied directly beneath the test specimen for 60 seconds at a location 14 inches (35.6 cm) from one end of the cable and between the supports separated by a 16 inch (40.6 cm) distance. For the splices made from the splice kits, the tip of the inner cone from the flame of a gas burner is applied for 60 seconds beneath the midpoint of the splice jacket. After subjecting the test specimen to the external flame for the specified time, the burner flame is removed from beneath the specimen while simultaneously turning off the heating current. The amount of time the test specimen continues to burn is recorded after the flame from the burner has been removed. The burn time of any material that falls from the test specimen after the flame from the burner has been removed is added to the total duration of flame. The length of burned (charred) area of each test specimen is measured longitudinally along the cable axis. The procedure is repeated for the remaining two specimens. For a cable or splice kit to qualify as flame resistant, the three test specimens must not exceed a duration of burning of 240 seconds and the length of the burned (charred) area must not exceed 6 inches (15.2 cm). The flame test of an electric cable is shown in Fig. 13.4 – the electric cable did not meet the test criteria.

13.4 Electric cable flame test (CFR 30, Part 7, Subpart K).

13.4.4   Signal cable test

The flame test of signal cables is conducted by using the same test apparatus as used for tests of electric cables and splice kits. However, no electric current is passed through the signal cable sample. Three samples of signal cable each 2 ft (0.61 m) long are tested. The test sample is centered horizontally in the test chamber on the three rods. The gas burner is adjusted to give an overall blue flame 5 inches (12.7 cm) high with a 3-inch (7.6 cm) inner cone. The tip of the inner cone from the flame of the gas burner is applied for 30 seconds directly beneath the sample centered between either of the end supports and the center support. The time the sample continues to burn after the burner flame is removed is recorded. The duration of burning includes the burn time of any material that falls from the test specimen after the igniting flame has been removed. The length of burned (charred) area along the axis of each test specimen is also measured and recorded.

For a signal cable to qualify as flame resistant the duration of burning must not exceed 60 seconds and the length of the burned (charred) area must not exceed 6 inches (15.2 cm) for any of the three samples tested.

13.4.5   Hydraulic fluid tests

The tests for determining the fire resistance of hydraulic fluids are specified in Title 30, Part 35 (CFR 30, 2005). The tests include an auto-ignition test, an evaporation for flammability test (wick test), and a spray flammability test. A hydraulic fluid must pass all three of these tests to qualify as a fire-resistant hydraulic fluid. A short description of each test and the criteria for passing each test is presented below. A more detailed description of each test may be viewed in 30 CFR, Part 35, which is available on MSHA's website.

Auto-ignition test

The auto-ignition test is conducted by heating a 200 ml (6.8 oz) Erlenmeyer glass flask in a heated furnace. The furnace is a refractory cylinder 5 inches (12.7 cm) high and 5 inches (12.7 cm) in internal diameter. Thermocouples are used to measure the temperature of the neck, mid-section and base of the glass flask. The test apparatus is shown in Fig. 13.5. A hypodermic syringe of 0.25 cc or 1 cc capacity equipped with a 2-inch (5.1 cm) No. 18 stainless steel needle is used to inject samples into the heated flask. To begin the test, the furnace is heated to the desired test temperature. A 0.07 cc sample of hydraulic fluid is then injected into the heated flask and an observation for flame is made. If no ignition of the sample occurs, the furnace temperature is raised 50 °F, (10 °C) and the procedure repeated. This process is repeated until ignition occurs, then the furnace temperature is lowered by 5 °F (approx 3 °C) until no ignition occurs. The process is performed with lower (0.03 cc) and higher (0.03 cc) increments of

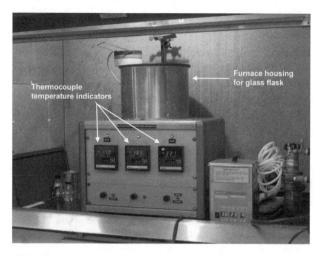

13.5 Test apparatus for determining the auto-ignition temperature of a hydraulic fluid (CFR 30, Part 35).

fluid and raising and lowering the furnace temperature until the lowest temperature at which ignition occurs is obtained.

Evaluation of test results

To pass the auto-ignition test, the test samples must yield an ignition temperature of 600 °F (315.6 °C) or greater.

Wick test

The wick test is used to determine the effect of evaporation on the reduction of fire resistance of a hydraulic fluid. Standard laboratory Petri dishes, approximately 90 mm diameter by 16 mm depth (3.5 inches diameter by 0.63 inches depth), are used to contain the test samples. Three uncovered Petri dishes are filled with 30 cc (1 oz) of the hydraulic fluid test sample. Two of these samples are placed in a gravity convection air oven that has been heated to 150 °F (65.6 °C). One sample is kept in the heated oven for two hours and the other sample is kept in the oven for four hours. The third sample remains at room temperature. Each of the three Petri dish samples of hydraulic fluid is used to soak five smoker's pipe cleaners. The five soaked pipe cleaners are then removed from each of the Petri dishes and drained of excess fluid.

An electrically operated cycling device, such as an automobile windshield wiper mechanism, is used to test the soaked pipe cleaners taken from each Petri dish. A pipe cleaner is attached to the cycling device so that it will enter and leave a flame from a Bunsen or equivalent laboratory natural gas burner. A square sheet metal box is built around the electrically operated cycling device and gas burner. An open metal tube, 1-1/4 inch in diameter, and located to one

13.6 Wick flame test using a pipe cleaner soaked in hydraulic fluid (CFR 30, Part 35). The pipe cleaner cycles through a Bunsen burner flame that is offset from the center of the metal platen.

side of the metal box is used to contain the gas burner. The burner flame exits through the metal tube out of the top of the metal box. The burner is adjusted to provide a non-luminous flame approximately 4 inches (10.2 cm) in height without forming a sharp inner cone. Each soaked pipe cleaner is attached to the cycling device and oscillated in a horizontal plane at 25 ± 2 cycles per minute through the burner flame. The pipe cleaner is observed for a self-sustaining flame as it enters and leaves the burner flame.

Figure 13.6 shows a wick test of a hydraulic fluid; the hydraulic fluid soaked pipe cleaner has caught fire after cycling through the burner flame. The number of cycles need to obtain a self-sustaining flame on each pipe cleaner is noted and averaged for each of the five soaked pipe cleaners tested from each of the three Petri dishes.

Evaluation of wick test results

To pass the wick test, the following test results are needed:

- For the five pipe cleaners soaked in hydraulic fluid at room temperature, the average number of cycles before attaining a self-sustaining flame must be 24 or more.
- For the five pipe cleaners soaked in hydraulic fluid evaporated in the oven for two hours, the average number of cycles before attaining a self-sustaining flame must be 18 or more.
- For the five pipe cleaners soaked in hydraulic fluid evaporated in the oven for four hours, the average number of cycles before attaining a self-sustaining flame must be 12 or more.

Spray flammability test

The spray flammability test is used to determine the fire resistance of a hydraulic fluid when it is subjected to three separate sources of ignition. The hydraulic fluid is heated in a mixing chamber to 150 °F (65.6 °C) and pumped under a pressure of 100 psi (6.9 bars) through an atomizing nozzle. The atomizing nozzle has a discharge orifice of 0.025-inch (0.064 cm) diameter, capable of discharging 3.28 gallons (12.42 liters) of water per hour with a spray angle of 90 degrees at a pressure of 100 psi (6.9 bars). A large rectangular metal chamber that is open at the front is used to contain the spray of the hydraulic fluid during a test and to hold each ignition source at a specified distance from the spray nozzle.

The three ignition sources used in the test of the hydraulic fluid spray are: (i) an electric arc produced across an electrode gap by a 12,000-volt transformer device, (ii) flame from a propane torch – Bernzomatic or equivalent, (iii) flame from ignition of cotton waste soaked in kerosene. The cotton waste is held in a metal trough (20 inches long by 2-1/4 inches wide by 2 inches deep) (50.8 cm by 5.7 cm by 5.1 cm) with a metal cover.

Prior to beginning a test, one of the three sources of ignition is placed into the metal test chamber. Each ignition source is placed at least 18 inches from the atomizing nozzle. Distances of 24 and 36 inches from the atomizing nozzle are also used for placing the ignition sources. The hydraulic fluid is sprayed at each of the ignition sources for one minute or until the flame or arc is extinguished (if less than one minute). A spray flammability test of a hydraulic fluid is shown in Fig. 13.7; the hydraulic fluid was ignited by the flame from kerosene soaked cotton waste and continued to burn.

Spray test evaluation

A hydraulic fluid passes the spray flammability test, if the tests do not result in an ignition of any sample of fluid or if an ignition of a sample does not result in flame propagation for a time interval not exceeding six seconds at a distance of 18 inches or more from the nozzle tip to the center of each igniting source.

13.4.6   Coatings/sealant test

Coating/sealants which are used to coat stoppings and other ventilation controls are discussed in the section on evaluated materials. The ASTM E-162 test, which is used to determine the flame resistance (flame spread index) for coatings/sealants is described in this section (ASTM E-162, 1987). This test uses a gas-fired panel 12 by 18 inches (30.5 cm by 45.7 cm) to radiate heat to a sample placed in front of the panel. The radiant panel test apparatus is shown in Fig. 13.8. The radiant heat output of the panel is adjusted to 1238 °F by the use of a radiation pyrometer. The radiation pyrometer is calibrated by the use of a blackbody enclosure. The test sample is six inches wide by 18 inches long

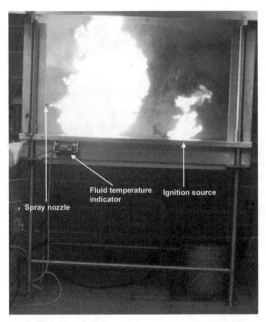

13.7 Spray flammability test of a hydraulic fluid ignited by the flame from cotton waste soaked in kerosene (CFR 30, Part 35).

(15.2 cm by 45.7 cm) and held by a metal holder at an incline of 30 degrees from the panel. The flame from an acetylene burner is used to ignite the upper edge of the sample.

A small metal stack with eight thermocouples of equal resistance and connected in parallel is placed above the top portion of the sample holder to

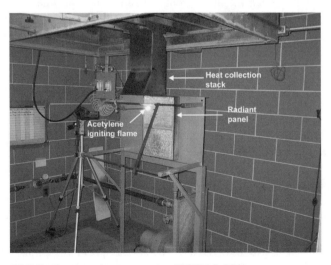

13.8 Radiant panel test apparatus (ASTM E-162).

13.9 Radiant panel test (ASTM E-162) of rigid urethane foam sealant.

collect the heat (fire products) generated by a sample during a test. A larger metal hood with an exhaust blower is used above the small metal stack. The exhaust blower is used to produce an air velocity of about 250 feet per minute (76 m/min) at the top of the small metal stack when the panel is at proper operating temperature. During a test, the progress of the flame is measured at 3 inch (7.6 cm) intervals along with its arrival time. The stack temperature is also recorded during a test and the maximum stack temperature is used in the calculation of a flame spread index for the test specimen. The test standard (ASTM E-162, 1987) requires four specimens of a given product to be tested. A test is completed if either the flame front from a burning specimen has traveled to the 15-inch mark or if exposed to the heated panel for 15 minutes provided the maximum stack thermocouple temperature is reached. A radiant panel test of a spray-applied, rigid urethane foam sealant is shown in Fig. 13.9. The progress of the flame from the foam sealant is approaching the 12-inch mark in Fig. 13.9.

Evaluation of test results
The flame spread index is determined by the use of an empirical formula. The empirical formula relates the flame travel time and the heat output created by the specimen in calculating a flame spread index. A flame spread factor is calculated arithmetically using the times that flame arrives at each 3-inch interval along the specimen. A heat evolution factor is calculated from the maximum observed stack thermocouple temperature along with a beta constant determined for each test apparatus by the use of a multi-port diffusion burner. An arbitrary constant for metric unit consistency is also used as a multiplier in calculating the heat evolution factor. The flame spread factor and the heat evolution factor are multiplied to obtain a flame spread index for each specimen

tested. An average flame spread index is then determined for the set of four specimens tested.

For sealants/coatings used on mine stoppings and other ventilation controls, a flame spread index of 25 or less is required. Sealants/coatings are discussed further in the section on evaluated materials.

13.4.7   Stoppings and other ventilation controls

Stoppings and other devices such as regulators, overcasts, undercasts and shaft partitions are used for the control of ventilation in mine operations. Discussion of these controls is provided in the section on evaluated materials. A test incorporating an ASTM E-119-88, or equivalent time/temperature heat input is used to determine the fire endurance rating of stoppings and certain ventilation controls (ASTM E-119, 1988). The ASTM E-119 test consists of a furnace heated to achieve specified temperatures at specified times. Figure 13.10 illustrates the time-temperature furnace heating curve used with respect to the ASTM E-119 test method. The ASTM E-119 test method has been used for many years to evaluate the performance of building constructions such as walls and partitions under fire exposure conditions.

A stopping construction or specified ventilation control product is exposed to a standard fire exposure that follows the standard time-temperature heating curve specified in the ASTM-E-119 test method. The size of the construction or unit tested may vary depending on its intended purpose. The area of exposure to the fire shall not be less than 100 square feet (9 square metres) with neither

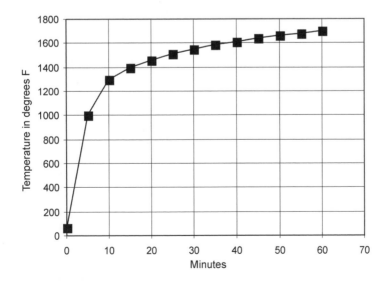

13.10 Furnace heat curve, fire endurance test, ASTM E-119.

dimension less than 9 feet (2.7 metres). The test exposure is 60 minutes in duration, but may be less if a condition of failure of the construction is observed. Nine thermocouples placed at specified locations according to the ASTM E-119 test method are used to measure the temperature on the unexposed surface of the construction. Temperature readings are taken at intervals not exceeding 15 minutes until a reading of 212 °F (100 °C) occurs. Then the readings may be taken more frequently, but not less than a five-minute interval.

Evaluation of test results
The construction meets the requirements for fire endurance if:

- it remains in place during the one hour period of the heat exposure
- the control does not permit the passage of visible flames during the one hour period of the exposure
- openings in excess of two square inches do not develop in the construction during the one hour period of the heat exposure.

13.5   Evaluated materials/products

This section presents a description of materials/products for which requirements for flame resistance are specified. The requirements for flame-resistant materials primarily encompass products used in underground coal mines.

13.5.1   Conveyor belts and system components

Conveyor belts for use in the underground coal mines of the U.S. must be flame resistant. The test for flame resistance is the SSFT as mentioned in preceding sections. Generally, conveyor belts are made from a rubber or plastic compound combined with one or more layers of fabric material or different diameters of steel cables. A conveyor belt may be made from polyvinyl chloride and fabric such as polyester or styrene-butadiene rubber and several layers of polyester or nylon fabric. Flame-retardant compounds used in the manufacture of conveyor belts, such as antimony oxide, enhance the ability of conveyor belts to pass the required flame test. Sometimes carbon black is added to the conveyor belt compounding and may enhance the undesirable feature of afterglow, depending on the amount in the formulation. Figure 13.3 shows the SSFT used to test for the flame resistance of conveyor belts.

More details on conveyor belts and their construction may be found in the Rubber Manufacturers Association Handbook (RMA, 1989). A list of conveyor belt manufacturers with assigned acceptance numbers for conveyor belts that has met the CFR30, 18.65 requirements is available on MSHA's website. The SSFT is also used to determine the flame resistance of some components of conveyor belt haulage systems. Components such as conveyor belt roller covers, belt scrapers and lagging material for use on belt drives are constructed in a different manner than conveyor belts. The formulation for these items can also be different. For example, belt scraper material may be made from hard plastic which is shaped in a manner to remove debris from a conveyor belt. These materials are recommended to be flame resistant for use in underground coal mines, although not mandated by regulation. The same flame test and criteria for flame resistance is applied to these materials as for conveyor belts. A list of the manufacturers of these materials and their assigned acceptance number is also available on MSHA's website.

13.5.2   Electric cables and splice kits

In underground coal mines, the electric trailing cable of equipment, high-voltage cables used on long walls, and the cables used within a mine battery are required to be flame resistant. Also, the splice kit (materials) used to make permanent repairs to flame-resistant electric cables are required to be flame resistant. The electric cable is an assembly of one or more insulated conductors of electric current under a common or integral jacket and may also contain one or more conductors that are not insulated. The cables are evaluated for flame resistance by the test and method specified in Title 30, Code of Federal Regulations, Part 7, Subpart K (CFR 30, 2005). In this particular test as indicated earlier in this chapter, an electric current is passed through the power conductors of the cable until a conductor temperature of 400 °F (204.4 °C) is reached. Then a flame from a natural gas type Tirrell burner is used to ignite the cable. The duration of burning is timed and the length of burned sample is measured. Three samples of cable are tested. To be considered as flame resistant, the duration of burning of each of the three cables must not exceed four minutes and the length of burned area must not exceed 15.24 cm (6 inches). The flame test of an electric cable that exhibits extensive burning is shown in Fig. 13.4. The Tirrell ignition burner can be seen in the bottom center of Fig. 13.4 between the two burning portions of cable. The cable obviously did not meet the flame test criteria.

Cable splice kits are used to make repairs in a flame-resistant electric cable. For cable splice kits, three samples of the same splice kit are also flame tested. Each splice kit is used to construct a splice on cable that has already been qualified as meeting the Subpart K flame test for electric power cables. The same flame test and criteria are applied to cable splice kits as for the electric trailing cables.

Signal cables are also evaluated for flame resistance under CFR30, Part 7, Subpart K. In contrast to an electric power cable, a signal cable may be a fiber optic cable or a cable containing electric conductors of cross-sectional area less than No.14 American Gauge Wire. The signal cable test method is similar to the electric cable test except electric power is not applied to the signal cable and the duration of burning must not exceed 60 seconds. The criterion for the length of burned area not to exceed 6 inches (15.24 cm) is the same as for the electric cable test method.

A list of the manufacturers of electric cables and cable splice kits and their assigned acceptance or approval number is available on MSHA's website. A list with respect to flame-resistant signal cables is also available on the MSHA website.

13.5.3 Hoses

There are several different types of hoses used in underground coal mine operations in the U.S. that are required to be flame resistant. These hoses are fire hose, fire suppression hose, and hose conduit. Fire hose is used for fire-fighting purposes in underground coal mines. Fire hose typically consists of a polyester cover and a rubber liner. Some fire hose products have a plastic impregnated hose cover and rubber liner. The fire hose is tested for flame resistance by using the SSFT method specified in Section 18.65 of Title 30 (CFR 30, 2005). However, only the liner (four samples) of the fire hose is tested for flame resistance. For underground coal mine use in the U.S., the cover of fire hose must be polyester, or other material with flame spread qualities and mildew resistance equal or superior to polyester. The criteria for fire hose to be considered as flame resistant are the same criteria as required for flame-resistant conveyor belts, i.e., average flame time not to exceed one minute and average afterglow not to exceed three minutes.

Fire suppression hose is used on equipment to convey an agent such as multipurpose dry chemical to suppress fire. The fire suppression hose may be made of rubber or plastic and may be reinforced with wire or fabric material. For use in U.S. underground coal mines, the cover of fire suppression hose is mandated to be flame resistant. The cover material (four samples) is tested for flame resistance by using the SSFT method specified in Section 18.65 of Title 30 (CFR 30, 2005). The criteria for fire suppression hose to be considered as flame resistant are the same criteria as required for flame-resistant fire hose, i.e., average flame time not to exceed one minute and average afterglow not to exceed three minutes.

Hose conduit, which is used for the mechanical protection of wiring on underground coal mine electrical equipment, is mandated by MSHA regulation to be flame resistant. In the case of hose conduit, four samples of the whole wall thickness (inside to outside) are tested for flame resistance by using the SSFT method specified in Section 18.65 of Title 30 (CFR 30, 2005). The sample size for the SSFT is 15.24 cm long by 1.27 cm wide by the thickness of the hose conduit (6-inches long by 1/2 inch wide by thickness). For hose conduit to meet the test criteria for flame resistance, the product must meet the same criteria as required for conveyor belts, i.e., average flame time not to exceed one minute and average afterglow not to exceed three minutes.

More details on hoses and their construction may be found in the Rubber Manufacturers Association Handbook (RMA, 2003). A list of the manufacturers of fire hose, fire suppression hose, hydraulic hose, and hose conduit and their assigned acceptance or approval number is also available on MSHA's website. Hydraulic hose used to convey fluid to power mining equipment is another product that is not mandated by U.S. MSHA regulation to be flame resistant. However, it is a recommendation that hydraulic hose should be flame resistant for use in underground coal mines. The SSFT is also used to determine the flame resistance of hydraulic hose in the same manner as fire-suppression hose. Hydraulic hose is typically a rubber compound reinforced by metal wire or fabric. The cover of hydraulic hose is tested and the same criteria for flame resistance are applied as for fire suppression hose.

13.5.4   Hydraulic fluids

Much of the mining equipment used in underground coal mines uses hydraulic fluid to power various components. The use of fire-resistant hydraulic fluid pertains to underground coal mines and is an option depending on the equipment and circumstances. Underground coal mine equipment which is electrically powered and attended (operated) by the presence of a miner is required to use fire-resistant hydraulic fluid or be equipped with a fire-suppression system. There are several options that involve the use of fire-resistant hydraulic fluid in unattended electrical powered equipment. The detailed requirements pertaining to the use of fire-resistant hydraulic fluids in attended and unattended electrical powered equipment in underground coal mines are specified in Title 30, Code of Federal Regulations, Part 75.1107 (CFR30, 2005). The flammability test requirements for hydraulic fluid are also specified in Title 30, Part 35.

There are three tests specified in Part 35 that a hydraulic fluid must meet to qualify as a fire-resistant hydraulic fluid. These tests are an auto-ignition test (Fig. 13.5), a hydraulic fluid saturated pipe cleaner (wick) test (Fig. 13.6), and a spray flammability test (Fig. 13.7). The details of these tests and the criteria to pass each of them are presented in the section on technical aspects and criteria. For a hydraulic fluid to qualify as fire resistant with respect to Part 35, it must exhibit an auto-ignition temperature of 600 °F (315.6 °C) or greater; it must not exhibit flame propagation for a time interval greater than six seconds; and it must not attain a self-sustaining flame on a fluid-soaked pipe cleaner before 24 cycles with the fluid at room temperature, before 18 cycles with the fluid after two hours of oven heating at 150 °F (65.6 °C), and before 12 cycles with the fluid after four hours of oven heating at 150 °F (65.6 °C).

There are different classes of hydraulic fluids that have met the fire-resistance requirements of Part 35. These classes are emulsions, glycols, and synthetics. The emulsion fluids typically contain water emulsified in oil. The oil molecule surrounds the water molecule, thereby giving lubricity to the fluid from the oil and fire resistance from the water. The percentage of water in emulsified fluids used in attended underground mine equipment is generally 40 to 45 percent. The emulsified fluids used in longwall mining roof supports contain about 95 to 98 percent water, since lubricity is not a major concern. Glycol fluids generally contain ethylene glycol mixed with about 45 percent water and are commonly known as an 'antifreeze' mixture. Synthetic fluids are chemical mixtures that may consist of phosphate esters or polymer compounds and generally cost more than emulsions or glycols. In comparison to regular petroleum oil, the lubricity of emulsions, glycols, and synthetics is not as good and their costs are much higher. The clear advantage of these classes of fluids over regular petroleum is fire resistance. An example of a hydraulic fluid which did not pass the spray flammability test is shown in Fig. 13.7. The flame propagation time of the fluid exceeded the required six seconds or less.

13.5.5   Mine equipment components

There are several different materials used as components of mine equipment. Packing material and rubber grommets are used to seal the entries through which electric cables pass or are routed, especially on permissible mining equipment. Decades ago, asbestos rope was used as packing material, but due to health issues it is no longer used. Non-asbestos substitutes are used today, which consists primarily of fiberglass rope. Packing material is typically available in both round and square configuration in diameters of 1/4-inch (0.635 cm) and larger. Both packing material and rubber grommets are required to be flame resistant and must meet the flame test requirements of Part 18, Section 18.65 (CFR 30, 2005). The packing material must also meet two additional test requirements. The one additional requirement is a compression test which is used to determine the compression ratio of the rope-like packing material. The other additional requirement involves subjecting the material which is packed in the steel glands of a closed test vessel to a series of natural gas–air explosions.

There are also materials used in the construction of underground coal mine battery boxes and used as insulation in battery boxes. These materials must meet the flame test requirements of Part 18, Section 18.65 (CFR 30, 2005). A list of the manufacturers of flame-resistant battery boxes and battery box insulation along with their assigned acceptance or approval number is available on MSHA's website.

13.5.6   Coatings (foam and other)

Coatings, also called sealants, are required to be flame resistant for certain mine applications and are available in several different types. These types are rigid urethane foams, cementatious materials, silicates, and epoxies. Urethane foam is typically applied by spray application; the others are typically applied by hand trowel or pressurized application. Sealants which are used to coat ventilation controls, such as underground coal mine stoppings, overcasts, and undercasts, must meet the requirements for flame resistance by having a flame spread index of 25 or less. Sealants that are also used to coat at least the first 200 ft (61 m) of intake air openings and exhaust air openings that are designated as escape-ways in underground metal-nonmetal mines must also meet a flame spread rating of 25 or less. The flame spread index is determined by ASTM E-162, Standard Test Method for Surface Flammability of Materials Using a Radiant Heat Energy Source, 1987 edition (ASTM E-162, 1987). The E-162 test (radiant panel test) is described in the Technical Aspects and Criteria section.

Figure 13.8 shows the radiant panel test apparatus. Figure 13.9 shows a radiant panel test of a urethane foam sealant as flame propagated down the foam sample. The achievement of a urethane or organic foam meeting a flame spread index of 25 or less is difficult to attain and more of a challenge than a cementatious or silicate sealant, which typically is composed of inorganic materials and thus would be noncombustible. The advantage of urethane foam as a sealant is its ability to expand and fill voids and its ability to withstand movement on an applied surface. Cementatious sealants are easy to apply, less expensive than foam, but do not withstand movement well and do not adhere as well as foam. Notwithstanding, these types continue to have application in mining operations and the choice may depend on local conditions and economics. A list of sealants that meet the flame spread index of 25 or less along with their manufacturers is available on MSHA's website.

13.5.7   Noise control materials

There are many applications for noise control materials in mining operations. Noise control products can be in the form of flexible foam, rubber-type barrier materials, and composites consisting of fiberglass or other substances encased in a covering such as vinyl. Noise control materials may present a flammability hazard, depending on the composition and manner of use (Litton *et al.*, 2003). There is no requirement that noise control materials must be flame resistant for use in mining operations. Based on the large-scale fire test data and ASTM E-162 flame spread index data of selected noise control materials, a flame spread index of 50 or less is recommended for surface mine equipment applications and 25 or less for underground mine equipment applications (Litton *et al.*, 2003).

13.5.8   Rib and roof support materials

Plastic and steel mesh-like materials are used in underground mine operations to retain material and prevent the spalling of the roof and ribs. The mesh material has different size openings and the thickness of the webbing that surrounds the openings varies. Plastic mesh is lighter than the steel mesh and provides the strength needed, but the flammability hazard is a concern. However, no requirement is mandated for the plastic rib and roof mesh to be flame resistant. Because of the large amount of surface area exposed after the material is put in place, it is recommended that it should be flame resistant especially for used in underground coal mines. The test for flame resistance of rib/roof support material is the same test that is used to determine the flame resistance of brattice cloth. However, in addition to suspending the rib/roof material in the same manner as brattice cloth, the rib/roof material is also held across the roof of the brattice test apparatus during a test. The test criteria and evaluation of results for the rib/roof material are the same as used for brattice cloth.

13.5.9   Ventilation controls (brattice, ventilation tubing, stoppings)

The discussion in this section pertains to ventilation controls which include brattice cloth and ventilation tubing and mine stoppings. Ventilation curtain, commonly known as brattice cloth, is used to direct the flow of air in underground coal mines. Several decades ago, jute and cotton duck were commonly used materials for brattice cloth. Today, nearly all brattice cloth used in underground coal mines is made of plastic-type materials. The plastic brattice cloth can be lightweight from about four ounces per square yard to heavier weight of around 44 ounces per square yard. The plastic brattice cloth is not subject to moisture collection from the mine atmosphere as was jute and cotton duck. The plastic brattice cloth essentially is more versatile and replaced jute and cotton duck. Some brattice cloths are transparent and thus provide visibility through the material and some brattice cloths used reflective material. Both of these features provide a safety advantage.

Ventilation tubing may be rigid or flexible. The rigid ventilation tubing ranges from about 16 inches (40.6 cm) in diameter and larger. The rigid ventilation tubing may be of metal or plastic composition. The plastic ventilation tubing may be composed of fiberglass and resin material to give strength to the product. Flexible ventilation tubing is typically constructed from brattice cloth material with a wire or plastic supporting structure. The flexible ventilation tubing may be easier to use in mine operations since it is collapsible, thus requiring less storage space than rigid ventilation tubing.

Because of the fire hazards in underground coal mines, brattice cloth and ventilation tubing are required to be flame resistant. Brattice cloth or ventilation tubing that is not flame resistant, can catch on fire and result in a large propagating flame. The flaming brattice cloth or ventilation tubing can cause coal and other combustible materials also to catch on fire, creating a potentially serious coal mine fire.

MSHA developed a test which was described previously in this chapter to evaluate the flame resistance of brattice cloth and ventilation tubing. The test, known as the standardized small-scale flammability test (SSFT) is laboratory-scale in size. The test apparatus is about three and a half feet high, two and a half feet wide and 5 feet long (1.07 m high by 0.76 m wide by 1.52 m long). The test was based on large-scale fire studies conducted by MSHA on many different brattice cloth and ventilation tubing products. Additionally, the NIOSH Pittsburgh Research Laboratory conducted a flammability study of rigid plastic ventilation tubing using the ASTM E-162 radiant panel test, the ASTM E-2863 oxygen index test and the MSHA SSFT apparatus and several different large-scale test setups. The NIOSH test results showed good agreement between the SSFT and their large-scale tests for the rigid ventilation tubing (Perzak *et al.*, 1987).

Brattice cloth and ventilation tubing used in underground coal mines and underground gassy metal-nonmetal mines is required by MSHA regulations to be flame resistant. The test (SSFT) to determine the flame resistance is specified in 30 CFR, Part 7, Subpart B (CFR 30, 2005). A synopsis of the test and criteria for flame resistance brattice cloth and ventilation tubing was presented earlier in this chapter. A list of the manufacturers of flame-resistant brattice cloth and ventilation tubing along with their assigned acceptance or approval number is available on MSHA's website.

Stoppings are used for ventilation control in underground coal and metal-nonmetal mines. The purpose of a stopping is to form an airtight wall across a mine passage. They may be constructed from concrete or cinder block, metal panels, cementatious materials combined with other components such as fly ash, and lightweight block products composed of inorganic materials. Some of these materials may also be used to construct overcasts, undercasts, shaft partitions, and regulators which are also used for ventilation control. Overcasts and under-casts are enclosed airways which serve as air bridges and allow two airstreams to cross without mixing. A shaft partition is a wall that generally divides a vertical mine passageway. A regulator is used to control the flow of mine ventilation in a particular area which may be a sliding door in a stopping.

These ventilation control devices, stopping, overcast, undercasts, shaft partitions, and regulators must met a mandatory requirement for a one hour fire endurance which is specified in Title 30, Part 75, Subpart D (CFR 30, 2005). The test used for the fire endurance determination is ASTM E-119-88 as discussed in section 13.4.

13.6   Future applications/trends

Some of the new materials developed that are used in mining operations are spin-offs or adopted from other industrial uses. The resulting increase in the use of plastic materials in the mining industry should continue to grow. Lightweight and stronger materials that are readily available and reasonable in cost will continue to be selected for use in mining operations. For some of the applications and types of materials that may be introduced, flammability could become an issue.

It is anticipated that modeling with respect to fire dynamics will play more of a role in materials selected for mine use. New tests or modification of existing tests for evaluating the flammability characteristics of newer materials take a lot of time to develop, validate and become standardized and accepted. In addition, the high costs of test and development programs and the lack of resources can impede new developments. The use of fire modeling will assist in a better understanding of the role of material combustion and assessing the fire safety of materials.

In another area, examining materials at the atomic level (nanotechnology) could result in new products. Science and technology on a nanoscale is quickly emerging. New advances in nanotechnology are anticipated to have future implications in a wide variety of scientific and engineering disciplines. Nano-technology is expanding worldwide. This expansion is evidenced in part by the announcement of U.S. National Nanotechnology Initiative. This initiative identifies nanotechnology as an emerging area of national interest, and there is provision for substantial increases in funding levels from U.S. Federal agencies. Other major scientific organizations such as the American Physical, Chemical, and Materials Research Societies will sponsor several sessions on nanotechnology-related areas during future annual meetings.

As an aspect of the future regarding some ongoing work involving nano-technology, a tiny amount of carbon nanotubes dispersed into polypropylene greatly reduces the polymer's flammability (Kashiwagi *et al.*, 2002). This research work by NIST scientists show the nanotubes outperformed existing environmentally friendly flame retardants. Better materials with lower flammability hazards will continue to be a future goal of developers and users. The benefits of better and more economical flame-resistant materials in the years to come will eventually find their way into the mining industry.

13.7 Summary

A discussion and description is provided on flammability tests for evaluation of certain materials used in U.S. mining operations. Most of the materials that are required to be flame resistant are those materials used in underground coal mines. Regulations to address materials with respect to flame resistance generally result from the incidents of mine fires. Mine fires prevalent with conveyor belts resulted in tests to determine flame resistance and instituting the requirement that conveyor belts must be flame resistant for use in underground coal mines. Other materials involved in underground coal mine fires such as electric cables and brattice cloth were also required to be flame resistant. The use of flame-resistant material in mining operations has, no doubt, resulted in a decrease in disastrous and serious fires. However, from past and present results, flammability tests do not necessarily describe the fire hazard of materials under actual fire conditions. Work continues in various organizations to develop more meaningful tests to determine the flammability hazard of materials in their use environment.

The role of tests for defining the flammability hazard of material used in an environment has been discussed in depth (Roux, 1976). Roux proposed that the fire hazard of a product might be defined as FH = PH × E, where FH is the fire hazard, PH is the potential harm, and E is the exposure. His proposed procedure was an attempt to characterize the fire hazard as it relates to the environment and the exposure in that environment.

Much work exists regarding the fire hazard of materials and the many tests designed to relate the material hazard to the work environment. For instance, the National Institute for Occupational Safety and Health (NIOSH) has conducted extensive research work on the flammability hazards of materials used in the mining industry. The NIOSH fire research work has been utilized by MSHA in addressing and enhancing fire safety and disseminating information on the flammability hazards of different materials used in the mining industry.

The dissemination of information on fire test methods and a host of other topics on fire continues through conferences and seminars held nationally and internationally. The information presented in this chapter adds to the body of information with respect to flammability tests for materials use in the mining industry and their requirements for flame resistance.

New materials continue to be developed which should be formulated with the flame-resistant qualities to provide the desired fire safety performance for use in mining operations. Over 50,000 materials can be utilized for the design and manufacture of engineered products, including metals, polymers, ceramics, and composites which may be used in the mining industry. Although metals and polymers are dominant materials for engineering applications, the use of composite materials is growing due to superior strength, low weight, and improved thermal and electrical performance characteristics. Sometimes newer materials such as the chlorinated polyethylene outer jacket for electric cables are introduced into the mining industry as a spin-off from other industrial developments. In addition, developments in nanotechnology may provide future advances in fire-resistant materials.

13.8 Sources of information

There are numerous sources of information on the subject of the flammability of materials and fire test methods. With the advent of the internet and the worldwide web, access to such information is much easier and quicker. Some of the more notable sources are listed. Reference material and publications are available through the websites of well established sources such as the U.S. National Institute for Science and Technology (NIST), the ASTM, NFPA, NIOSH, and the SFPE.

- National Institute for Science and Technology, Gaithersburg, MD, www.nist.gov

- FIREDOC, a bibliographic database available through NIST.
- American Society for Testing and Materials, West Conshohocken, PA, www.astm.org
- National Fire Protection Association, Quincy, MA, www.nfpa.org
- Society of Fire Protection Engineers, Bethesda, MD, www.sfpe.org
- National Institute for Occupational Safety and Health, www.cdc.gov/niosh/mining
- Mine Safety and Health Administration, www.msha.gov
- Fire Science and Technology, Inc., www.doctorfire.com
- Fire Research Information Services, NIST, www.bfrl.nist.gov/fris
- The International Organization for Standardization (ISO), www.iso.ch/
- Worchester Polytechnic Institute, Center for Fire Safety, www.wpi.edu/
- University of Maryland, Fire Protection Engineering, www.enfp.umd.edu/
- ATF Fire Research Laboratory, www.atf.treas.gov/labs/frl/index.htm

13.9   References

ASTM E-162, 1987, *Standard Test Method for Surface Flammability of Materials Using a Radiant Heat Energy Source*, 1987 edition.

ASTM E-119, 1988, *Standard Test Methods for Fire Tests of Building Construction and Materials*, 1988 edition.

Babrauskas, V., 1996, *Ten Years of Heat Release Research with the Cone Calorimeter*, Fire Science and Technology, Inc., http://www.doctorfire.com/cone.html

Babrauskas, V., 2003a, *Cone Calorimeter Bibliography*, Fire Science Publishers, Issaquah WA.

Babrauskas, V., 2003b, *Ignition Handbook*, Fire Science Publishers, Issaquah, WA.

CFR 30, 2005, Code of Federal Regulations, Title 30, *Mineral Resources*, Parts 1–199, U.S. Government Printing Office, July 1, 2005.

Clark, F.R.S., 1981, 'Fire Spread Tests – A Critique,' *Fire Technology*, Vol. 17, No. 2, May 1981.

Creswell, 1952, *Accident at Creswell Colliery, Derbyshire, on September 26, 1950*, by Sir Andrew Bryan, Her Majesty's Stationery Office, London, England.

CSA, 1987, Canadian Standards Association, *Fire-Performance and Antistatic Requirements for Conveyor Belting*, CAN/CSA-M422-M87, Ontario, Canada, May 1987.

Emmons, H.W., 1967, 'Fire Research Abroad,' *Fire Technology*, Vol. 3, No. 3, August.

FCMS&H, 1969, Federal Coal Mine Health and Safety Act of 1969, Public Law 91-173, 91st U.S. Congress, S. 2917, December 30.

Kashiwagi, T., Grulke, E., Hilding, J., Harris, R., Awad, A., and Douglas, J.F., 2002, 'Thermal Degradation and Flammability Properties of Polypropylene/Carbon Nanotube Composites,' *Macromolecular Rapid Communications*, Vol. 23, No.13, pp. 761–765.

Litton, C.D., Mura, K.E., Thomas, R.A. and Verakis, H.C., 2003, 'Flammability of Noise Abatement Materials used in Cabs of Mobil Mining Equipment,' *Proceedings of the Fire and Materials 2003 Conference*, San Francisco, California, USA, pp. 297–306, January 27–28.

McGuire, J.H. and Campbell, H.J., 1980, 'Surface Flammability Assessment, Part II – Validity and Application of Major Current Test Methods,' *Fire Technology*, Vol. 16, No. 2, May.

National Bureau of Standards, 1988, *Fire Hazard Comparison of Fire-Retarded and Non-Fire-Retarded Products*, U.S. Department of Commerce, National Bureau of Standards (NBS) Special Publication 749, July.

NFPA, 2003, *Fire Protection Handbook*, 19th edition, National Fire Protection Association, Quincy, MA 02269, U.S.A.

O'Neill, T.J. and Sweet, G.C., 1989, 'Life threat Hazard of Burning Materials,' *Polymers in Mining*, 3rd International Conference, The Plastics and Rubber Institute, University of Lancaster, United Kingdom, September 26–27.

Perzak, F.J., Lazzara, C.P., and Kubala, T.A., 1987, *Fire Tests of Rigid Plastic Ventilation Ducts*, U.S. Bureau of Mines Report of Investigations, RI 9085.

Pollack, S.P., 1956, *Research to Develop a Schedule for Testing Conveyor Belts for Fire Resistance*, U.S. Bureau of Mines, presented at the Ninth International Conference of Directors of Safety in Mines Research, Heerlen, Netherlands and Brussels, Belgium, June 28–July 4.

Roux, H.J., 1976, 'The Role of Tests in Defining Fire Hazard: A Concept,' in *Fire Standards and Safety*, ASTM Special Technical Publication 614, pp. 194–203.

RMA, 1989, Rubber Manufacturers Association, *Conveyor and Elevator Handbook*, 3rd edition, 98pp. (*www.rma.org*).

RMA, 2003, Rubber Manufacturers Association, *Hose Handbook*, 7th edition, 111pp. (*www.rma.org*).

Schlagel and Eisen, 1977, *Report on the Mine Disaster at the Schlagel & Eisen Colliery in Herten, Recklinghausen*, Mines Inspectorate, Nordrhein-Westfalen, Germany, October 27.

SPFE, 2002, *Handbook of Fire Protection Engineering*, 3rd edition, National Fire Protection Association, Quincy, MA 02269, U.S.A.

Flammability tests for railway passenger cars

R D PEACOCK and R W BUKOWSKI,
NIST Building and Fire Research Laboratory, USA

14.1   Introduction and history

Fire safety of passenger trains in the United States has been approached by regulating the flammability of furnishing and finish materials since early in the 20th century. In the past decade advances in technology associated with the development of high-speed trains has led to increasing use of light-weight synthetic materials whose end-use performance is not well predicted by small-scale tests. Considerable research has been conducted to develop quantitative methods to assess the fire performance of the entire system, starting with modern metrics for the flammability of materials. This chapter will summarize this transition and the research on which it is based.

Interest in improving the fire safety of passenger train vehicles is not new. From 1906 to 1928, the Pennsylvania Railroad undertook an ambitious program to replace their wooden passenger car fleet with all-steel passenger train cars due to a concern for safety and fire prevention.[1] A total of 5501 all-steel passenger train cars including baggage, mail, express, and dining cars were involved, representing an investment of approximately one hundred million dollars. More recently, emphasis on passenger comfort and aesthetic appeal led to the increased use of synthetic materials.[2] Plastic use in rail car interiors started in the early 1950s.[3,4] In the past decade global interest in high-speed trains has led to ever-increasing use of synthetics to reduce weight and increase energy efficiency. However, concern has been raised over the flammability and impact on fire hazard of plastics in their end-use configuration, even though they may be judged to be acceptable in small-scale tests.[5]

While nonmetallic materials have traditionally been used in seat cushioning and upholstery, their use in other system components such as coverings for floors, walls and ceilings, window glazing and window or door gasketing and nonstructural storage compartments have increased the fire load within the vehicles. In addition to the flammability characteristics of the interior furnishing materials, the size and design of the vehicle are all factors in determining the ultimate hazard to passengers and crew as a result of a fire.

Specific requirements for the flammability of materials in rail transportation vehicles first appeared in the U.S. in 1966.[3] These rail car specifications dictated 'flame tests' for seat foam materials before the material use would be approved for the original Amtrak Metroliner passenger rail cars. The U.S. National Academy of Sciences[6] provided general guidelines in 1979 for the use of flammable materials in rail transit vehicles. These guidelines recommended the use of only those polymeric materials that by testing and comparison, are judged to be the most fire retardant and that have the lowest smoke and toxic gas emission rates. Further, they suggested these be used sparingly, consistent with comfort and serviceability.

Rakaczky[7] examined the available literature on fire and flammability characteristics of materials which could be used in passenger rail transportation vehicles. With the exception of some documents published by the U.S. Federal Transit Administration (FTA), limited information was available for materials that related specifically to passenger rail vehicles. Much of the literature reviewed related more to other transportation applications (primarily aircraft) than to rail transportation. Key in the Rakaczky study, however, was a prevailing concern of many researchers of the ability of small-scale tests to predict large-scale burning behavior. Hathaway and Litant[8] provided an assessment of the state-of-the-art of fire safety efforts in transportation systems in 1979. Without annotation, they provide a bibliography of literature from 1970 to 1979. Peacock and Braun[9] studied the fire behavior of Amtrak passenger cars for the U.S. Federal Railroad Administration (FRA). They provide a review of material testing requirements and a comparison of small- and large-scale testing of vehicle interior materials. Schirmer Engineering Corporation studied the fire safety of railroad tunnels and stations in New York City, including the impact of passenger train flammability requirements on the fire load in tunnels and stations.[10] Peacock et al.[11] present a review and comparison of test requirements for passenger rail vehicles in several countries. The strengths and weaknesses of available methods for measuring the fire performance of rail transportation systems are evaluated.

14.1.1 U.S. requirements

Within the United States, the FRA, Amtrak, FTA, and National Fire Protection Association (NFPA) documents contain similar requirements covering the fire safety of materials used in passenger vehicles. German requirements include test methods used by the U.S. Federal Aviation Administration (FAA) for aircraft. A report to the Office of the Secretary of Transportation recognized the potential for similar requirements in multiple modes of transportation.[12] The review in the report was organized into topics, including fire protection and control, material controls, engine components, structural components, procedures, and buildings. Numerous areas were identified for potential cooperation and common requirements between different transportation modes. To date, the overlap is

Table 14.1 Test procedures and performance criteria for the flammability and smoke emission characteristics of materials used in passenger rail vehicles

Category	Function of material	Test method[1]	Performance criteria[1]
Cushions, mattresses	All	ASTM D 3675 ASTM E 662	$I_s \leq 25$ $D_s(1.5) \leq 100$ $D_s(4.0) \leq 175$
Fabrics	All	14 CFR 25, Appendix F, Part I (vertical test) ASTM E 662	Flame time ≤ 10 s Burn length ≥ 152 mm $D_s(4.0) \leq 200$
Interior vehicle components	Seat and mattress frames, wall and ceiling lining and panels, seat and toilet shrouds, trays and other tables, partitions, shelves, opaque windscreens, and combustible signage	ASTM E 162 ASTM E 662	$I_s \leq 35$ $D_s(1.5) \leq 100$ $D_s(4.0) \leq 200$
	Flexible cellular foams used in armrest and seat and mattress padding	ASTM D 3675 ASTM E 662	$I_s \leq 25$ $D_s(1.5) \leq 100$ $D_s(4.0) \leq 175$
	Thermal and acoustical insulation	ASTM E 162 ASTM E 662	$I_s \leq 25$ $D_s(4.0) \leq 100$
	HVAC ducting	ASTM E 162 ASTM E 662	$I_s \leq 25$ $D_s(4.0) \leq 100$
	Floor covering	ASTM E 648 ASTM E 162	CRF ≤ 5 kw/m^2 $D_s(1.5) \leq 100$ $D_s(4.0) \leq 200$

Category	Description	Standard	Criteria
	Light diffusers, windows and transparent plastic windscreens	ASTM E 162, ASTM E 662	$I_s \le 100$, $D_s(1.5) \le 100$, $D_s(4.0) \le 200$
Elastomers	Window gaskets, door nosings, intercar diaphragms, and roof mats	ASTM C 1166, ASTM E 662	Average flame propagation ≤ 101.4 mm, $D_s(1.5) \le 100$, $D_s(4.0) \le 200$
Exterior vehicle components	End caps, roof housings, articulation bellows, exterior shells, and component boxes and covers	ASTM E 162, ASTM E 662	$I_s \le 35$, $D_s(1.5) \le 100$, $D_s(4.0) \le 175$
Wire and cable	All	UL 1581, CSA C22.2, UL 1685, ANSI/UL 1666, NFPA 262, ASTM E 662	See note 2
	Control and low voltage	ICEA S-19/NEMA WC3, UL 44, UL 83	See note 2
	Fire alarm cable	IEC 60331-11	See note 2
Structural components	Flooring, other	ASTM E 119	Pass

1. Details of test methods and acceptance criteria are included in the NPFA standard.
2. All wires and cables shall be listed as being resistant to the spread of fire and shall have reduced smoke emissions.

primarily limited to material controls. Similar requirements in multiple rail transportation sectors are evident in the review below. In this chapter, the detailed review will be limited to the U.S. requirements, and will concentrate on material testing requirements. A more detailed report including requirements in France and Germany is available.[13]

The NFPA flammability and smoke emission requirements[14] for passenger train cars are summarized in Table 14.1. The Amtrak and FRA requirements are nearly identical to the NFPA version. The requirements are based in large part on two small-scale test methods – ASTM E 162, 'Surface Flammability of Materials Using a Radiant Energy Source'[15] (with a variant, ASTM E 3675 for cellular materials[16]) and ASTM E 662, 'Specific Optical Density of Smoke Generated by Solid Materials'.[17] Several additional standards are specified for individual material applications. With exceptions, the test methods are small-scale tests designed to study aspects of a material's fire behavior in a fixed configuration and exposure.

14.1.2   Regulation based on heat release rate

Recently the National Institute of Standards and Technology (NIST) examined the pivotal nature of heat release rate (HRR) measurements in detail.[18] Not only is heat release rate seen as the key indicator of large-scale fire performance of a material or construction, HRR is, in fact, the single most important variable in characterizing the 'flammability' of products and their consequent fire hazard. Examples of typical fire histories illustrate that even though fire deaths are primarily caused by toxic gases, HRR is the best predictor of fire hazard and the relative toxic potency of combustion gases plays a smaller role. The delays in ignition time, as measured by various Bunsen burner type tests, also have only a minor effect on the development of fire hazard.

There are at least two approaches to utilizing HRR data in material selection for any application:

- Use the heat release rate with appropriate limiting criteria for the selection of materials and constructions for the application. This is similar to the traditional approach of using the results of test methods to guide the selection of individual materials. The key limitation to this approach is the inability to judge a material in the context in which it is used and in conjunction with other materials in their end-use configuration.
- Use the heat release rate in a hazard analysis of the actual application. This removes the limitations of the traditional approach above. However, it requires consideration of how materials are combined in an application and thus is more difficult for individual material suppliers to judge the adequacy of their product.

Both these approaches are appropriate for passenger trains.

As noted above, appropriate acceptance criteria for application of HRR-based tests to passenger trains had not been developed. Widespread small-scale heat release rate test results were not yet available for materials in current use in passenger trains, although several organizations were testing materials and are developing such databases. Actual acceptance criteria must consider not only the desired level of protection, but also the current state-of-the-art in materials design for the application. Some testing is still required to establish equivalent criteria for current materials.

Once these test results became available, some large-scale testing of materials was required to establish or verify the predictive ability of the small-scale tests. This serves two purposes: (i) to provide a level of verification of the small-scale testing, and (ii) to minimize future large-scale testing needs for suppliers and manufacturers of passenger trains.

HRR is considered to be a key indicator of fire performance and is defined as the amount of energy that a material produces (over time) while burning. For a given confined space (e.g., rail car interior), the peak air temperature resulting from a fire increases as the HRR of the burning material increases. Even if passengers do not come into direct contact with the fire, they could be injured from exposure to high temperatures, heat fluxes, and toxic gases emitted by materials involved in the fire. Accordingly, the fire hazard to passengers and crew of these materials can be directly correlated to the HRR of a real fire.

14.1.3 NIST research

Beginning in the mid-1990s the FRA began funding NIST to develop a systematic approach to the quantification of fire hazards in passenger trains that could form the basis for regulatory reform. At the time the research began the FRA issued guidelines that were generally utilized by the U.S. rail industry. During the work a major train accident in Silver Spring, Maryland[19] occurred in 1996 that eventually resulted in the FRA guidelines being converted to federal regulations.

14.1.4 Related European passenger train studies

Several European countries have active programs to improve passenger train fire safety evaluation. A great deal of effort is being expended to relate small-scale and large-scale performance by the use of fire modeling. This work is being conducted by individual countries (France, Germany, Sweden, United Kingdom) and in coordinated activities under the sponsorship of the European Railway Research Institute (ERRI) and the Commission for European Standardization (CEN).

The British Rail (BR) small-scale test program was targeted at developing a database of HRR data for all rail materials in current use.[20] BR's cone calorimeter work was supplemented by large-scale assembly tests in a furniture calorimeter and included seat assemblies, sidewall and ceiling panel assemblies, catering refuse bags and contents, plastic towel dispenser units, and vending machines. No other test method data were available for the materials. The furniture calorimeter testing used the methods specified in the British Standards Institute (BSI) documents for the fire evaluation of mock-up upholstered furniture. These methods use small wood cribs as the ignition source. The British government's trend toward privatization of its rail industry has led to an increase in the rehabilitation of older equipment instead of the complete replacement of rolling stock. This has limited the availability of newer materials and assemblies for testing.

BR has also conducted several large-scale test burns of existing coaches and sleeping cars. While much of this work has been performed for internal use, some tests have been performed in connection with the ERRI activities. All of these tests relate to rail car fires on open trackways. Other large-scale fire tests of rail cars located in tunnels have been conducted as part of the Channel Tunnel safety work leading up to the operation of shuttle trains carrying passengers and motor vehicles between England and France.[21] In the process of testing representative materials using a cone calorimeter, the London Underground Limited (LUL) selected an exposure of $50\,kW/m^2$ for 20 minutes as a suitable exposure for material evaluation consistent with testing exposures and fire experience in the United Kingdom.[22]

In 1990, Göransson and Lundqvist studied seat flammability in buses and rail transit cars using material tests and large-scale tests.[23] All of the seats used high-resilience foam, covered with a variety of fabrics. Wall panels consisted of fabric-covered wood or metal panels. In the small-scale tests, the cone calorimeter was selected to provide ignition and HRR information. In large-scale tests, the maximum HRR of a seat assembly, about 200 kW, was not sufficient to ignite the panels or the ceiling 'quickly' (unfortunately, 'quickly' was not defined). However, ignition of adjacent seats was noted in large-scale mock-up tests.

In 1992, ERRI published a report that recommended supplementary studies be conducted to account for smoke opacity and toxicity hazards of materials.[24] Later in 1992, ERRI proposed that computer software be used to model half-scale and full-scale tests already carried out in order to compare computer results with actual results.[25] ERRI considered the use of the cone calorimeter to be the only small-scale apparatus suitable for providing useful data for computer modeling. A series of reports document the completion of ERRI rail coach tests.[26–30] In a test application, ERRI used the HAZARD I model[31] developed by NIST to simulate a fire in the British 3-meter test cube and concluded that the use of the model to simulate fires in a railway vehicle was feasible. Additional cone calorimeter and furniture calorimeter tests were conducted and numerous model simulations of fires within passenger rail coaches were performed. The results of the simulations were primarily aimed at comparing the model prediction to full-scale experiments and evaluating the ability of the model to be used in a rail environment. Use of fire models to validate the design of a passenger rail car in terms of passenger evacuation was proposed.

Numerous international conferences have been held and a very large research project was conducted in Norway under the auspices of EUREKA (European Research Coordination Agency) by nine Western European nations.[32] A 1995 EUREKA test report reviewed 24 fire incidents over 20 years (1971–1991), and presented the results of a series of tests in a tunnel utilizing aluminum and steel-bodied German (DB) Inter-City and Inter-City Express rail cars. An extensive series of full-scale fire tests were conducted and HRR values were developed. Although the primary focus of these tests was the effect of a burning vehicle on the environment within the tunnel, the results provide guidance on the burning properties of passenger rail car materials appropriate for fire hazard analysis that can be compared to the data used for this interim report. In addition to HRR, information on gas concentrations and smoke emission are included for a range of European passenger rail and transit cars. Temperatures within rail vehicles in the tests typically approached 1000 °C for fully-involved fires.

As part of the standardization efforts in the European Union, the FIRESTARR project examined the fire behavior of component materials used in passenger railway vehicles in small- and full-scale testing.[33] The program included small-scale testing of 32 materials in the cone calorimeter, along with additional ignition and flame spread tests on some materials. The same materials were tested in full-scale using the ISO 9705 room-corner test, furniture calorimeter, and compartment tests in a single 10 m^3 compartment. Results for small-scale tests and full-scale tests are available.[34] It was noted that the cone calorimeter allows products to be separated into categories of non-ignitable, difficult to ignite, or easy to ignite according to ignition times. The cone calorimeter also proved to be an appropriate tool for assessing heat release and dynamic smoke generation. Full-scale test results correlated well with small-scale tests for wall and ceiling linings, but less well for seating products due to the low number of seats ignited in the full-scale tests.

14.2   Small-scale testing

The cone calorimeter (ISO 5660-1/ASTM E 1354)[35] is a test method which provides measurements of HRR, specimen mass loss, smoke production, and combustion gases in a single test. Cone calorimeter tests were conducted at NIST on selected passenger rail car materials in current use in Amtrak vehicles.[36] These measurements included ignitability, HRR, and release rate for smoke, toxic gases, and corrosive products of combustion. With the use of a single test method for all materials, measured properties, such as HRR and smoke generation rate, were obtained under identical fire exposure conditions. Since each of the materials tested were in use, they had been tested by the methods specified in the

FRA Regulations, and copies of the test results were obtained from Amtrak. These were used to produce the comparisons presented below, that can be used to benchmark HRR performance for such materials. However, the data obtained from the test methods cited in the regulations, although providing relative ranking of materials under the exposure conditions of the test methods, do not provide quantitative data which can be used for a fire hazard analysis. HRR and other measurements generated from the cone calorimeter can provide similar performance rankings and can also be used as an input to fire modeling and hazard analysis techniques to evaluate the contribution of the individual components and materials in the context of their use.

14.2.1   Comparison of cone calorimeter with FRA regulatory test data

Ignition time, time-to-peak HRR, peak HRR, and several other values were measured in the cone calorimeter for each of the materials provided. A detailed report of the test results from both the FRA Regulatory test methods and the cone calorimeter is available.[37] In this section, the cone calorimeter test data are compared to the test data from the test methods specified in the FRA regulations. Although the primary use of the HRR data is as input to a fire hazard analysis, this comparison is also intended to provide a better understanding of the relationships and limitations of test data from the cone calorimeter relative to test data from FRA-specified test methods.

14.2.2   FRA regulatory test data

Several test methods cited in the FRA regulations include measures of material flammability flame spread (ASTM E 162, D 3675, and E 648), or ignition/self-extinguishment (FAR 25.853 and ASTM C 542). ASTM E 162 and D 3675 measure downward flame spread on a near vertically mounted specimen (the specimen is tilted 30° from the vertical with the bottom of the specimen further away from the radiant panel than the top of the specimen). ASTM E 648 measures lateral flame spread on a horizontally mounted specimen. Since ASTM E 648 was specifically designed to measure fire performance of flooring materials, it is the only test method that attempts to replicate end-use conditions. FAR 25.853 and ASTM C 542 are small burner tests which measure a material's resistance to ignition and burning for a small sample of material.

Because of specific end-use applications, not all materials required evaluation by the same test methods. Table 14.2 shows the materials tested in the study. Details of test results as reported by material manufacturers for the FRA-specified test methods are available.[37] Twenty-three materials were found to require ASTM E 162 or D 3675 testing. Test data were available for 21 of these materials. Although not specified in the FRA regulations, I_s values were also

Table 14.2 List of passenger train materials used in this study

Category	Sample number[1]	Material description
Seat and bed assemblies	1a, 1b, 1c, 1d	Seat cushion, fabric/PVC cover (foam, interliner, fabric, PVC)
	2a, 2b, 2c	Seat cushion, fabric cover (foam, interliner, fabric)
	3	Graphite-filled foam[2]
	4	Seat support diaphragm, chloroprene
	5	Seat support diaphragm, FR cotton
	6	Chair shroud, PVC/Acrylic
	7	Armrest pad, coach seat (foam on metal support)
	8	Footrest cover, coach seat
	9	Seat track cover, chloroprene
	10a, 10b, 10c	Mattress (foam, interliner, ticking)
	11a, 11b, 11c	Bed pad (foam, interliner, ticking)
Wall and window surfaces	12	Wall finishing, wool carpet
	13	Wall finishing, wool fabric
	14	Space divider, polycarbonate
	15	Wall material, FRP/PVC
	16	Wall panel, FRP
	17	Window glazing, polycarbonate
	18	Window mask, polycarbonate
Curtains, drapes, and fabrics	19	Privacy door curtain and window drape, wool/nylon
	20	Window curtain, polyester
	21	Blanket, wool fabric
	22	Blanket, modacrylic fabric
	23a, 23b	Pillow, cotton fabric/polyester filler
Floor covering	24	Carpet, nylon
	25	Rubber mat, styrene butadiene
Misc.	26	Café/lounge/diner table, phenolic/wood laminate
	27	Air duct, neoprene
	28	Pipe wrap insulation foam
	29	Window gasketing, chloroprene elastomer
	30	Door gasketing, chloroprene elastomer

1. Letters indicate individual component materials in an assembly. Individual component materials are listed in order in parentheses following the material description.
2. All foam except sample 3 is the same type.

available for the window and door gasketing. Of the materials currently in use, two did not meet FRA flammability performance criteria, a space divider and a window mask. Polycarbonate is used both as window glazing and as an interior space divider. As a window glazing, the material meets the FRA regulations. However, when used as an interior space divider, the recommended performance criteria are stricter. The Amfleet II window mask is an older material which has been in use since before adoption of the FRA regulations.

Floor covering materials are evaluated using ASTM E 648. Data was available for only one Amtrak material, floor carpeting, which met the FRA regulation performance criteria. No data were available for the sheet rubber flooring.

FAR 25.853 was applicable to 11 samples or 16 unique component materials. Test data on burn length was available for five of the 16 materials. Flame time was available for two of the 16 materials. The five tested materials met the FRA regulations for burn length. Data for the ASTM C 542 small burner test was not considered because ASTM C 542 is a simple pass-fail test and not appropriate for comparison to HRR test data.

The FRA regulations require ASTM E 662 testing for all materials. The 30 samples represent 40 unique component materials. Test data was available for 25 components at the $D_s(1.5)$ level and 27 components at the $D_s(4.0)$ level of performance. At $D_s(1.5)$, five materials were found not to meet FRA regulations. At $D_s(4.0)$, 7 materials were found not to meet FRA regulations. Most of these materials (seat support diaphragm, armrest pad, footrest pad, seat track cover, window gasketing, and door gasketing) represent a small portion of the fire load in a typical vehicle interior. The test data for these components should not be of great concern for fire safety. The last component, a window mask is an older material which has been in use since before adoption of the FRA regulations. Materials in newer cars are well within the FRA recommended performance criteria. Taken together, it is unclear whether the contribution from all these materials would be significant.

14.2.3   Cone calorimeter test method evaluation

All cone calorimeter tests in this study were conducted at a heat flux exposure of $50\,kW/m^2$. This level represents a severe fire exposure consistent with actual train fire tests. With the high performance level typical of currently used materials, levels higher than $50\,kW/m^2$ are unlikely. A spark ignitor was used to ignite the pyrolysis gases. All specimens were wrapped in aluminum foil on all sides except for the exposed surface. A metal frame was used and where necessary a wire grid was added to prevent expanding samples from entering into the cone heater. Data obtained from the cone calorimeter tests are shown in Table 14.3. Included in the table are ignition time, peak HRR, peak specific extinction area (SEA), and average HRR and SEA for the first 180 s of each test. More extensive tabulations of data are available.[37]

Table 14.3 Summary of cone calorimeter heat release rate data for individual materials

Sample no.	Ignition time (s)	Time to peak HRR (s)	Peak HRR (kW/m^2)	HRR 180s average (kW/m^2)	Peak SEA (m^2/kg)	SEA 180s average (m^2/kg)
1a	15	25	75	40	210	30
1b	5	15	25	5	n.a.	420
1c	10	20	425	30	420	230
1d	5	10	360	30	1040	780
2a	15	25	80	40	210	30
2b	5	15	25	5	n.a.	420
2c	10	35	265	50	600	390
3	10	10	90	45	430	50
4	30	55	295	115	1780	1390
5	5	10	195	10	1350	490
6	30	350	110	95	1420	490
7	15	170	660	430	1130	780
8	25	100	190	95	1420	490
9	20	40	265	205	1250	1140
10/11a	10	20	80	20	280	80
10/11b	5	10	25	<5	n.a.	70
10/11c	5	10	150	5	140	80
12	30	95	655	395	860	510
13	20	35	745	90	460	260
14	110	155	270	210	1960	1010
15	25	40	120	100	1330	700
16	55	55	610	140	930	530
17	95	245	350	250	1170	1000
18	45	70	400	110	720	680
19	15	20	310	25	480	380
20	20	30	175	30	1090	800
21	10	15	170	10	2440	560
22	15	25	20	<5	n.a.	n.a.
23	25	60	340	110	660	570
24	10	70	245	95	770	350
25	35	90	305	180	1600	1400
26	45	55	245	130	250	80
27	30	55	140	70	1100	810
28	5	10	95	40	1190	690
29	30	330	385	175	1390	1190
30	40	275	205	175	1470	1200

Peak HRR varied over more than an order of magnitude from 25 kW/m^2 for a thin fabric interliner (Sample 10b) to 745 kW/m^2 for a wall fabric (Sample 13). In general, Table 14.3 shows lower peak HRR rates for the seat and mattress foams, ranging from 65 to 80 kW/m^2 and higher values for wall surface materials, ranging from 120 to 745 kW/m^2. Other fabric and thin sheet materials display intermediate values between these two extremes. This performance is consistent with the current FRA regulations which provide strictest flame spread index requirements for seat foam (for example, $I_S = 25$ in ASTM E 162), intermediate requirements for most other materials ($I_S = 35$ in ASTM E 162), and least stringent requirements for window materials ($I_S = 100$ in ASTM E 162). The HRR for the window mask (Sample 19), is one of the highest of the materials tested and certainly does not fit into the 'intermediate' group as would be expected from the FRA criteria. This result is consistent with ASTM E 162 test data above.

Cone calorimeter smoke data is usually presented in terms of a 'specific extinction area,' which is a measure of the smoke production of a material. Like the specific optical density measurement in ASTM E 662, the specific extinction area is a measure of the attenuation of light by soot particles. The cone calorimeter smoke data show trends similar to the HRR data. The lowest values were noted for the seat and mattress foams (Samples 1, 2, 3, and 10). Highest values were noted for several thin materials: seat support diaphragm (Sample 4), seat track cover (Sample 9), rubber flooring (Sample 25), and gasketing (Samples 29 and 30). The thicker polycarbonate space divider and window glazing (Samples 15 and 18) also had high smoke values. Several materials showed elevated HRR and smoke values over an extended period of time. For example, the following materials showed HRR values greater than 100 kW/m^2 for more than 500 s: space divider (Sample 15), wall material (Sample 17), window glazing (Sample 18), window gasketing (Sample 29), and door gasketing (Sample 30). Smoke values generally paralleled the HRR results. Although the peak HRR of these materials fall into an intermediate range, the extended duration of the HRR curve makes these materials important for study in future fire hazard analysis efforts.

14.2.4   Observations from small-scale testing

For many of the materials, the cone calorimeter results were strong indicators of results from the FRA regulation tests. Equally, there were cone calorimeter results which were not indicative of the FRA regulation test results. For example, several materials which had low I_S values in the ASTM E 162 test had higher HRR values in the cone calorimeter. One material had a low HRR value and a high I_S. The following rationale was used in comparing the cone calorimeter test data with data from the FRA regulation tests:

- The comparison between ASTM E 162/D 3675 and the cone calorimeter shows that peak HRR in the cone calorimeter is predictive of an upper bound on I_S. With one exception for a unique seating foam, materials which have a low HRR values have a correspondingly low I_S.
- The Bunsen-burner test specified in FAR 25.853 is a self-extinguishment test which assesses a material's resistance to small ignition sources. For the cone calorimeter, a comparable value is based upon the ratio of the ignition time to

the peak HRR. A simple linear regression resulted in a high correlation coefficient of $r^2 = 0.98$. The char length comparison is based on a limited amount of data.

- Only two flooring materials were available for cone calorimeter testing in the current study. Thus, there are too few data for a meaningful comparison between the test methods for passenger train applications.

- For equivalence to ASTM E 662, an optical density measure was derived as an integrated value based upon the smoke extinction coefficient from the cone calorimeter. Comparing values from the cone calorimeter and ASTM E 662 for this calculated smoke density showed an appropriate comparison for the four-minute E 662 values in 17 of the 22 cases where data were available. A simple linear regression resulted in a good correlation coefficient of $r^2 = 0.87$. No appropriate comparison was apparent for the 1.5 minute values. Since the main purpose of using the D_S values derived from cone calorimeter data is to demonstrate the comparability of cone calorimeter data to ASTM E 662 data, the four-minute values provide a sufficient comparison.

14.3   Tests on assemblies

To aid in the development of realistic-scenario fires for use in the fire hazard analysis conducted for this study, large-scale furniture calorimeter tests were conducted on large-scale assemblies of rail car materials currently used in intercity passenger train service. Like the small-scale cone calorimeter, the primary measurement from this large-scale test is the HRR of the burning assembly sample when exposed to an ignition source.

14.3.1   Results

Table 14.4 summarizes the test results. Total peak HRR ranged from 30 kW to 920 kW, including any contribution from the ignition sources used. After subtracting the HRR of the ignition source, these values ranged from 15 kW to 800 kW. Trash bags taken from overnight, intercity train services were also characterized as a representative severe ignition source that could be present. The actual trash bag peak HRR from an Amtrak overnight train ranged from 55 to 285 kW. Heavier and more densely packed trash bags had lower HRR values than lighter bags since the dense packing prevents the heavier bags from burning completely. A newspaper-filled trash bag representative of the lighter trash-filled bags used as the ignition source for many of the assembly tests had a peak HRR of 200 ± 35 kW based on one standard deviation. In the assembly tests, the HRR of seat, bed, wall, and ceiling carpet, window drape/privacy curtain, and window assemblies were characterized. Total peak HRR ranged from 30 kW for a seat assembly exposed to a 17 kW gas burner ignition source to 920 kW for a lower and upper bed assembly exposed to a newspaper-filled trash bag and gas burner ignition source.

Table 14.4 Peak HRR measured during furniture calorimeter assembly tests

Material/test assembly	Ignition source[1,2] (kW)	Peak HRR[2] (kW)
Trash bags ranging in mass from 1.8 kg to 9.5 kg	25	30–260
Coach seat assemblies (foam cushion, wool/nylon upholstery, PVC/acrylic seat shroud)	17–200	15–290
Lower bed with bedding and pillow	200	550–640
Upper and lower beds with bedding and pillow	200	720
Wall carpet on a wall or a wall and ceiling, wool	50	290–800
Window drape/door privacy curtain assemblies, wool/nylon	25	40–170
Wall/window assemblies, FRP and polycarbonate	50–200	80–250

1. Does not include contribution of ignition source.
2. Uncertainty in heat release rate measurement is estimated to be 2% to 17% of the peak HRR.

All the assemblies tested were extremely resistant to ignition and required an initial fire source ranging from 17 to 200 kW to ignite. Materials and products that comply with the current FRA-cited fire tests and performance criteria are difficult to ignite, requiring ignition strengths of double to ten times those used for similar materials and products found outside of the passenger rail car operating environment. Some of the materials do not contribute to the fire even with these ignition sources.

These assemblies are typical of intercity passenger rail cars in the U.S. While commuter rail cars or rail transit vehicles may have different levels of furnishings, results for some of the assemblies (such as the seat assemblies) may be appropriate for these applications as well. Since the focus of this study is primarily passenger rail car interior design, all of these results apply to interior ignition scenarios. Exterior ignition sources, which may be important in some environments, particularly in the design of tunnel ventilation systems, were not considered. Such scenarios have been considered elsewhere.[38]

14.3.2   Observations from assembly testing

The results of the assembly tests showed:

• The net peak HRR from actual trash bags from an Amtrak overnight train ranged from 30 to 260 kW. Heavier and more densely packed trash bags had lower peak HRR values than lighter bags.

• All the assemblies tested were extremely resistant to ignition. They required an ignition source ranging from 17 to 200 kW to ignite. Some of the materials did not contribute to the HRR of the fire even with the largest ignition source.

• For the assembly tests, the HRR of seat, bed, wall and ceiling carpet, window drape and door privacy curtains, and wall/window assemblies were quantified. Total peak HRR ranged from 30 kW for a seat assembly exposed to a 17 kW gas burner ignition source to 920 kW for a lower and upper bed assembly exposed to a newspaper-filled trash bag ignition source. Although difficult to ignite, the wall carpeting and window glazing produce high HRR values once ignited.

- For the sleeping compartment economy bedroom tests, the small enclosed geometry allowed a much larger HRR to develop than for the seat assembly tests, although the materials are similar.

As in the 1983 tests NIST conducted on large-scale mock-ups of Amtrak coach cars, wall carpet and wall/window assemblies, although slow to ignite, are seen as the most important materials for fire growth once ignited. In addition, the effect of geometry also noted in the Amtrak report is confirmed by the sleeping car compartment bed tests.

14.4   Full-scale tests

Amtrak donated an Amfleet I passenger rail coach car to FRA for the research program. Materials present in the test car reflect a cross-section of typical interior component materials used in current Amtrak passenger trains. The seat assemblies, wall and ceiling lining materials, and floor coverings represent the greatest mass of interior fire load found in the test car and in most passenger rail cars.

14.4.1   Test car

The interior length of the car is 22.1 m. The interior width of the car, at the floor level, is 2.7 m. The center aisle ceiling height is 2.2 m in the seating area and 2 m at each end of the car for the first 2.7 m from each end of the car. Ten rows of seat assemblies were installed on both sides of the center aisle.

The exterior of the car is constructed of corrugated stainless steel. One end of the car had significant structural damage, including a roof penetration. However, there was very little damage to the interior of car on the opposite end. The car was equipped with a vestibule area at each end of the car; each end also had two side doors (one on each side) and interior end doors. The total car length is 26 m.

The interior of the car was divided into two main sections by a bulkhead with a 2 m high by 0.75 m wide doorway. This doorway had the same dimensions as the interior doorways on either end of the car. The fire test area was on one side of the bulkhead while the other side of the rail car, the damaged end, was used as a smoke collection area.

In addition to the center bulkhead, steel frame walls covered with gypsum board and calcium silicate were used to create a fire-resistant bulkhead in the area where the handicapped rest room module had been removed from the end of the car. A smoke curtain consisting of the steel and gypsum board and calcium silicate construction from the ceiling to half the height of the interior was added to the smoke collection area of the car. These bulkheads were added to protect the fire end of the car during repeated fire tests and to allow for the measurement of HRR by oxygen consumption in the smoke collection area. Calcium silicate board was also installed on the ceiling above the gas burner to protect the end of the car from repeated fire tests. The area between the end bulkhead and the seat assemblies provided a location for the gas burner used in some of the tests. Figure 14.1 shows the interior configuration of the test vehicle.

Starting with the upper portion of the car, the center ceiling panels of the car consist of a laminated sandwich of melamine and aluminum plywood (plymetal). The curved portions of the ceilings and walls are sheathed with wool carpet (Sample 12), glued to perforated metal. The carpet is covered by rigid polyvinyl chloride acrylic (PVC) panels (Sample 6). The window masks consist of fiberglass-reinforced plastic (FRP) polycarbonate (Sample 18). A layer of vinyl fabric covers a thin layer of foam on the underside of the luggage rack. PVC/acrylic rigid panels are attached over the vinyl. The top of the

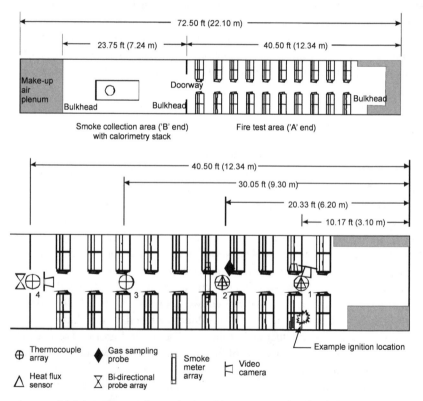

14.1 Interior configuration and instrumentation for full-scale tests in a passenger train car.

luggage rack has metal sheeting. Wool carpet is used to cover the lower portions of the wall and the full height of the permanent end of the car interior bulkhead (Sample 12) while nylon carpet over foam padding covers the floors (Sample 24). The seat cushions are composed of neoprene/polyurethane foam, covered with a cotton fabric interliner, with a fabric/vinyl upholstery (Samples 1a through 1c). The seat support diaphragm (flat 'spring') is made of chloroprene elastomer (Sample 4). The seats have steel frames with PVC acrylic shrouds (Sample 6). The armrest pad is chloroprene elastomer over a steel support. The windows in the car are composed of polycarbonate (Sample 17) and they are held in place by a chloroprene elastomer gasket (Sample 29). Wool/nylon window drapes used to line the windows in some premium and longer-distance services were also included in the test program (Sample 20). Table 14.5 includes a summary of data from tests conducted in the car. Additional details of the tests are available.[39]

14.4.2   Comparison of small- and full-scale test data

The NIST research included testing of the same set of passenger rail car materials in small-scale, full-scale assembly tests, and full-scale tests in an actual rail coach car. In addition, earlier research included testing of similar materials and train car geometries in small and full scale. This section compares the various test results for passenger rail car materials in small and full scale. The comparison discussed in this section is intended to aid in the development of appropriate criteria for material screening and to place the current test results in context with earlier research.

To realize the maximum benefits of performance-based designs, a low-cost method to screen materials is an important complement to an overall system fire safety analysis. As small-scale tests are significantly cheaper than full-scale tests, use of a small-scale screening method would minimize costs both to the manufacturer and the end-user.

Table 14.6 shows a comparison of passenger rail car materials tested in the cone calorimeter with assembly test results. With the exception of the seat cushion assembly, the relative ranking of materials in the cone calorimeter is similar to the rank order in the furniture calorimeter. For the seat cushion assembly, the cone calorimeter result ranks higher than the furniture calorimeter result. This is likely a result of testing with several different ignition sources in the furniture calorimeter, ranging from the small TB 133 burner to a 400 kW gas burner. In contrast, the cone calorimeter results with a 50 kW/m^2 incident flux represents only more severe ignition scenarios. Considering only the most severe ignition source in the furniture calorimeter data would bring this result in line. It is important to note that this comparison was limited to only five different materials. Additional material data would help refine the comparison.

Table 14.5 Summary of test results for large-scale tests conducted in a single-level passenger rail coach car

| | Gas temperature (°C) | | | | Heat flux (kW/m²) | | Gas concentration (percent volume fraction) | | | | | |
| | Upper layer | | Lower layer | | | | O₂ | | CO₂ | | CO | |
	Peak	At time	Peak	At time	Peak	At time	Min	At time	Peak	At time	Peak	At time
Slow t² gas burner	398	600	106	600	19	625	16	630	3.0	620	0.02	–[1]
Medium t² gas burner	331	320	81	315	16	317	17	330	2.4	325	0.01	–
Fast t² gas burner	376	155	79	155	15	155	16	170	2.8	190	0.01	–
Ultra-fast t² gas burner	372	80	73	80	14	80	17	95	2.3	95	0.03	–
Window drape with 25 kW burner	53	510	32	600	0.31	540	20	600	0.31	–	0.01	–
Corner test (wall carpet and FRP panel) with trash bag	183	300	61	320	9	270	17	290	3.7	300	0.2	280
TB 133 ignition on seat	47	600	32	365	0.24	560	21	–	0.23	–	0	–
25 kW burner ignition on seat	53	565	31	255	0.46	505	21	–	0.32	–	0	–
Trash bag on seat	363	270	131	260	27	90	12	285	6.6	290	1.4	285

1. Measured condition is at an ambient or near-ambient value and roughly constant throughout the test.

Table 14.6 Cone calorimeter and furniture calorimeter material rankings

↑ Increasing fire hazard →	Cone calorimeter ranking	Furniture calorimeter ranking
	Wall carpet	Wall carpet
	Window components	Window components
	Privacy door curtains	Seat assembly
	Drapes	Privacy door curtains
	Seat assembly	Drapes

However, it is important to understand the limitations of the comparisons. The comparisons support a fire protection engineer's intuition: low HRR materials are inherently less hazardous than high HRR materials which ignite easily and facilitate flame spread. However, physical phenomena that are not evaluated using the simple peak HRR comparison, such as geometry, burnout time, or smoke and toxic gas production may have a significant impact upon actual burning behavior of passenger rail car materials. Additional research is appropriate to fully understand the comparison between small- and full-scale testing. For example, Janssens has developed a simple flame spread model for application of cone calorimeter data in fire hazard analysis of commuter rail vehicles.[40] Thus, small-scale testing is most appropriate as a screening tool for alternative material selection.

14.4.3   Comparison with earlier research

The 1984 NIST study includes test results on several mock-up configurations of Amtrak passenger cars.[9] That study includes several of the same materials used in this current study. The Eureka tests included temperature measurements inside rail vehicles in several tests.[32] Fire growth in WMATA subway vehicles have been previously studied by NBS (now NIST).[41] Table 14.7 shows test data

Table 14.7 Comparison of selected full-scale test results from several test series

Test series	Peak upper gas temperature (°C)
Current study	50–360
1984 NBS study[12]	114–825
Eureka study[21]	270–900
NBS subway study[24]	55–290

from these three studies along with comparable data from the current study derived from Table 14.5.

In Table 14.7, results from each test series show a considerable range of values due to different materials and configurations included in the tests. The three earlier studies all included older materials such as untreated urethane foam seating that would not meet current FRA requirements. In these three studies, the higher peak temperatures are noted for configurations including these older materials. The Eureka study shows particularly high temperatures since the tests were full burnout tests intended to study the fire environment inside a tunnel, rather than in the car. The WMATA data show lower temperatures for one of the tests compared to the fully-furnished intercity rail cars. This may be due to either the more limited furnishing of the subway car mockup tests or the small 28 g ignition source for the test. The expected high performance of FRA-compliant materials is evident in the lower peak temperatures from the current study compared to other fully furnished rail cars in the 1984 FRA/Amtrak and Eureka tests.

14.4.4   Key observations from full-scale tests

The gas burner tests served two primary purposes: verification of fire modeling results obtained from the hazard analysis and estimation of the uncertainty of the measurements. The replicate measurements from the gas burner tests proved to be very repeatable. As an example, the average uncertainty of the upper layer temperature measurements for the slow, medium, fast and ultra-fast t-squared fires ranged from 3.1 percent to 10.8 percent.

The flame spread and growth tests clearly supported the conclusion from the full-scale assembly tests that a significant ignition source was necessary to sustain significant flame spread. The three tests which used small ignition sources (25 kW burner on seat, TB 133 burner on seat, and 25 kW burner on drapes), each yielded temperature and species levels near to slightly above ambient after six minutes. The tests that used the trash bag as an ignition source (trash bag in corner and trash bag on the seat) exhibited sustained flame spread and extension, producing temperatures and species concentrations sufficient to render the main compartment untenable in about 100 s.

For the five flame spread and growth tests, the range of ignition source strengths indicated that an ignition source size between 25 kW and approximately 200 kW is necessary to promote significant fire spread, which is consistent with the conclusion that the ignition source strength of passenger rail car materials is two to ten times greater than those of typical office furnishings. Given an ignition source of the magnitude of a large trash bag, however, significant flame spread is observed. For the largest ignition source tests, conditions within the rail car can become untenable.

14.5   Future trends

It is clear from this work and that of others that the future of material flammability regulation will be through fire hazard analysis where the safety of the whole system is evaluated in context.[42,43] This parallels the global transition of standards and building regulations from prescriptive to performance-based, also utilizing fire hazard analysis to assess whether the fire safety objectives are met. For example, both ASTM International and the Society of Fire Protection Engineering have resources for the application of fire hazard analysis techniques to passenger train fire safety.[44,45] This transition has a significant impact on materials testing procedures since fire hazard assessment requires material performance data on reaction to fire and cannot use data on relative rankings under a single set of conditions. Thus, traditional test methods are making way for modern, property measurement methods like the cone calorimeter.

14.6   References

1. *The Passing of the Wooden Passenger Car from This Railroad*, Pennsylvania Railroad Information (1928).
2. Troy, J. J., 'Fire Protection Provisions for Rapid Transit Systems', *Fire J. 70*, 13–17 (1976).
3. Wilkins, J. J. and Cavanaugh, R. R., *Plastics and Rail Transportation, Polymeric Materials and Their Use in Transportation*, April 27–29, 1977, The Polytechnic Institute of New York (1977).
4. Cavanaugh, R. R., 'Selection of Materials for Ground Transportation Vehicles', in *Proceedings of the Sixth International Conference on Fire Safety*, January 12–16, 1981, Menlo Park, CA (1981).
5. *Passenger Fire Safety in Transportation Vehicles*, Arthur D. Little Report No. C-78203 to Transportation Systems Center, U.S. Department of Transportation (1975).
6. 'Fire Safety Aspects of Polymeric Materials', Volume 8, *Land Transportation Vehicles*, National Materials Advisory Board Pub. NMAB 318-8, National Academy of Sciences (1979).
7. Rakaczky, J. A., *Fire and Flammability Characteristics of Materials Used in Rail Passenger Cars*. A Literature Survey, U.S. Army Ballistic Research Laboratory, Report No. ARBRL-MR-03009 (1980).
8. Hathaway, W. T. and Litant, I., *Assessment of Current U. S. Department of Transportation Fire Safety Efforts*, U.S. Department of Transportation Report No. UMTA-MA-06-0051-79-4 (1979).
9. Peacock, R. D. and Braun, E., *Fire Tests of Amtrak Passenger Rail Vehicle Interiors*, Natl. Bur. Stand. (U.S.), Technical Note 1193 (1984).
10. Schirmer Engineering Corporation, *Life Safety Study and Computer Modeling Analysis for New York City Railroad Tunnels and Pennsylvania Station*, contract to The National Railroad Passenger Corporation, Sec No. 15-890-29-04-00.
11. Peacock, R. D., Reneke, P. A., Jones, W. W., Bukowski, R., W., Babrauskas, V., 'Concepts for Fire Protection of Passenger Rail Transportation Vehicles: Past, Present, and Future'. *Fire and Materials*, Vol. 19, No. 2, 71–87, March/April 1995.
12. Hathaway, W. T., *Commonalities in Transportation Fire Safety: Regulations,*

Research and Development, and Data Bases, U.S. Department of Transportation, Transportation Systems Center, Report DOT-TSC-OST-80-5 (1980).

13. Peacock, R. D., Bukowski, R. W., Jones, W. W., Reneke, P. A., Babrauskas, V., and Brown, J. E., *Fire Safety of Passenger Trains: A Review of Current Approaches and of New Concepts*, Natl. Inst. Stand. Technol., Tech. Note 1406 (1994).

14. National Fire Protection Association, NFPA 130, *Standard for Fixed Guideway Transit Systems*, 2003 edition, NFPA, Quincy, MA (2003).

15. ASTM International, 'Standard Test Method for Surface Flammability of Materials Using a Radiant Heat Energy Source', ASTM E162-02a, *Annual Book of ASTM Standards*, Vol. 04.07 (2002).

16. ASTM International, 'Standard Test Method for Surface Flammability of Cellular Materials Using a Radiant Heat Energy Source', ASTM D3675-01, *Annual Book of ASTM Standards*, Vol. 09.02 (2001).

17. ASTM International, 'Standard Test Method for Specific Optical Density of Smoke Generated by Solid Materials', ASTM E662-03, *Annual Book of ASTM Standards*, Vol. 04.07 (2003).

18. Babrauskas, V. and Peacock, R. D., 'Heat Release Rate: The Single Most Important Variable in Fire Hazard', *Fire Safety J.*, Vol. 18, No. 3, pp 255–272 (1992).

19. National Transportation Safety Board (NTSB). *Railroad Accident Report: Derailment and Collision of Maryland Rail Commuter MARC Train 286 with National Railroad Passenger Corporation Amtrak Train 29, near Silver Spring, Maryland, on February 16, 1996*. Adopted: July 3, 1997. Washington, DC. Report No. NTSB/RAR-97/02.

20. Young, R. 'Behaviour of Rail Vehicles Components in Fires.' *Proceedings of Materials in Mass Transport*, May 26, 1988. Rapra Technology Limited, Shawbury, UK. 1988.

21. Malhotra, H.L. 'Fire Protection in Channel Tunnel Car Carrying Shuttles.' *Proceedings of Interflam '93*, 6th International Fire Conference. Interscience Communications, Ltd. London, 1993.

22. Young, Roy. 'Development of Heat Release Techniques Within European Railways.' *Proceedings of the Fire Safety by Design Conference*. University of Sunderland, UK. Volume 3, 1995.

23. Göransson, U. and Lundqvist, A. *Fires in Buses and Trains, Fire Test Methods.* Swedish National Testing and Research Institute, Sweden. SP Report No. 1990: 45, 1990.

24. ERRI. Working Party B 106.2. *Coaches; Reasons for Undertaking Supplementary Studies on Improvement of Protection of Coaches Against Fire*. ERRI B 106 RP 22. Utrecht, July 1992.

25. *Fires in Railway Vehicles with a View to Improving Passenger Safety*. ERRI B 106 RP 25. Utrecht, November 1992.

26. ERRI. Young, R. and Metral, S. Edited by A. Kaminski, ERRI Technical Advisor. Improvement of the Protection Against Fire of Passenger Rolling Stock; Progress Report on the Tests Carried Out Using A cone calorimeter and the Calculations with HAZARD 1.1. Software Package. ERRI C 204/RP 1. Utrecht, June 1994.

27. ERRI. Improvement of the Fire Protection of Passenger Rolling Stock; Progress Report on the Tests Carried Out Using A cone calorimeter and the furniture calorimeter and on the Calculation with the HAZARD 1.1 Software Package. ERRI C 204/RP 2. Utrecht, 1995.

28. ERRI. Metral, S., J. Wolinska, and G. Barbu. Improvement of the Protection Against Fire of Passenger Rolling Stock; Results of Laboratory Tests using a cone calorimeter to Provide Data for Computation. ERRI C204.1/DT301. Utrecht, 1994.

29. ERRI. Young, R. Scientifics/British Railway Board. Edited by A. Kaminski, ERRI Technical Advisor. *Improvement of the Fire Protection of Passenger Rolling Stock; Results of Additional Cone Calorimeter Tests on Seat Materials and Furniture Calorimeter Tests on Mock-Up Seats.* ERRI C204.1/DT319. Utrecht, June 1995.

30. ERRI. Metral, S., Wolinska, J. and Barbu, G. *Improvement of Fire Protection of Passenger Rolling Stock; Computer Simulation of the Fire Process Within Railway Passenger Coaches Using the HAZARD 1.2 Software.* ERRI C204.1/DT358. Utrecht, November 1997.

31. Peacock, R. D., Jones, W. W., Bukowski, R. W., and Forney, C. L., 'Technical Reference Guide for the HAZARD I Fire Hazard Assessment Method,' *NIST Handbook 146*, Vol. II, 1991.

32. *Fires in Transport Tunnels, Report on Full-Scale Tests*, EUREKA Project 499: FIRETUN. Editor Studiengesellschaft Stahlanwendung e.V., D-40213 Düsseldorf, Germany, November 1995.

33. Le Tallec, Y., Sainrat, A. and Le Sant, V. 'The FIRESTARR Project – Fire Protection of Railway Vehicles'. *Proceedings of Fire and Materials 2001.* 7th International Conference and Exhibition. Interscience Communications Limited. January 22–24, 2001, San Francisco, CA, pp. 53–66, 2001.

34. Final Report, Report C/SNCF/01001, SNCF, Direction Matérial et de la Traction, Paris. April 2001.

35. ASTM International. 'Standard Test Method for Heat and Visible Smoke Release Rates for Materials and Products Using an Oxygen Consumption Calorimeter'. E1354-03. *Annual Book of ASTM Standards.* Vol. 04.07, 2003.

36. Peacock, R. D., Bukowski, R. W., Markos, S. H., 'Evaluation of Passenger Train Car Materials in the Cone Calorimeter'. *Fire and Materials*, Vol. 23, 53–62, 1999.

37. Peacock, R. D. and E. Braun. *Fire Safety of Passenger Trains Phase I: Material Evaluation (Cone Calorimeter).* Natl. Inst. Stand. Technol. NISTIR 6132, March 1999.

38. Kennedy, W. D., Richard, R. E. and Guinan, J. W. 'A Short History of Train Fire Heat Release Calculations,' 1998 Annual American Society of Heating, Refrigeration, and Air-Conditioning Engineers Conference, Toronto, Canada, June 20–24, 1998.

39. Peacock, R. D., Averill, J. D., Madrzykowski, D., Stroup, D. W., Reneke, P. A., and Bukowski, R. W., *Fire Safety of Passenger Trains; Phase III: Evaluation of Fire Hazard Analysis Using Full-Scale Passenger Rail Car Tests*, Natl. Inst. Stand. Technol., NISTIR 6563, December 2003.

40. Janssens, M. and J. Huczek, 'Fire Hazard Assessment of Commuter Rail Equipment', *Fire Risk & Hazard Assessment Research Application Symposium*, Baltimore, MD, July 24–26, 2002, pp. 404–433.

41. Braun, E. 'Fire Hazard Evaluation of the Interior of WMATA Metrorail Cars'. *Natl. Bur. Stand.* (U.S.). NBSIR 75-971, December 1975.

42. Peacock, R. D., Bukowski, R. W., Reneke, P. A., Averill, J. D., and Markos, S. H. 'Development of a Fire Hazard Assessment Method to Evaluate the Fire Safety of Passenger Trains'. *Proceedings of Fire and Materials 2001.* 7th International Conference and Exhibition. Interscience Communications Limited. January 22–24,

2001, San Francisco, CA, pp. 67–78, 2001.

43. Peacock, R. D., Reneke, P. A., Averill, J. D., Bukowski, R. W., and Klote, J. H. 'Fire Safety of Passenger Trains. Phase 2. Application of Fire Hazard Analysis Techniques'. Natl. Inst. Stand. Technol., NISTIR 6525; p. 185. December 2002.

44. ASTM International, 'Guide for Fire Hazard Assessment of Rail Transportation Vehicles', ASTM E2061-03, *Annual Book of ASTM Standards*, Vol. 04.07 (2003).

45. Bukowski, R. W. 'Fire Hazard Assessment for Transportation Vehicles', Section 5, Chapter 15, *SFPE Handbook of Fire Protection Engineering*, 3rd edition, Phillip J. DiNenno, P.E., editor-in-chief, Society of Fire Protection Engineers, Boston, MA. March 2002.

15

Ships and submarines

B Y   L A T T I M E R , Hughes Associates, Inc., USA

15.1   Introduction

Ships and submarines are generally constructed using a limited amount of combustible materials. This approach to fire safety onboard marine vessels partially originated from a fire in 1934 on the U.S. passenger ship *Morro Castle*, which resulted in 124 lost lives.[1] The ship was constructed of wood, which allowed the fire to spread rapidly throughout the vessel. Following this tragedy, a U.S. congressional subcommittee was formed to develop recommendations on improving fire protection features on ships. Based on results from large-scale testing conducted onboard the *S.S. Nantasket*, the subcommittee recommended using noncombustible construction materials.[2] This led to more restrictive use of combustible materials throughout ships.

Current codes and standards for ships and submarines still use noncombustible construction as the cornerstone for most ship and submarine designs. However, the use of combustible materials is allowed in a variety of applications for different types of vessels. The focus of current codes and standards is to protect passengers from loss of life. Similar to buildings, marine vessels use a variety of fire protection features to achieve passenger safety. This can include fire detection, fire suppression, fire resistance boundaries, and limited combustible materials with regulated flammability, smoke production and toxic gas generation. This chapter focuses on the flammability requirements of combustible materials permitted for use on ships and submarines.

Flammability tests and requirements discussed in this chapter are from the most commonly used marine vessel codes. It should be noted that most materials used onboard ships and submarines may also be regulated for smoke and/or toxic gases. Requirements for smoke and toxic gas production is not described in detail in this chapter, but some smoke production requirements are provided when they are being measured in the same test used to evaluate flammability. Following a review of the flammability requirements for different construction features on ships and submarines, some expected future trends are provided on using fire models to assist in designing and evaluating fire-safe materials for future vessels.

15.2 Ships

Ships have flammability requirements for division construction, construction in egress routes, interior finish on boundaries, insulation, furnishings, wires and cables, and pipes and ventilation ducts. The flammability requirements for these different construction features will depend on the adopted code as well as the size and function of the ship. A brief overview is provided on the different codes and standards that are used to regulate material flammability on ships. Material flammability requirements in these codes are then presented for the different ship construction features listed above.

15.2.1   Codes and standards

Codes and standards have been developed for ships expected to be used internationally, locally, or for war purposes. The most frequently used codes and standards have been developed by the International Maritime Organization (IMO) and the National Fire Protection Association (NFPA). Codes for local and war purposes may vary from country to country. Often governments will adopt IMO and NFPA codes for these purposes. Therefore, this section will focus on the flammability requirements in the IMO and NFPA codes.

IMO codes include the Safety of Life at Sea (SOLAS)[3] and High Speed Craft (HSC) Code.[4] SOLAS includes regulations for passenger ships, cargo ships, and tankers. The HSC Code includes requirements for international passenger craft that do not travel more than four hours to reach a place of refuge or for international cargo craft of more than 500 gross tons that do not travel more than eight hours to reach a place of refuge. Both SOLAS and HSC Code do not include craft of war or troop craft, craft not mechanically propelled, wooden primitive craft, pleasure craft not engaged in trade, and fishing craft. The Fire Test Procedures (FTP) Code[5] contains the flammability tests and performance criteria for combustible materials allowed by SOLAS and HSC Code.

NFPA codes include NFPA 301 'Safety to Life from Fire on Merchant Vessels'[6] and NFPA 302 'Fire Protection Standard for Pleasure and Commercial Motor Craft'.[7] NFPA 301 contains requirements for tanker and cargo vessels, towing vessels, and various passengers carrying vessels. It does not cover pleasure craft or commercial craft of less than 300 gross tons, which are covered by NFPA 302.

All codes except NFPA 302 promote usage of noncombustible materials for all ship construction. In these codes, a noncombustible material is a material that will not ignite or contribute heat to a fire. IMO further defines a noncombustible material as a material that neither burns nor gives off flammable vapors when heated to 750 °C. The IMO FTP code requires noncombustible materials pass the ISO 1182:1990 test[8] using the following criteria[5]:

- the average furnace and surface thermocouple rise as calculated in 8.1.2 of ISO 1182 is less than 30 °C

- the mean duration of sustained flaming does not exceed 10 s as calculated in 8.2.2 of ISO 1182
- the average mass loss does not exceed 50% as calculated in 8.3 of ISO 1182.

NFPA 301 considers a material to be noncombustible if it passes ISO 1182 with the above IMO criteria or 46 CFR 164.009.[9]

Where combustible materials are allowed, the combustible material or assembly must be demonstrated to meet the flammability requirements. The combustible material flammability test methods and criteria are provided in the following sections.

15.2.2   Division construction

Divisions are bulkheads and decks that divide the ship into compartments. Divisions are typically constructed of steel or an equivalent noncombustible material. The IMO and NFPA both classify divisions as A, B, and C Class. These divisions are used to classify the range of fire resistance performance of a bulkhead or overhead, but also provide restrictions on the use of combustible materials. The A Class divisions must be constructed of steel or an equivalent material. Equivalent material would be a noncombustible material that by itself or due to insulation has the structural integrity of steel after a specified fire exposure (e.g., aluminum with fire insulation). The B and C Class divisions are constructed of noncombustible materials but are allowed to have combustible veneers on their surface. Flammability requirements for these combustible veneers will be discussed in the interior finish section.

Both IMO and NFPA allow combustible construction on specific types of ships. The IMO HSC Code has fire resisting class divisions which may be constructed of fire restricting materials. In IMO Resolution MSC.40(64),[10] fire restricting materials are required to pass the ISO 9705 room corner fire test[11] shown in Fig. 15.1. In this test, the back wall, two side walls and ceiling are

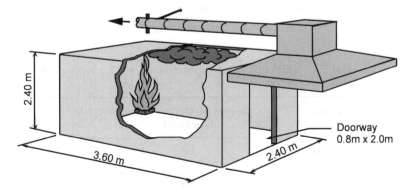

15.1 ISO 9705 room corner fire test (ref. 11).

Table 15.1 IMO HSC fire restricting material flammability ISO 9705 test criteria (ref. 10)

Variable		Criteria
Heat release rate*	Test average Peak 30 s average	$\leq 100\,kW$ $\leq 500\,kW$
Smoke production	Test average Peak 60 s average	$\leq 1.4\,m^2/s$ $\leq 8.3\,m^2/s$
Flame spread on walls		Must not reach closer than 0.5 m from the floor when further than 1.2 m from the corner with the initiating fire
Flaming droplets or debris		None may reach the floor further than 1.2 m from the corner with the initiating fire

* Heat release rate due to the lining only, burner contribution excluded.

lined with the material. The initiating fire in the back corner is a 100 kW fire for ten minutes followed by a 300 kW fire for ten minutes, with propane as the fuel. The test criteria provided in Table 15.1 require a fire restricting material to have limited heat release rate, limited flame spread, and no flaming droplets or debris 1.2 m from the initiating fire.[10]

NFPA 301 allows passenger vessel Group IV and V to use combustible construction. Flammability requirements for these vessels are provided in Table 15.2. A Group IV ship is allowed to use combustible construction, fiber reinforced plastic (FRP), when the ship has no more than 150 passengers a day and no more than 12 overnight passengers. In this case, the FRP must have a flame spread index (FSI) of less than 100 when tested in accordance with NFPA 255 'Standard Method of Test of Surface Burning Characteristics of Building Materials'[12] or ASTM E84 'Standard Test Method for Surface Burning Characteristics of Building Materials'.[13] A Group V ship, which has no more than 450 passengers per day and no overnight passengers, can follow the HSC Code, which as described above allows fire restricting materials for division construction.

Table 15.2 NFPA 301 flammability requirements for combustible division construction

Ship group	Passengers per day		Test method	Criteria
	Total	Overnight		
Group IV	150	12	NFPA 255 or ASTM E84	$FSI < 100$
Group V	450	0	ISO 9705	See Table 15.1

15.2.3   Construction in egress routes

There are flammability requirements for construction materials, interior finish and furnishings in egress routes. The flammability requirements for interior finish and furnishings are the same as those described in sections 15.2.5 and 15.2.6. Materials used to construct egress features, such as stairways, platforms, ladders, and landings, are generally required to be of noncombustible construction. SOLAS requires stairways and lift trunks in accommodation spaces, service spaces, and control stations to be constructed of steel or equivalent material. A specified number of steel ladders are also required. In the HSC Code, stairways can be constructed of a noncombustible or fire restricting material. Flammability requirements for fire restricting materials are provided in Table 15.1. In NFPA 301, stairs, platforms and landings used to connect more than three decks must be noncombustible.

15.2.4   Interior finish

Interior finish are materials that cover the division construction materials and include finish materials on bulkheads, overheads, and decks. Bulkhead and overhead materials are tested with the same test method and have the same requirements. Decks, however, have slightly different requirements and are sometimes tested in a horizontal orientation.

Bulkheads and overheads

SOLAS and the HSC Code require that interior finish materials be 'low flame spread' and meet the surface flammability requirements when tested in accordance with IMO Resolution A.653(16).[14] This is the lateral flame spread test shown in Fig. 15.2, and it is equivalent to the test method described in ASTM E1317 'Standard Test Method for Flammability of Marine Surface Finishes'.[15] In this test, the vertical test sample is exposed to a heat flux that varies along the length of the sample as shown in Fig. 15.2, with a peak heat flux of $50\,kW/m^2$. Measurements made during the test include the time to ignition, flame front location on the sample with time, and the gas temperature in the stack. The gas stack temperature is a measure of the heat release rate of the sample. Flammability criteria for bulkheads and overheads are provided in Table 15.3. In addition to these requirements, materials cannot produce flaming droplets or have jetting combustion due to adhesives or bonding agents. The test is terminated when the material does not ignite after a ten-minute exposure, the material burns out, or after 40 minutes of exposure. Combustible veneers are permitted if the gross calorific value measured in ISO 1716[16] is less than 45 MJ/m^2.

The surface flammability of interior finish materials in NFPA 301 in general may be evaluated either through IMO Resolution A.653(16) test method or

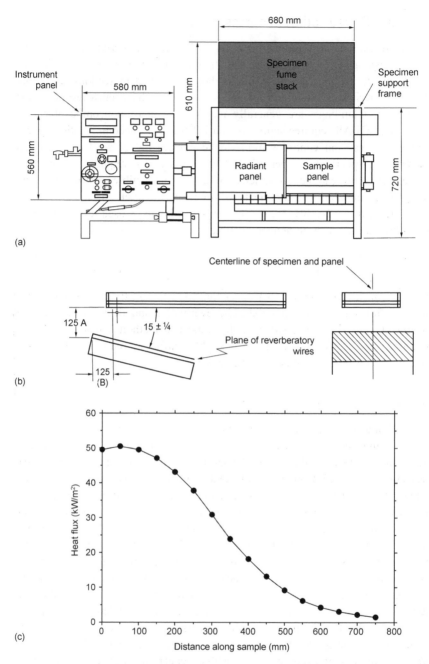

15.2 IMO Resolution A.653(16) and ASTM E1317 lateral flame spread test (a) front view showing entire apparatus (adapted from ref. 13); (b) top view of burner and sample (adapted from ref. 14); (c) heat flux variation along length of sample.

Table 15.3 Bulkhead and overhead interior finish material flammability criteria in the IMO Resolution A.653(16) (ref. 14)

Critical heat flux at extinguishment (kW/m^2)	Heat for sustained burning (MJ/m^2)	Total heat release (MJ)	Peak heat release rate (kW)
≥ 20.0	≥ 1.5	≤ 0.7	≤ 4.0

Table 15.4 Bulkhead and overhead interior finish of flammability test method options and criteria in NFPA 301

Test method option	Criteria
IMO Resolution A.653(16)	See Table 15.3.
NFPA 255/ASTM E84	FSI $\leq$ 20 and SDI $\leq$ 10
NFPA 255/ASTM E84*	FSI $\leq$ 75 and SDI $\leq$ 450
NFPA 286*	No flashover, total smoke released not to exceed 1,000 m^2

*Only for passenger vessel accommodation spaces with sprinklers.

NFPA 255/ASTM E84 test. As shown in Table 15.4, the performance criteria for the IMO resolution A.653(16) test are the same as those in Table 15.3. The NFPA 255/ASTM E84 performance criteria are flame spread index (FSI) not exceeding 20 and smoke developed index (SDI) not exceeding 10. The only exception to this is for interior finish, other than textile wall coverings, in passenger vessel accommodation areas with sprinklers. In these areas, interior finish must have a FSI not exceeding 75 and a SDI not exceeding 450 when tested in accordance with NFPA 255/ASTM E84. In addition, the material must be tested in accordance with NFPA 286 'Standard Methods of Fire Tests for Evaluating Contribution of Wall and Ceiling Interior Finish to Room Fire Growth'.[17] In this test, the material must not cause the room to reach flashover and the total smoke released must not exceed 1,000 m^2.

Deck coverings

Decks on ships may be covered with multiple layers of combustible materials. The material applied directly onto the deck, including any paint primers, is called the primary deck covering or deck overlay. Layers of material above the primary deck covering are referred to as either the floor covering or deck finish. Deck overlays and deck finishes that are required in SOLAS and the HSC Code to be 'low flame spread' must be tested in accordance with the test method in IMO Resolution A.653(16), which is shown in Fig. 15.2. Paints may be

Table 15.5 Deck overlay and deck finish material flammability criteria in the IMO Resolution A.653(16)

Critical heat flux at extinguishment (kW/m^2)	Heat for sustained burning (MJ/m^2)	Total heat release (MJ)	Peak heat release rate (kW)
≥ 7.0	≥ 0.25	≤ 1.5	≤ 10.0

*Deck overlay no flaming droplets, deck finish no more than 10 flaming droplets.

Table 15.6 Flammability test options in NFPA 301 for deck finishes and overlays

Test method option	Criteria
IMO Resolution A.653(16)	See Table 15.5
NFPA 253/ASTM E648	Critical radiant flux not less than 4.5 kW/m^2

exempted. Deck overlays and finishes are tested in a vertical orientation and must meet the requirements provided in Table 15.5. In addition to these requirements, deck overlays cannot produce flaming droplets and deck finishes cannot produce more than ten flaming droplets. Materials also must not have jetting combustion due to adhesives or bonding agents. The test is terminated when the material does not ignite after a ten-minute exposure, the material burns out, or after 40 minutes of exposure. For deck coverings having multiple layers of different materials, each individual layer may be required to be tested.

NFPA 301 also requires deck finishes and overlays to meet flammability requirements. Deck finishes or overlays that consist of 100% wool are assumed to meet the flammability requirements. Deck finishes are to be tested in accordance with 16 CFR 1630[18] and must also pass one of the tests provided in Table 15.6. One of these tests is IMO Resolution A.653(16), which has been previously described and is shown in Fig. 15.2. The other flammability test that can be conducted is NFPA 253/ASTM E648 'Standard Test Method for Critical Radiant Flux of Floor-Covering Systems Using a Radiant Heat Energy Source',[19,20] see Fig. 15.3. During this test, the sample is exposed to the heat flux shown in Fig. 15.3 and is ignited by a pilot on the end exposed to the maximum heat flux of 10 kW/m^2. The flame front location during the test is recorded until the flame front stops progressing. The critical radiant heat flux is the heat flux from the heater at the furthest point of flame spread along the sample. Materials must have a critical radiant flux not less than 4.5 kW/m^2.

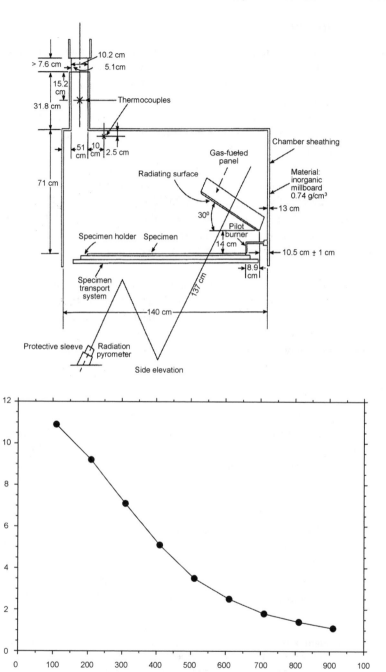

15.3 NFPA 253/ASTM E648 floor radiant panel test.

15.2.5   Insulation

Insulation that is applied to boundaries may also have flammability requirements. In SOLAS and the HSC Code, insulation applied to boundaries must
have 'low flame spread' as measured using IMO Resolution A.653(16). The
performance criteria are the same as those for interior finish materials, see Table
15.3. Insulation for cold service piping must also qualify as a 'low flame spread'
material when tested using IMO Resolution A.653(16) and with the
requirements in Table 15.3.

NFPA 301 requires that insulation used on structural boundaries meet 46
CFR 164.007.[21] Insulation on cold service piping must be tested in accordance
with NFPA 255/ASTM E84. In NFPA 255/ASTM E84, the material must have
a FSI not exceeding 25, a SDI not exceeding 50, and no flaming droplets.
Insulation on ventilation ducts and pipes must have a FSI not exceeding 20 and
a SDI not exceeding 10 when tested in accordance with NFPA 255/ASTM
E84.

NFPA 302 has an interior finish flammability requirement in the machinery
spaces. Thermal and acoustic insulation applied in a compartment or enclosure
containing an internal combustion engine or heater must be tested in accordance
with NFPA 255/ASTM E84. As seen in Table 15.7, the thermal and acoustic
insulation must have a flame spread index (FSI) not exceeding 75.

15.2.6   Furnishings

Several types of furnishings on ships require flammability testing. These include
case furniture, upholstered furniture, other furniture, bedding, and draperies. The
flammability tests and criteria in SOLAS, HSC Code, and NFPA 301 are
presented in following sections for each of these furnishings.

Case furniture

SOLAS requires case furniture be constructed of noncombustible material. HSC
Code allows case furniture to be constructed of noncombustible or fire restricting
material (see Table 15.1 for requirements). Combustible veneers are permitted if
the gross calorific value measured in ISO 1716 is less than 45 MJ/m^2.

NFPA 301 allows case furniture constructed of noncombustible or com-

Table 15.7 Flammability criteria in NFPA 302 for thermal and acoustic insulation
applied in enclosures with an engine or heater

Method	Criteria
NFPA 255/ASTM E84	FSI $\leq$ 75

Table 15.8 ISO 9705 test requirements for combustible construction in case furniture to be in accordance with NFPA 301

	Variable	Criteria
	Flashover	Room cannot reach flashover
Heat release rate*	Test average	$\leq 100\,kW$
	Peak 30 s average	$\leq 500\,kW$
Smoke production	Test average	$\leq 1.4\,m^2/s$
	Peak 60 s average	$\leq 8.3\,m^2/s$

*Heat release rate due to the lining only, burner contribution excluded.

bustible materials. For noncombustible case furniture with a combustible veneer no more than 3 mm thick, the veneer must have a FSI not exceeding 20 and SDI not exceeding 10 when tested in accordance with NFPA 255/ASTM E84. For combustible furniture, the material of construction must pass the ISO 9705 room corner fire test with the material lining the side walls, back wall, and ceiling of the room. The test must be conducted with an initiating fire of 100 kW for ten minutes followed by a 300 kW fire for an additional ten minutes. Test criteria provided in Table 15.8 are similar to those in Table 15.1 for a fire restricting material except there are no extent of flame spread requirements, no flaming droplet/debris requirements, and there is a requirement that the room cannot reach flashover.

Upholstered furniture

SOLAS and HSC Code have flammability requirements for the frame and the upholstered part of the upholstered furniture. SOLAS requires that frames be constructed of a noncombustible material, while in the HSC Code frames can be a noncombustible or fire restricting material (see Table 15.1 for requirements). The upholstered part of the furniture must be tested in accordance with IMO Resolution A.652(16). In this test a sample of the upholstered part of the furniture is placed on the back and bottom of a test seat frame. The sample is exposed to a cigarette exposure and a butane flame. If progressive flaming or smoldering is observed during the one-hour test period after either the cigarette or butane ignition exposure, then the upholstered sample fails.

NFPA 301 requires that upholstered furniture comply with NFPA 266 'Standard Test Method of Upholstered Furniture Exposed to Flaming Ignition Source',[22] ASTM E1537 'Standard Method for Fire Testing of Upholstered Furniture Items',[23] or UL 1056 'Fire Test of Upholstered Furniture'.[24] Furniture must meet the limited heat release rate criteria in Table 15.9.[25]

Table 15.9 Upholstered furniture fire tests and criteria

Code	Test method	Criteria
SOLAS, HSC Code	IMO Resolution A.652(16)	No progressive flaming or smoldering one hour after exposure
NFPA 301	NFPA 266	Maximum heat release rate no more than 80 kW
	ASTM E1537	Maximum total heat released during initial ten minutes less
	UL 1056	than 25 MJ

Other furniture

SOLAS requires all other furniture that is not considered upholstered (chairs, sofas, tables) to be constructed of noncombustible materials. In the HSC Code, all furniture not upholstered can be constructed of noncombustible materials or fire restricting materials (see Table 15.1 for requirements).

NFPA 301 does not have flammability regulations on other furniture that is not upholstered, except for stacked chairs. Stacked chairs are defined in the standard as chairs that can be stacked more than three chairs high. Stacked chairs must individually meet the requirements for an upholstered chair provided in Table 15.9. In addition, it must be demonstrated that the stacked chairs cannot cause the room of use to reach flashover. This could be done either through a large-scale test or through modeling that is approved by the authority having jurisdiction.

Bedding

SOLAS and HSC Code require flammability testing on bedding be conducted in accordance with IMO Resolution A.688(17).[26] This includes mattresses, mattress pads, pillows, quilts, blankets, and bedspreads. Tests are performed on bedding samples that are 450 mm by 350 mm, except for pillows which are tested full-size. In tests on mattress pads, the test sample includes the mattress covered in the mattress pad. Samples are placed on a test frame platform and an ignition source is applied to the top surface of the sample. Bedding is classified as not readily ignitable if it does not progressively smolder or exhibit flaming ignition, as defined in the test method, over the one-hour test period.

NFPA 301 has flammability requirements for mattresses, mattress pads, and mattresses with foundations, but no requirements for other types of bedding. Mattresses, mattress pads, and mattresses with foundations must comply with 16 CFR 1632 'Standard for Flammability of Mattresses'.[27] In addition, they need to be tested in accordance with NFPA 267,[28] ASTM E1590,[29] or UL 1895[30] and meet the requirements in Table 15.10.

Table 15.10 Bedding fire tests and criteria

Code	Test method	Criteria
SOLAS, HSC Code	IMO Resolution A.688(17)	No progression in smoldering or flaming ignition one hour after exposure
NFPA 301	16 CFR 1632 NFPA 267 ASTM E1590 UL 1895	Pass Maximum heat release rate no more than 100 kW Maximum total heat released during initial ten minutes less than 25 MJ

Drapes

Materials used in vertically hanging drapes also have flammability requirements. SOLAS and HSC Code require that these materials be tested in accordance with IMO Resolution A.563(14)[31] and meet the test criteria in Table 15.11. In this test, a small swatch of material is exposed to a small flame either at the bottom of the swatch or in the center. The sample must not continue to burn for more than five seconds after burner application, must have limited char length, must not burn through to the edge of the sample, and cannot ignite cotton below the sample by flaming drops/debris. Tests must be conducted on samples after appropriate aging and/or washing.

NFPA 301 requires that draperies either pass NFPA 701 'Standard Methods of Fire Tests for Flame Propagation of Textiles and Films'[31] or the IMO Resolution A.563(14)[32] with the criteria in Table 15.11.

Table 15.11 Flammability tests and criteria for materials used in vertically hanging drapes

Code	Test method	Criteria
SOLAS, HSC Code	IMO Resolution A.563(14)*	No flaming 5 s after exposure. Char length no more than 150 mm. No burn-through to sample edge. No ignition of cotton below sample. Surface flash propagation no more than 100 mm from point of ignition.
NFPA 301	IMO Resolution A.563(14) NFPA 701	Same as above See NFPA 701

*Tests conducted on samples after appropriate aging and/or washing.

15.2.7   Wires and cables

There are no flammability requirements for wire and cables in SOLAS or the HSC Code, but NFPA 301 has an extensive set of flammability requirements for various types of wires and cables. Refer to NFPA 301 for a list of the flammability tests and requirements for wires and cables.

15.2.8   Pipes and ventilation ducts

Pipes and ventilation ducts are permitted to be made out of combustible materials. SOLAS and HSC Code require combustible ventilation ducts and pipes to qualify as a 'low flame spread' material, and the material must be tested in accordance with IMO Resolution A.653(16) and meet the criteria in Table 15.3. If the interior and exterior of the ventilation duct are constructed of different materials, both sides of the duct must be tested in IMO resolution A.653(16).

NFPA 301 requires that combustible pipes and ventilation ducts have a FSI not exceeding 20 and a SDI not exceeding 10 when tested in accordance with NFPA 255/ASTM E84.

15.3   Submarines

Guidelines and codes for the construction of submarines contain some flammability requirements, but not in the detail provided for ships. Two commonly used standards are the IMO MSC/Circ.981 'Guidelines for the Design, Construction, and Operation of Passenger Submersible Craft'[33] and American Bureau of Shipping (ABS) 'Rules for Building and Classing Underwater Vehicles, Systems and Hyperbaric Facilities 1990'.[34] Flammability requirements in both standards are discussed below.

IMO MSC/Circ.981 states that the construction should minimize the potential smoke and fire hazards. The design should consider material toxicity and materials with low flame spread characteristics. In addition, all materials and equipment within the craft must be noncombustible for the oxygen level expected, which will typically be close to ambient levels.

The ABS rules have different flammability requirements for interior and exterior materials. Interior materials must be noncombustible or a fire restricting material, as defined in IMO Resolution MSC.40(64).[10] This includes materials used to make linings, deck coverings, ceilings, insulation, partial bulkheads, and seating. External materials, which include decks, deck coverings, skins and fairings, must not readily ignite or produce toxic or explosive hazards. Manned compartments inside chambers, dive training centers, and dive simulators must contain a minimal amount of combustible materials. In these areas, linings, deck coverings, ceilings, insulation, partial bulkheads, seating and bedding are to be constructed of fire restricting materials as defined in MSC.40(64).[10]

15.4   Future trends

Future ships and submarines will likely include more fiber-reinforced plastic (FRP) composite materials. As these materials become more advanced, the potential applications for composite materials will increase. Fire models are being used to assist in the development of composite materials, especially to evaluate the performance of composite materials in full-scale tests like the ISO 9705 room-corner fire test using small-scale test data on the material.

The U.S. Coast Guard sponsored a study[35] to investigate methods for predicting material performance in the ISO 9705 room-corner test using small-scale data from the ASTM E1354 cone calorimeter.[36] In this study, two different approaches were used to predict whether a material is expected to meet the fire restricting material performance criteria in Table 15.1. The first approach was based on flame spread theory and an empirical correlation that would simply indicate whether a material was expected to pass or fail the ISO 9705 test. The other approach evaluated was the use of state-of-the-art fire models to predict the behavior of materials in the ISO 9705 fire test. These methods were evaluated for composite materials that had undergone small-scale cone calorimeter testing as well as full-scale ISO 9705 room-corner fire testing.[37,38] A list of materials that have been included in the model validation are provided in Table 15.12 along with cone calorimeter data for each material.

15.4.1   Flammability parameter

The likelihood of flashover and the heat release rate due to the lining in the ISO 9705 test were found to correlate with the flammability parameter. The flammability parameter is based on the flame spread equations developed in ref. 39 and is a measure of whether a material will spread flame. The flammability parameter was originally developed for the U.S. Navy to evaluate the use of passive fire protection linings such as insulations.[40,41] The flammability parameter is calculated using data from the cone calorimeter data at $50\,kW/m^2$ with the following relation,

$$F = 0.01\dot{Q}'' - (t_{ig}/t_{burn}) \qquad\qquad 15.1$$

where $\dot{Q}''$ is the test average heat release rate (kW/m^2), t_{ig} is the time to ignition (sec) and t_{burn} is the burning duration (sec).

Theoretically, a value of $F < 1.0$ indicates that the material will not spread flame while flame spread is expected with $F > 1.0$. Beyler et al.[35] conducted an analysis comparing ISO 9705 test average and peak heat release rates with the flammability parameter. Materials with $F < 0.0$ were measured to pass the fire restricting material requirements; however, materials did not consistently fail the fire restricting requirements until $F > 0.5$. Materials with $0.0 < F < 0.5$ had variable performance with some passing and some failing. The deviation from

Table 15.12 Cone calorimeter data of composite materials. Data for materials 1–8 is from ref. 37 while other data is ref. 38

No.	Material/ incident heat flux	Critical heat flux (kW/m²)	Time to ignition (s)	Burning duration (s)	Test average heat release rate, $\dot{Q}''$ (kW/m²)	Total heat released (kJ/m²)	Test average HOC, $\Delta H_{c,eff}$ (kJ/kg)
1	FR Phenolic	49					
	50 kW/m²		324	241	19	4.6	4.9
	75 kW/m²		78	385	50	19.4	9.1
	100 kW/m²		16	604	41	24.6	9.0
2	FRM	54					
	50 kW/m²		NI	–	–	–	–
	75 kW/m²		78	270	26	7.1	11.3
	100 kW/m²		14	261	35	9.1	8.0
3	FR Polyester	18					
	25 kW/m²		249	189	59	11.2	10.9
	50 kW/m²		65	703	64	44.7	11.4
	75 kW/m²		27	704	72	51.0	11.5
4	FR Vinyl Ester	18					
	25 kW/m²		306	203	75	15.2	12.8
	50 kW/m²		75	983	67	65.5	12.9
	75 kW/m²		34	782	85	66.4	14.5
5	FR Epoxy	20					
	50 kW/m²		123	90	60	5.4	10.0
	75 kW/m²		59	436	41	17.7	8.8
	100 kW/m²		36	419	35	14.7	5.8

6	Coated FR Epoxy	34						
	50 kW/m²		68	37	24	0.9	7.9
	75 kW/m²		30	410	32	13.0	8.0
	100 kW/m²		20	310	34	10.4	7.1
7	Polyester	17						
	25 kW/m²		123	558	109	60.9	22.9
	50 kW/m²		30	355	193	68.5	21.3
	75 kW/m²		16	292	189	55.1	20.6
8	FR Modified Acrylic	19						
	25 kW/m²		462	426	50	21.3	11.5
	50 kW/m²		93	1013	47	47.6	13.0
	75 kW/m²		62	1057	51	54.4	12.3
9	FR Phenolic 1391	34						
	50 kW/m²		615	587	26	15.0	5.0
	75 kW/m²		211	691	41	28.0	10.8
	100 kW/m²		163	781	36	27.8	10.9
10	FR Phenolic 1407	32						
	50 kW/m²		340	350	43	14.1	11.6
	75 kW/m²		163	363	48	20.0	12.7
	100 kW/m²		62	325	60	20.0	13.9

FR = fire retardant FRM = fire restricting material HOC = heat of combustion

Table 15.13 Correlation between the flammability parameter and ISO 9705 full-scale test heat release rate and occurrence of flashover (ref. 38)

Material no.	Time to flashover (s)	ISO 9705 heat release rate (kW)			Flammability parameter, F
		Test average	Max 30 s avg.	Pass/fail[1]	
1	NR	62	159	Pass	−1.15
2	NR	31	112	Pass	NC
3	342	203	677	Fail	0.54
4	300	224	798	Fail	0.60
5	1002	125	454	Fail	−0.07
6	NR	31	82	Pass	−0.41
7	102	170	402	Fail	1.39
8	682	127	657	Fail	0.42
9	NR	60	104	Pass	−0.79
10	NR	59	123	Pass	−0.54
U.S. Navy	–	≤ 100	≤ 500		

NR = not reached
NC = not calculated due to no ignition at 50 kW/m^2
1. Heat release rate requirements only.

the theoretical value of 1.0 is attributed to the hot gas layer that develops inside of the ISO 9705 room, which pre-heats material and enhances flame spread. Therefore, materials are observed to spread flame in the specified ISO 9705 test at a lower value of F. Changing the room size, initiating fire, or room door size could all affect the value of F where flames will spread over the material.

Lattimer and Sorathia[38] calculated the flammability parameter for the materials in Table 15.12. Table 15.13 contains a comparison of the flammability parameter with the occurrence of flashover and the heat release rate results in the ISO 9705 test. Materials with F less than approximately zero passed the fire restricting material heat release rate requirement and did not cause the room to reach flashover. Results from analysis of Beyler *et al.*[35] considering a broader range of materials were consistent with those in Table 15.13.

Assuming that the threshold for meeting the fire restricting material ISO 9705 heat release rate requirements is $F < 0$, eqn 15.1 is transformed into

$$0.01\dot{Q}'' < (t_{ig}/t_{burn}) \tag{15.2}$$

Noting that the total heat released, Q'', is the average heat release rate, $\dot{Q}''$, multiplied by the burning duration, t_{burn}, eqn 15.2 becomes,

$$Q'' < 100t_{ig} \tag{15.3}$$

Therefore, the total heat released per unit area (kJ/m^2) must be less than 100 times the time to ignition (s) to meet the heat release rate requirements for a fire restricting material in the specified ISO 9705 test. This relation indicates that materials having higher times to ignition can release more heat and still meet the requirements. For example, consider a material tested at 50 kW/m^2 in the cone calorimeter that has an ignition time of $t_{ig} = 150$ s. According to eqn 15.3, this material would meet the ISO 9705 heat release rate requirements if the total heat released per unit area at 50 kW/m^2 was less than 15,000 kJ/m^2. The allowed total heat released per unit area would increase as the time to ignition of the material increased. Note that this relation has only been evaluated for use in the ISO 9705 test with an initiating fire of 100 kW for ten minutes followed by 300 kW for ten minutes. If the initiating fire, room size, or room door opening size are changed, then eqn 15.3 may not be appropriate.

15.4.2 Fire growth modeling

Fire growth models[35,38,42–45] have also been used to predict material performance in the ISO 9705 test using small-scale test data to develop model input. These models vary in level of complexity, and all required validation with large-scale test data on a variety of material linings to demonstrate predictive capability. They can predict the time variation in heat release rate and, in some cases, the smoke production. Such models are typically general enough that they can predict material fire performance in other spaces besides the ISO 9705 room and with other types of fires besides the 100/300 kW standard initiating fire.

Fire growth models are becoming more sophisticated, and predictions are getting more consistent with test data. In the study conducted by Beyler et al.[35] for the U.S. Coast Guard, three different fire growth models were evaluated. These models were determined to predict trends generally but were not sufficiently accurate to discriminate whether a material would meet the fire restricting performance criteria in Table 15.1. More recent versions of fire growth models have shown more promise. Lattimer and Sorathia[38] conducted predictions on the ten materials in Table 15.12 using the model described in ref. 45. A comparison of the model results and test data is provided in Table 15.14. The model is generally able to determine whether a material will meet the heat release rate requirement but overestimates the smoke production rate. A comparison of time predictions of heat release rate and smoke production rate with test data is provided in Fig. 15.4. As seen in these figures, model results are in good agreement with test data. As shown in Table 15.14 and Fig. 15.4, there are some fire growth models that are now sufficiently accurate to screen material performance for meeting the fire restricting material requirements.

15.5 Conclusions

Codes and standards for ships and submarines emphasize the use of a limited amount of combustible materials for construction materials and furnishings. Where combustible materials are allowed, these materials generally have to meet a flammability requirement.

Table 15.14 Comparison of fire growth model results (ref. 38) with data (refs 37, 38)

Material no.	Time to flashover (s)[1]		Heat release rate (kW)				Smoke production rate (m²/s)			
			Test average		Max 30 s avg.		Test average		Max 60 s avg.	
	Data	Model	Data	Model	Data	Model	Data	Model	Data	Model
1	NR	NR	62	75	159	202	1.5	3.1	5.4	5.2
2	NR	NR	31	23	112	84	0.2	2.6	0.5	4.2
3	342	475	203	248	677	971	9.4	16.1	21.7	43.9
4	300	370	224	298	798	1020	10.2	22.1	26.3	56.2
5	1002	978	125	135	454	940	6.7	5.0	26.4	18.2
6	NR	NR	31	45	82	190	1.4	3.3	3.5	6.7
7	102	44	170	167	402	241	2.3	3.8	2.7	3.8
8	682	332	127	216	657	770	0.5	2.1	4.8	4.3
9	NR	NR	60	63	104	180	0.3	3.3	0.7	5.9
10	NR	NR	59	78	123	208	0.4	3.7	0.6	7.8
IMO	—		≤100		≤500		≤1.4		≤8.3	

NR = not reached
1. Flashover assumed to occur when heat release rate exceeds 1000 kW.

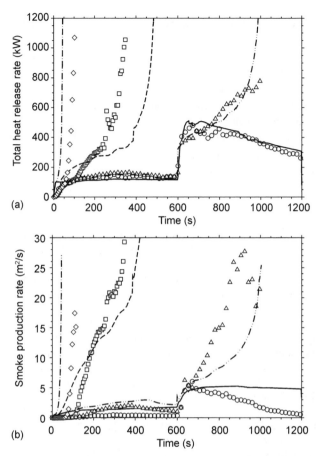

15.4 Measured and predicted (a) total heat release rate and (b) smoke production rate in ISO 9705 tests on different materials. Material No. 1 (○ (data), — (model)), Material No. 3 (□ (data), - - (model)], Material No. 5 (△ (data), -..- (model)), Material No. 7 (◇ (data), -.- (model)]. Data from ref. 37.

Flammability requirements for combustible materials on ships are contained in SOLAS, HSC Code, FTP Code, NFPA 301, and NFPA 302. The majority of the fire test procedures and performance criteria for SOLAS and HSC Code are provided in the FTP Code. In SOLAS, flammability of most combustible surfaces is regulated by IMO Resolution A.653(16), which is a lateral flame spread test with the sample in the vertical orientation. HSC Code allows construction material and furnishings to be made of a combustible material as long as it meets the fire restricting material performance requirements. To be considered a fire restricting material, the material must be tested in the ISO 9705 room-corner fire test and have a limited heat release rate and smoke production rate. NFPA 301 uses a variety of flammability tests to regulate use of combustible materials. In general, NFPA 301 gives ship builders the option to meet the SOLAS flammability requirements or pass a NFPA/ASTM flammability test.

IMO and ABS both have standards that discuss the flammability require-ments of combustible materials on the inside and outside of submarines. IMO emphasizes the use of noncombustible materials, while the ABS standard allows materials such as linings and deck coverings to be constructed of fire restricting materials as defined in IMO MSC.40(64).

An increase in the use of composite materials onboard ships and submarines is likely. Fire models can be used to assist in the development of new materials and evaluate the material performance in actual locations of use.

15.6   References

1. Eberly, R. and Colonna, G.R., 'Marine Vessels,' Section 14, Chapter 6, *Fire Protection Handbook*, 19th edn, Volume II, National Fire Protection Association, Quincy, MA, 2003.
2. US Senate Report 184, 75th Congress, 1st Session, 17 March 1937, pp. 8–9.
3. IMO, *SOLAS*, Consolidated Edition, International Maritime Organization, London, 1997.
4. IMO Resolution MSC.36(63), *International Code of Safety for High-Speed Craft (HSC Code)*, International Maritime Organization, London, 1995.
5. IMO Resolution MSC.61(67), *International Code for Application of Fire Test Procedures (FTP Code)*, International Maritime Organization, London, 1998.
6. NFPA 301, *Code for Safety to Life from Fire on Merchant Vessels*, 2001 edn, National Fire Protection Association, Quincy, MA, 2001.
7. NFPA 302, *Fire Protection Standard for Pleasure and Commercial Motor Craft*, 1998 edn, National Fire Protection Association, Quincy, MA, 1998.
8. ISO 1182:1990, *Fire Test-Building Materials-Non-combustibility Test*, International Organization for Standards, Geneva, Switzerland, 1990.
9. 46 CFR 164.009, *Noncombustible Materials for Merchant Vessels*, U.S. Code of Federal Regulations, 2005.
10. IMO Resolution MSC.40(64), *Standard for Qualifying Marine Materials for High-Speed Craft as Fire-Restricting Materials*, International Maritime Organization, London, 1998.
11. ISO 9705:1993(E), *Fire Tests – Reaction to Fire – Full-Scale Room Test for Surface Products*, International Organization for Standards, Geneva, Switzerland, 1993.
12. NFPA 255, *Standard Method of Test of Surface Burning Characteristics of Building Materials*, 2000 edn, National Fire Protection Association, Quincy, MA, 2000.
13. ASTM E84, *Standard Method of Test of Surface Burning Characteristics of Building Materials*, American Society of Testing and Materials, West Conshohocken, PA, 1997.
14. IMO Resolution A.653(16), *Recommendation on Improved Fire Test Procedures for Surface Flammability of Bulkhead, Ceiling and Deck Finish Materials*, International Maritime Organization, London, 1998.
15. ASTM E1317, *Standard Method for Flammability of Marine Surface Finishes*, American Society of Testing and Materials, West Conshohocken, PA, 2000.
16. ISO 1716:1973, *Building Materials – Determination of Calorific Potential*,

International Organization for Standards, Geneva, Switzerland, 1973.

17. NFPA 286, *Standard Methods of Fire Tests for Evaluating Contribution of Wall and Ceiling Interior Finish to Room Fire Growth*, 2000 edn, National Fire Protection Association, Quincy, MA, 2000.

18. 16 CFR 1630, *Standard for Surface Flammability of Carpets and Rugs*, U.S. Code of Federal Regulations, 2005.

19. NFPA 253, *Standard Test Method for Critical Radiant Flux of Floor-Covering Systems Using a Radiant Energy Source*, 2000 edn, National Fire Protection Association, Quincy, MA, 2000.

20. ASTM E 648-03, *Standard Test Method for Critical Radiant Flux for Floor-Covering Systems Using a Radiant Heat Energy Source*, American Society of Testing and Materials, West Conshohocken, PA, 2003.

21. 46 CFR 164.007, *Structural Insulations*, U.S. Code of Federal Regulations, 2005.

22. NFPA 266, *Standard Test Method of Upholstered Furniture Exposed to Flaming Ignition Source*, 1998 edn, National Fire Protection Association, Quincy, MA, 1998.

23. ASTM 1537, *Standard Method for Fire Testing of Upholstered Furniture Items*, American Society of Testing and Materials, West Conshohocken, PA,1996.

24. UL 1056, *Fire Test of Upholstered Furniture*, Underwriters Laboratories, Inc., Northbrook, IL, 1989.

25. IMO Resolution A.652(16), *Recommendation on Fire Test Procedures for Upholstered Furniture*, International Maritime Organization, London, 1998.

26. IMO Resolution A.688(17), *Fire Test Procedures for Ignitability of Bedding Components*, International Maritime Organization, London, 1998.

27. 16 CFR 162, *Standard for Flammability of Mattresses*, U.S. Government.

28. NFPA 267, *Standard Method of Test for Fire Characteristics of Mattresses and Bedding Assemblies Exposed to Flaming Ignition Source*, 1998 edn, National Fire Protection Association, Quincy, MA, 1998.

29. ASTM 1590, *Standard Method for Fire Testing of Mattresses*, American Society of Testing and Materials, West Conshohocken, PA, 1999.

30. UL 1895, *Fire Test of Mattresses*, Underwriters Laboratories, Inc., Northbrook, IL, 1995.

31. NFPA 701 *Standard Methods of Fire Tests for Flame Propagation of Textiles and Films*, 1998 edn, National Fire Protection Association, Quincy, MA, 1998.

32. IMO Resolution A.563(14), *Amendments to the Recommendation on Test Method for Determining the Resistance to Flame of Vertically Supported Textiles and Films*, International Maritime Organization, London, 1998.

33. IMO MSC/Circ.981 *Guidelines for the Design, Construction, and Operation of Passenger Submersible Craft*, International Maritime Organization, London, 20 January 2001.

34. ABS, *Rules for Building and Classing Underwater Vehicles, Systems and Hyperbaric Facilities 1990*, Notice No. 3, American Bureau of Shipping, Houston, TX, 1 January 2002.

35. Beyler, C.L., Hunt, S.P., Lattimer, B.Y., Iqbal, N., Lautenberger, C., Dembsey, N., Barnett, J., Janssens, M., Dillon, S., and Greiner, A., *Prediction of ISO 9705 Room/ Corner Test Results*, Report No. R&DC21599, U.S. Coast Guard, (1999).

36. ASTM E1354, *Standard Test Method for Heat and Visible Smoke Release Rate for Materials and Products Using an Oxygen Consumption Calorimeter*, American Society of Testing and Materials, West Conshohocken, PA, 1997.

37. Janssens, M., Garabedian, A., and Gray, W., *Establishment of International Standards Organization (ISO) 5660 Acceptance Criteria for Fire Restricting Materials Used on High Speed Craft*, Final Report, CG-D-22-98, U.S. Coast Guard, (1998).
38. Lattimer, B.Y. and Sorathia, U., 'Composite Fire Hazard Analysis Tool for Predicting ISO 9705 Room-Corner Fire Test', *SAMPE 04*, Long Beach, CA, 2004.
39. Mowrer, F., and Williamson, R., *Fire Safety Science – Proceedings of the Third International Symposium*, pp. 689–698, 1991.
40. Williams, F., Beyler, C., and Iqbal, N., 'Flame Spread Evaluation for the U.S. Navy Passive Fire Protection (PFP) Test Materials,' *NRL Ltr Rpt Ser* 6180/0216A.2, 1995.
41. Williams, F., Beyler, C., Hunt, S., and Iqbal, N., *Upward Flame Spread on Vertical Surfaces*, NRL/MR/6180-97-7908, 1997.
42. Quintiere, J.G., 'A Simulation for Fire Growth on Materials Subject to a Room-Corner Test,' *Fire Safety Journal*, Vol. 20, pp. 313–339, 1993.
43. Goransson, U. and Wickstrom, U., 'Flame Spread Predictions in the Room/Corner Test Based on the Cone Calorimeter,' *Interflam 90*, pp. 211–218, 1990.
44. Karlsson, B., 'Calculating Flame Spread and Heat Release Rate in the Room Corner Test, Taking Account of Pre-Heating by the Hot Gas Layer,' *Interflam 93*, pp. 25–37, 1993.
45. Lattimer, B.Y., Hunt, S.P., Wright, M., and Beyler, C., 'Corner Fire Growth in a Room with a Combustible Lining,' *Fire Safety Science – Proceedings of the Seventh International Symposium*, pp. 689–698, 2002.

16

Tests for spontaneous ignition of solid materials

H W A N G , CSIRO Manufacturing and Infrastructure Technology, Australia,
B Z D L U G O G O R S K I and E M K E N N E D Y,
The University of Newcastle, Australia

16.1 Introduction

Some solid materials are susceptible to spontaneous heating; that is, they may increase their temperature without assistance from an external heat source. Usually, spontaneous or self-heating arises naturally as an outcome of low temperature ($< 100\,°C$) exothermic oxidation reactions, or condensation of moisture. Both phenomena can occur on pore surfaces, in the case of porous materials, or on the outside surfaces of micron-sized solid particles. Once the rate of heat generation exceeds the rate of heat dissipation to the environment, self-heating leads to self-ignition (or auto-ignition), and eventually to flaming or smouldering combustion. For this reason, we denote spontaneous combustion to include all three stages of this process, that is: (i) the initial self-heating; (ii) auto-ignition (i.e., in the absence of pilots and external heat sources); and, (iii) flaming or non-flaming combustion.

In nature, some materials may display the propensity to self-heating and auto-ignition, even at temperatures close to ambient. These materials are denoted as pyrophoric. Haystacks were perhaps the first materials observed to ignite spontaneously by early farmers, as noted by Cuzzillo (1997) in his literature survey. It is now appreciated that a variety of materials display similar pyrophoric properties, including coal, charcoal, carbon black, wood, grass, tree leaves, rags, and some other agricultural products such as cotton and grain. Some chemical products, such as alkali metals, white phosphorous, calcium oxide or quicklime acids, zinc powder, organic peroxides, plastics or solid foams, may also exhibit the tendency to combust spontaneously with or without being exposed to an oxidative environment. Other materials, such as solid benzoyl peroxide, may decompose exothermically in processes that eventually lead to fires and explosions (Bowes 1984).

Spontaneous combustion of solid materials has always been one of the major hazards in construction, transport, processing and mining industries, as extensively reported in the literature (Coward 1957; Kuchta *et al.* 1980; Bowes 1984;

Table 16.1 Factors that determine the tendency of coals to combust spontaneously (Singh *et al.* 1993)

Physical	Environmental
Rank of coal	Ambient temperature
Porosity	Available air
Petrographic composition	Amount of moisture
Pyritic content	Previous history
Particle size	Mining method
Catalyst mineral content	Cover
Hardness	Coal thickness
Thermal conductivity and specific heat	Geological conditions

Carras and Young 1994; Gouws and Knoetze 1995; Cuzzillo 1997). For instance, spontaneous combustion of freshly mined coal has been found to be one of the major reasons responsible for the fire incidents occurring in coal mines. Statistical data show that in the 1970s an average of eight cases of spontaneous combustion in coal mines occurred each year in France and Great Britain (Kuchta *et al.* 1980). In South African collieries, during the period from 1970 to 1990, more than a third of the 254 underground fires were caused by spontaneous combustion of coal (Gouws and Knoetze 1995).

The physical and chemical properties of a material and the conditions during its mining, transport, storage, and further processing in an industrial operation modify the material's propensity to ignite spontaneously (Coward 1957; Kim 1977; Bowes 1984; Singh *et al.* 1993; Cuzzillo 1997; Wang *et al.* 2003b). For instance, it is known that the rank of a coal, its moisture content, the percentage of fixed carbon and some other parameters, as shown in Table 16.1 (Singh *et al.* 1993), may dramatically alter the tendency of coal to self-ignite. In addition, the geometry of the solid as well as the access and type of the oxidation medium play significant roles in spontaneous combustion. For example, compacting the coal impedes oxygen and moisture access to a stockpile which can delay self-heating.

Since the beginning of the last century, the testing methods, together with the relevant pass/fail criteria, have been established to evaluate the liability of the materials to spontaneous combustion. This has been accomplished by characterising the hazardous properties of the pyrophoric materials, and to determine the conditions that facilitate spontaneous combustion. The first parameter introduced by early investigators to characterise the ignition process was the so-called auto-ignition temperature (Wheeler 1918; Parr and Coons 1925); the lower the auto-ignition temperature the more hazardous the material. Over the remaining part of the last century, it has been recognised that the propensity of a solid to combust spontaneously could be examined by observing the temperature rise within the material, its consumption of oxygen, formation of oxidation products both in gas and solid phases and change in the material's mass (van Krevelen 1993; Wang *et al.* 2003b).

Usually, a test method is intimately related with the construction and operational details of the testing apparatus, and involves the determination of relevant indexes. Such indexes are then used to assess the propensity of solid materials to combust spontaneously, in specific applications. In broad terms, the test methods can be classified as those based on calorimetric (i.e., involving the evaluation of the generated heat) and non-calorimetric measurements. Heat-based tests methods attempt to account for or to measure the heat released during the self-heating process. These test methods include adiabatic and isothermal calorimetry, differential thermal analysis (DTA), differential scanning calorimetry (DSC), basket heating, Chen's method, and the conventional crossing point temperature (CPT) (Ponec *et al.* 1974; Kim 1977; Kuchta *et al.* 1980; van Krevelen 1993; Carras and Young 1994; Wang *et al.* 2003b). The next paragraph summarises the non-heat-based measurements.

The measurement of mass change during the self-heating stage of spontaneous combustion is usually performed with a micro balance or a thermogravimetric analyser (TGA). The so-called oxygen adsorption and the isothermal flow reactor techniques examine the capacity of a material to consume oxygen and release gaseous products of oxidation at low temperatures (Ponec *et al.* 1974; Carras and

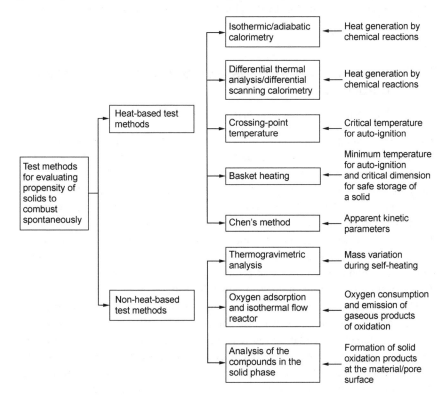

16.1 Summary of the test methods.

Young 1994; Wang *et al.* 2003b). The physical and chemical titration, infra-red spectroscopy (IR), Fourier transform infra-red spectroscopy (FTIR), X-ray photoelectron spectroscopy (XPS), secondary ion mass spectrometry (SIMS) and ^{13}C nuclear magnetic resonance (^{13}C NMR), electron spin resonance (ESR) and electron paramagnetic resonance spectroscopy (EPR) can determine, sometimes *in situ*, the type and the rate of formation of products of low-temperature exothermic reactions in the solid phase (Nelson 1989; Wang *et al.* 2003b). Figure 16.1 provides a summary of both heat- and non-heat-based methods.

The content of this chapter is organised as follows: Based on the classification of the test methods, sections 16.2–16.6 describes the calorimetric test methods adopted in practice for the examination of spontaneous combustion of solid materials. Furthermore, section 16.2 provides the details of the experimental set-ups, introduces the indexes established for quantification of the self-heating and self-ignition behaviour. This is followed, in sections 16.7–16.9, by a review of non-calorimetric-based methods adopted in the literature for the examination of these two phenomena. Section 16.10 discusses the advantages and drawbacks of the existing test methods and addresses the reliability of the existing indexes obtained from the experimental measurements. Conclusions, together with suggestions for the development of improved indexes, are given in section 16.11.

16.2   Heat-based methods: adiabatic and isothermal calorimetry

16.2.1 Experimental set-up and procedure

These two techniques measure the heat released by a solid undergoing self-heating under preset and controlled conditions. Although both techniques involve a calorimetric vessel to accommodate a sample, with the vessel surrounded by a protective jacket or shield, their operation is completely different. In an adiabatic calorimeter, one inhibits the exchange of heat between the vessel and the environment. As a result, the temperature rise in the sample can be directly applied to gauge the amount of heat generated by the sample. In an isothermal calorimeter, the reactor is immersed in a large bath held at a constant temperature, and the heat released by the exothermic reactions in the reactor is determined by measuring the heat dissipated to the bath.

The most important consideration in constructing an adiabatic calorimeter is to maintain the adiabatic condition of the reactor's wall. This can be achieved either by completely insulating the reactor or by controlling the temperature of the environment in such a way that it closely follows the temperature of the sample. In practice, the first approach is difficult to implement, especially for experimental systems that involve gases flowing in and out of the reactor. Thus, a carefully controlled external heating system is usually applied to maintain the

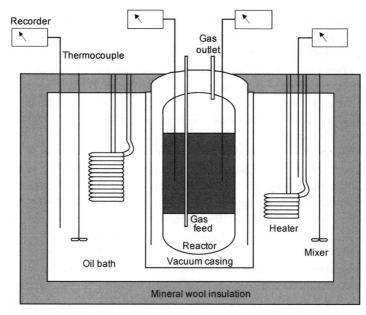

16.2 Schematic diagram of an adiabatic reactor (modified from Cygankiewicz (2000)).

temperatures of the environment and the incoming gases as closely as possible to the sample temperature. This prevents the transfer of heat across the reactor's walls.

Figure 16.2 provides a sketch of a typical adiabatic calorimeter. The oil bath surrounds the reaction vessel equipped with vacuum casing. Temperature sensors measure both the sample and oil bath temperatures, with the difference in the readings of the two sensors employed to control the bath temperature, by means of electric heaters immersed in the oil. The bath also preheats the incoming gases prior to their introduction into the reactor (although, Fig. 16.2 does not illustrate this detail). This prevents cooling down of the sample by convection.

The precise control of the power supplied to the heaters is the critical consideration in the construction of a reliable adiabatic calorimeter. An erroneous amount of energy supplied to the bath can lead to the bath temperature either over- or undershooting the reactor temperature, rather than following closely the temperature rise in the sample. As early as 1924, Davis and Byrne (Davis and Byrne 1924) designed an electric circuit to control the power supplied by the heating coil to the oil bath, which allowed very accurate temperature control. Successive investigators (Cudmore 1964; Guney and Hodges 1968; Humphreys *et al.* 1981; Vance *et al.* 1996) improved the original design by incorporating the photoconductive cell, transistor amplifier and auto-transformer. Modern designs involve computer-aided systems with sophisticated controls to maintain the adiabatic conditions at the reactor's walls (Ren *et al.* 1999; Cygankiewicz 2000).

In an isothermal calorimeter, the reaction vessel is placed in a large and sealed bath filled with a pure chemical of a melting point close to ambient temperature. Typical bath chemicals include ice, diphenylether, phenol or naphthalene with melting points of 0, 26.9, 41 and 80 °C, respectively (Sevenster 1961a; Ponec *et al.* 1974; Chen *et al.* 1990). Good experimental apparatus is designed to melt only a fraction of the chemical present in the bath during an experiment. The heat transferred to the bath is determined by measuring the pressure drift or volumetric expansion of the bath material using a mercury capillary (Sevenster 1961a; Ponec *et al.* 1974; Chen *et al.* 1990; Carras and Young 1994; Cygankiewicz 2000). The isothermal calorimetry is especially well suited to determine the minute amounts of heat generated during the self-heating of solids. However, the apparatus has the drawback of being able to be operated only at discrete temperature settings.

Although both adiabatic and isothermal calorimeters are suitable for measuring small amounts of heat given off by self-heating solids, the adiabatic calorimeters are more suitable for experiments that are several days in duration (Cygankiewicz 2000). In addition, the adiabatic measurements tend to mirror more closely the material behaviour in real situations, where temperature rise of a sample leads to thermal runaway. For these reasons, adiabatic calorimetry has become a popular testing methodology, as justified by a number of applications described in the literature (Davis and Byrne 1924; Cudmore 1964; Humphreys *et al.* 1981; Smith and Lazzara 1987; Miron *et al.* 1990; Cliff *et al.* 1996; Vance *et al.* 1996; Beamish *et al.* 2001).

From several procedures described in the literature, we quote the one employed by Beamish and co-workers (Beamish *et al.* 2001) for studying the self-heating behaviour of coals of various ranks in the adiabatic calorimeter:

1. A dried sample, 150 g in mass and consisting of particles less than 212 μm in size, is placed in the reactor.
2. The reaction vessel is inserted into the oven that has been already preheated to 40 °C. Oxygen-free nitrogen is then flowed into the reactor until the temperature of the sample stabilises at 40 °C.
3. The oven temperature is subsequently switched to the automatic control enabling the oven temperature to adjust itself to the sample temperature.
4. The experiment is started by introducing pure oxygen into the reactor at the rate of 50 cm^3/min.
5. The temperature history is recorded with a data logger.
6. The power and gas supply are shut down once the sample temperature reaches 180 °C.

The experimental procedures reported by other research groups are for the most part very similar to that described above. The differences relate to the experimental settings, as listed in Table 16.2.

Table 16.2 Typical experimental settings adopted by various authors in experiments involving adiabatic calorimeters

Reference	Reactor geometry and material	Sample mass	Oxidation medium	Start and finish temperatures
Davis and Byrne (1924)	Glass tube	$\sim$30 g	Dried oxygen at $\sim$150 cm^3/min	Various start temperatures but stopped at $\sim$140 °C
Guney and Hodges (1968)	Filter tube, 4.6 cm in internal diameter	100 g (fresh or dried)	Water vapour saturated air at 200 or 250 cm^3/min	$\sim$40 °C at the beginning
Humphreys et al. (1981)	Cylindrical vacuum flask, 250 cm^3 in volume	130 g (dried)	Oxygen at 100 cm^3/min	40 and 200 °C
Smith and Lazzara (1987)	Stainless steel cylinder, 12.1 cm in diameter and 10.8 cm in height	100 g (fresh)	Water vapour saturated air at flow rate of 200 cm^3/min	Various start temperatures but stopped at $\sim$140 °C
Vance et al. (1996)	Thermal vacuum flask, 500 cm^3 in volume	150 g (fresh)	Oxygen at $\sim$50 cm^3/min	40 and 150 °C
Beamish et al. (2001)	Not reported	150 g (fresh)	Oxygen at 50 cm^3/min	40 and 180 °C

16.2.2   Indexes for evaluating the self-heating behaviour and typical results

Figure 16.3 shows typical results for coal samples of various moisture content undergoing self-heating in an adiabatic reactor in the presence of oxygen flowing at a rate of 50 cm^3/min (Vance *et al.* 1996). It can be readily inferred from the figure that, the accumulation of heat within the sample leads to a significant temperature rise that accelerates the oxidation reactions and results in thermal runaway. These results correspond to the observations reported by others, for example, Davis and Byrne (1924), Cudmore (1964), Humphreys and co-workers (1981), Smith and Lazzara (1987); Miron and co-investigators (1990); and Beamish (2001). However, a number of exceptions to the thermal runaway behaviour have also been reported in the literature.

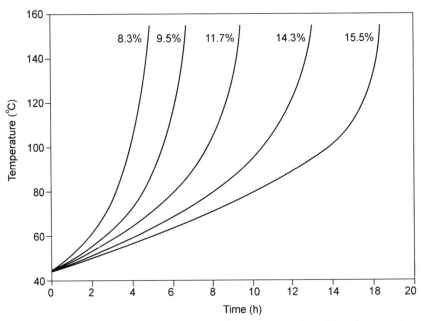

16.3 Effect of the moisture content of a New Zealand sub-bituminous coal on the self-heating behaviour of the material in the adiabatic reactor (Vance *et al.* 1996).

Following the same approach, some workers (Guney and Hodges 1968; Bhattacharyya *et al.* 1969; Singh and Demirbilek 1987; Ren *et al.* 1999) noted experimental measurements with no significant temperature rise and no thermal runaway. Rather, levelling off followed a slight temperature rise for several coals in experiments up to 12 h in duration, as illustrated in Fig. 16.4 (Ren *et al.* 1999). Ren *et al.*'s experiments were performed with air saturated and flowing at the rate of 200 cm^3/min. Obviously, the variation in the character of the temperature histories between the results of Ren *et al.* (1999) and those of Vance and co-workers (1999) may be simply attributed to the differences in the physical and chemical properties of the coal samples themselves. Unfortunately, in situations such as those just described, one can also question the effectiveness of the temperature control system itself for maintaining the adiabatic conditions at the reactor's walls. In our view, it is critically important that initial testing and debugging of a new adiabatic reactor always includes experiments with solid materials that display runaway behaviour.

Indexes have been derived from the adiabatic tests to rank the propensity of solids to combust spontaneously. Initially, Humphreys and co-workers (1981) observed that the temperature of coals in an adiabatic oven often increases linearly with time up to 70 °C, except for coal samples with no propensity to self-heating. Consequently, they proposed the gradient of the linear plots, denoted as R_{70} and presented in units of °C/h, as an index to rank the rate of heat

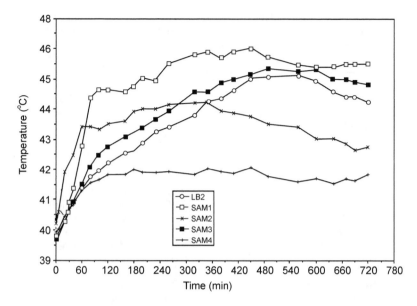

16.4 Temperature histories of five coal samples undergoing self-heating tests in the adiabatic reactor, with air saturated with moisture flowing through the reactor at 200 cm³/min (Ren *et al.* 1999).

generation during the early stage of coal oxidation. This index has been widely used in Australia to assess the risk posed by different coals as a consequence of self-heating. The index classifies coals as high, medium and low risk if their R_{70} is above 0.8, between 0.8 and 0.5 and below 0.5 °C/h, respectively (Beamish *et al.* 2001).

Ren and co-workers (Ren *et al.* 1999) studied the liability of 18 pulverised coals to heat spontaneously, following the approach of Humphreys *et al.* (1981), with the initial test temperature between 40 to 60 °C. Although Ren *et al.* observed no thermal runaway in their experiments, the results characterised spontaneous heating in terms of two new indexes: the initial rate of heating (IRH) and the total temperature rise (TTR); the latter inferred from the initial and final temperature readings in each test (Ren *et al.* 1999). Table 16.3 illustrates typical values of these two indexes with the corresponding risk assessment for a number of bituminous coals.

16.2.3   Minimum self-heating temperature (SHT)

Smith and Lazzara (1987) modified the adiabatic calorimetry method to develop a testing technique to determine the so-called minimum self-heating temperature (SHT), which the authors defined as the lowest initial temperature necessary for the appearance of the thermal runaway in the reactor operated under the

Table 16.3 The initial rate of heating and the total temperature rise for 18 highly volatile bituminous coals; from the investigation of Ren *et al.* (1999)

Coal sample	IRH (°C/h)	TTR (°C)	Risk classification
SAM4	1.35	2.11	Low
INDON2	1.46	2.02	Low to medium
INDON4	1.70	2.39	Low to medium
US1-1	1.42	2.61	Medium
US1-2	1.64	2.97	Medium
UK1	1.70	3.12	Medium
INDON3	2.22	3.60	Medium
SAM5-1	2.23	3.32	Medium
SAM5-2	2.85	3.64	Medium
SAM3	1.16	5.02	Medium to high
INDON1	1.64	4.24	Medium to high
INDON5	1.83	4.34	Medium to high
AUS1	1.89	4.23	Medium to high
UK2	1.63	5.61	High
SAM6	1.34	5.23	High
SAF1	2.71	4.76	High
SAM2	3.20	4.19	High
SAM1	2.96	6.21	Very high

prescribed experimental conditions. Although Smith and Lazarra introduced their method to test coals, the method can be applied to other materials as well. Smith and Lazzara (1987) reported the SHT values for various coals, as listed in Table 16.4. The authors classified coals according to their risk of spontaneous ignition as low, medium and high, taking the minimum self-heating temperature as the index.

The innovative investigation performed by Smith and Lazzara (1987) also included the measurement of the oxidation products at the gas exit with the typical results illustrated in Fig. 16.5. A strong dependence of the amount of CO produced on the temperature can be readily seen in the figure. In fact, the concentration of CO has been used as an indicator for detecting the self-heating phenomenon at mining sites, with the term 'CO make index' coined to denote this indicator. Note that several environmental factors, including the oxygen concentration in the environment, often affect the determination of the CO make index (Coward 1957; Cliff *et al.* 1996; Wang *et al.* 2002a,b).

Since the experimental procedure for the determination of the minimum self-heating temperature differs from that described in section 16.2.1, we outline it below:

1. A pulverised coal sample, 100 g in size and containing particle sizes between 74 and 150 μm, is placed in a reactor.
2. The sample is dried in the stream of dry nitrogen that flows into the reactor at the rate of 200 cm^3/min and at 67 °C.

Table 16.4 Typical values of the minimum self-heating temperature from the modified adiabatic calorimetry according to Smith and Lazarra (1987)

Coal sample	Rank	Minimum SHT, °C	Relative self-heating tendency
Lehigh bed	Lignite	35	High
No. 80-1	High volatile C bituminous	35	
No. 80-2	High volatile C bituminous	40	
F*	High volatile C bituminous	45	
Beulah-Zap	Lignite	60	
E-1	High volatile A bituminous	65	
E-2	High volatile A bituminous	65	
B-1	High volatile A bituminous	70	Medium
No. 6*	High volatile C bituminous	70	
B-2	High volatile A bituminous	75	
Clarion	High volatile A bituminous	75	
D-2	High volatile A bituminous	80	
Lower Sunnyside-2	High volatile A bituminous	80	
Lower Sunnyside-1	High volatile A bituminous	85	
D-1	High volatile A bituminous	90	
Pittsburgh*	High volatile A bituminous	90	
Lower Kittanning	High volatile A bituminous	100	
Pocahontas 3-2*	Low volatile bituminous	110	Low
Pocahontas 3-1*	Low volatile bituminous	115	
Coal Basin-1	Moderate volatile bituminous	120	
Coal Basin-2	Moderate volatile bituminous	120	
Blue Greek	Low volatile bituminous	135	
Mary Lee	Low volatile bituminous	135	
Anthracite	Anthracite	>140	

* Miron and co-workers later reproduced the measurements of the minimum SHT for these coals using the same equipment and the same experimental procedure (Miron *et al.* 1990). The values obtained by these authors differ by up to 10 °C from those presented in the table, suggesting an uncertainty of up to 10 °C in the determination of the minimum SHT for a specific coal.

3. The oven temperature is then equilibrated at ambient temperature. In subsequent experiments, this initial temperature is elevated in increments of 5 °C.

4. Once the sample temperature equilibrates, humid air is introduced into the reactor at the same flow rate.

5. A data logger records the temperature history, as measured by a thermocouple placed in the centre of the sample.

6. If the test results demonstrate no thermal runaway, the experiment is repeated with a fresh sample, with the initial temperature increased by 5 °C, as indicated in point 3.

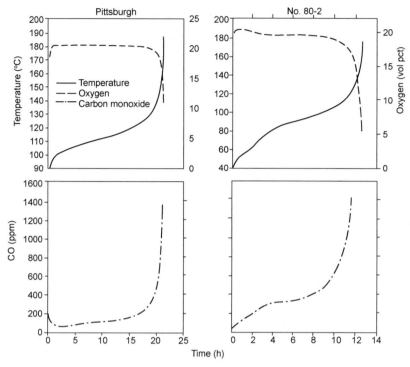

16.5 Consumption of O_2 and evolution of CO during the self-heating tests of Pittsburgh and No. 80-2 coals, with the initial test temperatures of 90 and 40 °C, respectively (Smith and Lazaara 1987).

16.2.4   Relationship between self-heating and material properties

Singh and Demirbilek (1987) examined the self-heating behaviour of 47 coals using the adiabatic calorimetry. By applying multiple regression analysis, these authors attempted to correlate the initial rate of heating (IRH) and total temperature rise (TTR) indexes with 13 material properties: calorific value (CV), moisture content on the air dry basis (MT), total moisture (TM), superficial moisture (SM), volatile matter (VM), density (DE), fixed carbon (FC), ash (ASH), total sulphur (TS), pyrite sulphur (PS), organic plus sulphate sulphur (OSS), total iron (TI), non-pyritic iron (NPI). Table 16.5 includes the empirical equations relating the indexes with the material properties. Note that the quality of fit deteriorates significantly once the correlation includes coals of different ranks. The results of Singh and Demirbilek confirm the existence of close relationships between material properties and the indexes that gauge propensity of materials to self-heat. Unfortunately, such relationships cannot yet be predicted accurately by coupling the material properties with the fundamental mass and heat transfer phenomena operating during the self-heating process.

Table 16.5 Correlations between IRH and TTR indexes and material properties (after Singh and Demirbilek (1987)). Abbreviations of material properties are defined in the text

Coal type	Equation derived by multiple regression	Correlation coefficient	Standard error (estimated)
High-volatile bituminous coal A	IRH = 17.8 + 1.5 ln(TI) − 0.211 ln(NPI) − 2.7 ln(TS) + 0.33 ln(PS) + 0.65 ln(OSS) − 0.4 (DE) − 0.0004 (CV) − 0.29 (MT) − 0.136 (VM) − 0.346 (ASH) + 0.463 (TM)	0.95	0.289
	TTR = 23.5 + 0.473 ln(NPI) + 0.765 ln(PS) − 2.07 ln(OSS) − 1.44 (DE) − 0.0003 (CV) − 1.4 (MT) − 0.135 (VM) − 0.245 (ASH) + 0.916 (TM)	0.913	0.685
High-volatile bituminous coal B	IRH = − 180.0 − 3.13 ln(TI) + 0.977 ln(NPI) − 0.092 ln(TS) + 2.24 ln(PS) + 0.27 ln(OSS) + 0.59 (DE) − 0.0001 (CV) + 1.91 (MT) + 2.03 (VM) + 1.92 (ASH) + 1.86 (FC) + 0.185 (TM)	0.970	0.298
	TTR = − 17.0 + 0.432 ln(NPI) + 1.15 ln(OSS) + 19.9 (DE) − 0.0004 (CV) − 0.509 (MT) + 0.109 (VM) − 0.273 (ASH) + 0.64 (TM)	0.96	0.933
Medium-volatile bituminous coal	IRH = − 4.38 + 4.55 ln(TI) − 4.36 ln(NPI) + 1.38 ln(TS) − 1.70 ln(PS) − 1.85 ln(OSS) − 6.75 (DE) + 0.0001 (CV) + 3.26 (MT) + 0.074 (VM) + 0.017 (ASH) + 0.15 (TM)	0.998	0.208
	TTR = 0.487 + 2.18 ln(TI) − 2.91 ln(NPI) + 2.36 ln(TS) − 1.91 ln(PS) − 2.84 ln(OSS) − 16.8 (DE) + 0.0001 (CV) + 10.7 (MT) + 0.3 (VM) − 0.1 (ASH) + 1.07 (TM)	0.997	0.780
All coals	IRH = − 12.4 + 0.707 ln(TI) − 0.229 ln(NPI) + 0.146 ln(TS) − 0.394 ln(PS) − 0.379 ln(OSS) + 4.22 (DE) + 0.0001 (CV) + 0.071 (MT) + 0.0155 (VM) + 0.0023 (ASH) + 0.058 (TM)	0.736	0.730
	TTR = − 35.6 + 1.46 ln(TI) − 0.268 ln(NPI) − 0.54 ln(TS) − 0.41 ln(PS) − 0.193 ln(OSS) + 16.5 (DE) + 0.0004 (CV) + 0.185 (MT) + 0.0112 (VM) − 0.0144 (ASH) + 0.173 (TM)	0.731	2.117

16.3   Heat-based methods: differential thermal analysis (DTA) and differential scanning calorimetry (DSC)

These two thermal analysis techniques quantify the heat released by a sample undergoing physical or chemical changes. Differential thermal analysis (DTA), involves heating of a small amount of a sample and an inert reference material, at a preset steady rate, and continuously monitoring the temperature difference between the sample and the inert reference material. The sample and the inert reference material are usually placed in one furnace. The output thermogram, a record of temperature difference between the sample and the reference material, provides a means of quantifying the evolution of heat during the spontaneous heating of a solid (Banerjee and Chakravorty 1967; Marinov 1977; Pope and Judd 1980; Clemens *et al.* 1991). The working principle of DSC differs slightly from that of DTA. The sample and reference material are usually placed in two separate furnaces. The heat required to maintain the sample and the reference material at identical temperatures throughout the measurement serves to quantify the heat released by the sample. The book by Pope and Judd (1980) provides a good introduction to the two techniques.

First investigations involving the application of the DTA and DSC techniques for the examination of the propensity of solids to combust spontaneously began to appear in the literature in 1960s. Initially, Banerjee and Chakravorty (1967) attempted to establish a standard procedure to test coals for their tendency to self-heat by analysing the factors affecting the thermogram traces during measurement in differential thermal analysers. These factors included the properties of the reference material, heating rate, type of the material of the sampling cell, particle size and crystallinity of the sample, packing density, sample mass and furnace atmosphere. Banerjee and Chakravorty recommended that coal samples tested in a DTA instrument should contain particles 72 mesh in size and be mixed with alumina, 0.6 g of the sample should be packed in the volume of 1 cm^3 and the heating rate should be set to 5 °C/min.

16.3.1   Auto-ignition temperature

For materials that auto-ignite in the presence of oxygen, the experimental results indicate the existence of three distinct stages in the heating behaviour: (i) in the first stage the endothermic reactions predominate as a consequence of the removal of the inherent moisture present in the pores; (ii) in the second stage, the exothermic reactions grow in significance and the rate of heat evolution commences to increase with temperature; (iii) a very steep rise in heat evolution, as reflected in the temperature readings, defines the third stage. Banerjee and Chakravorty (1967) suggested that the self-heating behaviour of coals could depend on the nature of heat evolution in the second stage, but proposed no quantitative criterion for evaluating the propensity of coals to spontaneous combustion.

Following the same approach, Davies and Horrocks (1983) investigated the spontaneous combustion behaviour of cotton cellulose. Using a DTA instrument with aluminium oxide as the reference material, they placed about 10 mg of Sudanese cotton mixed with 36 mesh inert powder in the measuring cell. Pre-treated air flowing at a rate of between 50 and 300 cm³/min created the oxidation environment in the apparatus. The experiments involved ramping the temperature from the ambient to 500 °C at heating rates between 1 and 20 °C/min. Davies and Horrocks observed three consecutive exothermic reactions to follow the initial pyrolysis of cellulose. These exothermic reactions correspond to the oxidation of volatiles at ~345 °C, the oxidation of solid char at ~460 °C and the oxidation occurring at ~500 °C, ascribed to a char-related gas reaction.

The intersection of the extrapolated baseline (just ahead of the peak) with the tangent to the slope of the leading shoulder of the peak, as denoted by T_i in Fig. 16.6, establishes the onset temperature for auto-ignition. In general, T_i depends on the heating rate but not on the gas flow rate of the oxidation medium introduced to the thermal analyser. Table 16.6 justifies this remark with a typical set of experimental measurements.

ASTM E537-02 provides a standardised methodology for the determination of auto-ignition temperatures using DSC instruments (ASTM 2002). The

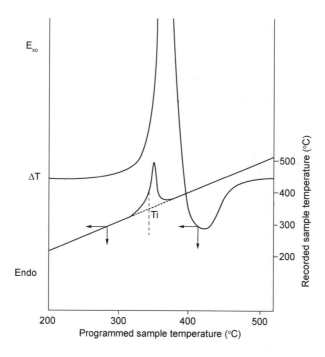

16.6 Typical DTA thermogram demonstrating the determination of the critical temperature for the onset of auto-ignition (Davies and Horrocks 1983).

Table 16.6 Auto-ignition temperatures of cotton cellulose determined from a DTA instrument operated at various heating and air flow rates (Davies and Horrocks 1983)

Heating rate (°C/min)	Spontaneous combustion temperature at various flow rates (cm³/min)			
	50	100	200	300
1	288	290	283	291
2	303	302	302	304
3	311	314	314	313
4	320	320	322	322
5	322	324	324	322
6	326	326	326	326
8	334	334	331	333
10	340	338	337	338
15	346	344	344	344
20	351	350	351	353

experimental conditions and procedure differ slightly from that described in the last paragraph. ASTM E537-02 recommends a sample size of about 5 mg, and a heating rate of between 2 and 20 °C/min, with the flow rate of inert or reactive purge gases set at 50 ± 5 cm³/min.

Another ASTM standard denoted as E698-05 covers the determination of the kinetic parameters from DSC measurements (ASTM 2005). However, Gray and Macaskill have recently objected to applying this standard to materials that heat spontaneously, arguing that this violates the assumption of low values of the Frank-Kamenetskii parameter (see section 16.4.1) implied in E698-05 (Gray and Macaskill 2004).

16.3.2   Other methodologies for assessing the propensity to spontaneous combustion

Huang and Wu (1994) applied the DTA technique to determine the auto-ignition temperature of some conventional explosives, such as RDX, HMX, Tetryl, PETN, TNT and nitrocellulose, based on the energy balance around the sample in the reactor. Huang and Wu realised that the rates of energy generated in the reactor and dissipated to the environment can be written, respectively, as

$$q_1 = VQA\exp\left(-\frac{E}{RT}\right)$$
(16.1)

and

$$q_2 = Sh(T_s - T)$$
(16.2)

By applying the sufficient and necessary conditions for the onset of criticality

$$q_1|_{T=T_i} = q_2|_{T=T_i} \quad \text{and} \quad \frac{dq_1}{dt}\Big|_{T=T_i} = \frac{dq_2}{dt}\Big|_{T=T_i} \qquad \text{16.3a,b}$$

Huang and Wu demonstrated that the auto-ignition temperature of an energetic material corresponds to

$$T_i = -\frac{E}{2R} + \sqrt{\left(\frac{E}{2R}\right)^2 + \frac{ET_{ib}}{R}} \qquad \text{16.4}$$

where T_{ib} is the temperature at the commencement of the steepest slope in the DTA or DSC thermogram. Table 16.7 lists the auto-ignition temperatures for several energetic compounds (Huang and Wu 1994). Note that T_{ib} exceeds the auto-ignition temperature calculated from eqn 16.4. We remark that this method applies for the determination of the auto-ignition temperature only when the rate of temperature rise of a sample significantly surpasses the heating rate set in a run.

Recently, Garcia and co-workers (1999) experimented with samples of three Colombian coals ground to less than 100 Tyler mesh. Samples were first dried by heating them at the rate of 10 °C/min up to 110 °C, in a stream of nitrogen. After the sample cooled down to 50 °C, the oxidation experiment commenced by flowing high purity oxygen into the reactor at the rate of 20 cm³/min and heating the sample at the rate of 10 °C/min up to 600 °C. Supplementary experiments performed by heating a dried sample from 50 to 600 °C under N_2 served to factor

Table 16.7 Auto-ignition temperatures of explosives calculated from the DTA measurements (Huang and Wu 1994)

Compound	Heating rate (K/min)	T_{ib} (K)	E (kJ/mol)	T_i (K)
RDX	6	479.9	187.9	470.1
	10	483.6	187.4	473.6
	15	489.5	195.8	479.7
HMX	6	549.6	216.3	537.8
	10	550.2	209.2	538.6
	15	553.0	241.8	542.9
Tetryl	6	444.7	136.8	433.3
	10	448.2	140.6	436.9
	15	449.3	130.5	437.1
PETN	6	439.5	196.6	431.4
	10	442.4	195.0	434.4
	15	453.1	197.9	444.9
TNT	6	554.3	92.0	529.0
	10	560.7	87.9	533.7
NC	10	460.5	180.7	451.1

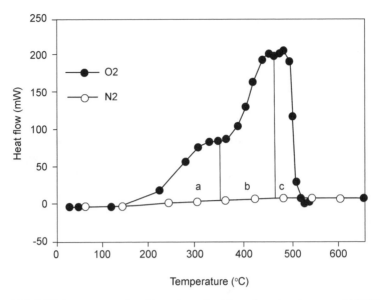

16.7 DSC thermograms for a Venecia coal subjected to a heating rate of 10 °C/ min under O_2 and N_2 (Garcia *et al.* 1999).

out the heat effects owing to the devolatilisation and condensation reactions occurring during pyrolysis.

Figure 16.7 demonstrates typical thermograms obtained by heating one of the Columbian coals under oxygen and nitrogen. The integration of the area under the curve in the thermograms yields the enthalpy of the coal oxidation. Two or three regions can be distinguished in the thermograms, depending on the coal type. The temperature at the onset of fast oxidation denotes the starting point of the first region. This onset temperature provides a convenient measure or an index for ranking the susceptibility of coals to combust spontaneously. Note, however, that this temperature does depend on the prior oxidation history of the material as shown in Table 16.8.

Table 16.8 Effect of weathering time on the onset temperature for three Columbian coals (Garcia *et al.* 1999)

Weathering period (day)	Onset temperature of heat evolution (°C)		
	Venecia coal	Amaga coal	Titirbi coal
0	189	203	204
15	198	207	209
30	199	209	209
45	205	211	210
75	205	213	210
105	212	214	211

16.4   Heat-based methods: basket heating

16.4.1 Mathematical considerations

This method was developed by Bowes and Cameron, based on the Frank-Kamenetskii model for thermal ignition of packed solids (Bowes and Cameron, 1971; Bowes, 1984). For a sample undergoing spontaneous combustion, the general energy conservation can be written as

$$\rho c_p \frac{\partial T}{\partial t} = k \left(\frac{\partial^2 T}{\partial x^2} + \frac{j}{x} \frac{\partial T}{\partial x} \right) + \rho Q A \exp \left(-\frac{E}{RT} \right) \qquad 16.5$$

where $j = 0$, 1, or 2 for the slab, cylinder or sphere, respectively, subject to the following boundary conditions:

$$\frac{\partial T}{\partial x} = 0 \qquad\qquad \text{at } x = 0 \qquad\qquad\qquad 16.6$$

$$\frac{\partial T}{\partial x} = \mp h(T - T_a) \qquad \text{at } x = \pm r, \qquad\qquad 16.7$$

where r represents the radius or the half-thickness of the sample, h denotes the convective heat transfer coefficient at the solid surface, and T_a signifies the temperature of the environment.

As summarised by Cuzzillo (1997), eqn 16.5 implies the following assumptions:

1. The material is homogeneous, isotropic, and possesses constant physical properties.
2. Heat evolution is a consequence of one-step Arrhenius-like overall reaction that does not depend on the consumption of oxygen.
3. The ratio of the activation energy to the universal gas constant, E/R, substantially exceeds the ambient temperature, i.e., $RT_a/E \ll 1$.
4. No other physical processes, such as water evaporation, make a significant contribution to the energy balance.

At steady state, eqn 16.5 assumes the following non-dimensional form

$$\frac{d^2\theta}{dz^2} + \frac{j}{z} \frac{d\theta}{dz} + \delta \exp(\theta) = 0 \qquad\qquad 16.8$$

where the non-dimensional quantity z represents x/r, and θ is defined as

$$\theta = \frac{E}{RT_a^2}(T - T_a) \qquad\qquad\qquad 16.9$$

and the symbol δ denotes the so-called Frank-Kamenetskii parameter, determined from

$$\delta = \frac{\rho Q A}{k} \cdot \frac{E r^2}{RT_a^2} \exp \left(-\frac{E}{RT_a} \right) \qquad\qquad 16.10$$

Table 16.9 Critical values of the Frank-Kamenetskii parameter for most common geometries (Bowes 1984)

Geometry	j (see eqn 16.5)	Critical dimension	$\delta_c(r)$	θ_0
Infinite plane slab	0	Thickness, $2r$	0.857	1.119
Cube	3.280	Side, $2r$	2.569	1.888
Infinite square rod	1.443	Side, $2r$	1.700	1.492
Infinite cylinder	1	Radius, r	2.000	1.386
Short cylinder	2.728	Radius, r	2.844	1.778
Sphere	2	Radius, r	3.333	1.622

The boundary conditions for eqn 16.8, deduced from eqn 16.9 and definition of z, correspond

$$\nabla\theta|_0 = 0 \quad \text{and} \quad \theta|_s = 0 \qquad\qquad 16.11$$

The lack of a solution to eqn 16.8 implies the appearance of auto-ignition behaviour, which occurs once the Frank-Kamenetskii parameter exceeds its critical value, δ_c. The shape of the solid material and the Biot number hr/k constitute the most important parameters affecting δ_c (Bowes 1984; Cuzzillo 1997). Table 16.9 summarises the critical values of the Frank-Kamenetskii parameter for various geometries.

Taking natural logarithm and rearranging eqn 16.10 results in

$$\ln \frac{\delta_c T_a^2}{r^2} = \ln \frac{\rho QAE}{kR} - \frac{E}{RT_a} \qquad\qquad 16.12$$

or

$$\ln \frac{\delta_c T_a^2}{\rho r^2} = \ln \frac{QAE}{kR} - \frac{E}{RT_a} \qquad\qquad 16.13$$

Experimental measurements obtained for a series of basket sizes, $\ln (\delta_c T_a^2/r^2)$ and plotted as a function of the reciprocal temperature, $1/T_a$, permit a least squares fitting of eqn 16.12 or 16.13, which produces an estimate of the activation energy.

16.4.2   Experimental set-up and procedures

A typical apparatus for basket heating comprises: (i) a pre-shaped container (basket) made of wire mesh for holding a sample, with each testing programme necessitating a number of basket sizes; (ii) an oven to provide a constant temperature environment; and, (iii) a thermocouple connected to a data logging system for recording the temperature inside the sample. Most commonly, experiments involve baskets shaped as cubes, cylinders and spheres, with a single shape selected for a testing programme.

The runaway temperature needs to be determined for each basket size, by adhering to the following procedure: (i) a basket is charged with a sample and a thermocouple is inserted in the middle of the basket; (ii) the basket together with the thermocouple is inserted into an oven, which is normally preheated to more than 100 °C; (iii) the data logger records the temperature to determine the presence or absence of the thermal runaway. Depending on the outcome of the initial test, one increases or decreases the oven temperature to bracket the runaway temperature in subsequent experiments. Tests with different basket sizes provide additional critical temperatures required for the determination of the activation energy and the intercept from eqn 16.13. Equation 16.13 can then serve to estimate the maximum size of a pile or a layer of a material as a function of the ambient temperature.

16.4.3   Examples of applications

Using the basket heating technique, Hill and Quintiere (2000) examined the critical temperature of auto-ignition of metal waste to determine the maximum dimensions for the safe storage of this material. The samples comprised aluminium residue collected from a polishing plant in the Midwestern United States. Hill and Quintiere performed experiments with cubic baskets with side lengths of 51, 76 and 102 mm. The typical temperature histories, shown in Fig. 16.8, correspond to the tests conducted at 148.3 and 150.0 °C, for a basket size of 76 mm. The researchers observed the presence and absence of the thermal

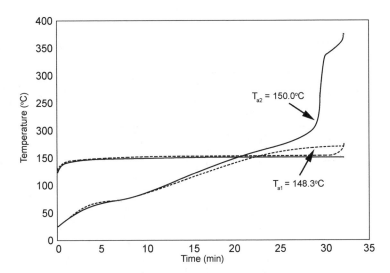

16.8 Temperature histories of the aluminium polishing residue stored in a cube basket, 76 mm in side length, being exposed to the ambient temperatures of 148.3 and 150.0 °C (Hill and Quintiere 2000).

Table 16.10 Critical temperatures for auto-ignition of aluminium polishing residue determined by Hill and Quintiere (2000) by the basket-heating technique

Basket size/ side of cube (mm)	Oven temperature with no thermal runaway (°C)	Oven temperature with thermal runaway (°C)	Mean critical temperature (°C)
51	165.2	165.8	165.5
76	148.3	150.0	149.2
102	129.3	134.9	132.1

runaway at 150 and 148.3 °C, respectively. This means that the critical temperature for this particular geometry lies between 148.3 and 150.0 °C. In addition to these results, Table 16.10 lists the experimental measurements for the two other basket sizes. Note that the increase in the sample size from 51 to 102 mm decreases the critical temperature by about 30 °C.

Bowes and Cameron reported the auto-ignition temperatures and ignition delay times for chemically activated carbon powders (Bowes and Cameron 1971). The samples were prepared in two batches under identical conditions, with the apparent density of between 360 and 380 kg/m^3. As shown in Table 16.11, the auto-ignition temperature decreases with the increase in the dimensions of the cubic sample. Similarly to the results of Hill and Quintiere, this behaviour reflects the magnitude of heat losses from the samples, which

Table 16.11 The auto-ignition temperatures for chemically activated carbon determined by heating the carbon in hexahedral baskets (Bowes and Cameron 1971)

Batch	Linear dimension (mm)	Ratios of sides	δ_c	Auto-ignition temperature (°C)	Time to ignition (h)
A	25.4	1:1:1	2.60	143	—
A	51	1:1:1	2.60	125	1.3
B	51	1:1:1	2.60	118	1.3
A	76	1:1:1	2.60	113	2.7
B	102	1:1:1	2.60	110	5.6
B	102	1:1:1	2.60	106	5.6
A	152	1:1:1	2.60	99	14
B	152	1:1:1	2.60	95	14
B	204	1:1:1	2.60	90	24
B	610	1:1:1	2.60	60	68
A	25.4	1:2:3	1.19	129	—
A	25.4	1:3:3	1.07	125	—
A	51	2:3:3	1.65	117	—

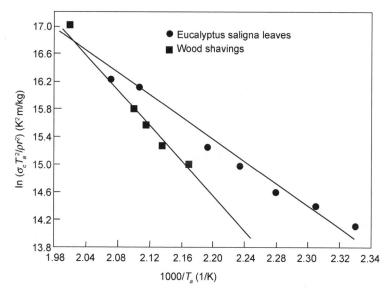

16.9 Application of eqn 16.13 to wood shavings and leaves of Sydney Blue Gum (*E. saligna*) (Jones and Raj 1988).

decrease for larger samples with smaller surface to volume ratios. In other words, larger amounts of solid materials, in storage or in transport, can self-ignite at lower ambient temperatures.

As another example, using cubic baskets with the side lengths ranging from 4 to 10 cm, Jones and Raj (1988) studied the critical ambient temperatures for auto-ignition of two organic materials, i.e. *Eucalyptus saligna* (Sydney Blue Gum) leaves collected from the wildland and wood shavings from an industrial operation. Their experimental measurements when plotted in the form of ln $(\delta_c T_a^2/r^2)$ versus $1/T_a$, following eqn 16.13, yielded the estimates of the activation energies of 71 and 95 kJ/mol for the leaves and the wood shavings, respectively. See Fig. 16.9 for the illustration.

16.5   Heat-based methods: Chen's method

Chen's method, introduced in the 1990s, avoids the labour intensity associated with the basket heating technique. The operation of the Chen's method relates to the observation that for a sample experiencing external heating and exothermic reactions, the heat flux vanishes some distance away from the centre of the sample. In other words, the temperature profile becomes locally flat. The application of the heat conduction equation over a control volume near the sample's centreline allows the elimination of the heat conduction term from eqn 16.5, leading to (Chen and Chong 1995, 1998)

$$\frac{\partial T}{\partial t}\Big|_{T=T_{cr}} = \frac{QA}{C_p} \exp\left(-\frac{E}{RT}\right)\Big|_{T=T_{cr}} \qquad\qquad 16.14$$

where both sides of eqn 16.14 are evaluated at the temperature for which the temperature profile becomes locally flat at the centreline.

Taking the logarithm of both sides of eqn 16.14 yields

$$\ln\frac{\partial T}{\partial t}\Big|_{T=T_{cr}} = \ln\left(\frac{QA}{C_p}\right) - \frac{E}{RT_{cr}}. \qquad\qquad 16.15$$

In practice, two thermocouples are placed at and a short distance away from the centreline of the sample, and the temperature histories are recorded at least until both temperature traces cross each other. For this reason, Chen and Chong (1995, 1998) named this technique as a crossing point temperature method. However, this introduces some confusion with the classical crossing point temperature method (CPT) described in section 16.6. To avoid the ambiguity, we prefer to apply the term Chen's method to the present technique.

In terms of experimental infrastructure, Chen's method necessitates a pre-shaped basket made of metal mesh to contain the solid particles, an oven to provide an environment with constant temperature, and two temperature sensors (rather than one for basket heating) together with the data logging system for monitoring the temperature rise within the sample. Usually, the basket is cylindrical or cubical in shape with cylindrical baskets being several diameters in length.

Normally, four to six experiments, performed under different ambient temperatures (or with different basket sizes), suffice to collect enough experimental measurements to determine the kinetic parameters. This significantly reduces the workload for testing in comparison to that of basket heating. For this reason, the method has been extensively applied to test solid materials for their propensity to self ignite. Each experiment yields an estimate of the natural logarithm of the temperature slope of the left-hand side of eqn 16.15 at T_{cr}. Subsequent plotting of $\ln(\partial T/\partial t)|_{T=T_{cr}}$ as a function of the reciprocal T_{cr} temperature, followed by the least squares fitting of the straight line, provides the kinetic parameters.

The steps enumerated below outline the experimental procedure (Chen and Chong 1995, 1998):

1. Solid particles are packed into a steel mesh basket that is covered with a lid.
2. Two thermocouples are inserted into the sample through two pre-drilled holes in the basket lid, with one thermocouple positioned at the centreline and the other a small distance away.
3. The assembly incorporating the basket and the thermocouples is placed into the oven, which is usually preheated to more than 140 °C.
4. The temperature logger commences to record both temperature readings, for convenience the temperature measurements may be displayed with data acquisition software on a computer screen in real time.

5. The test is terminated when the temperature difference between the two thermocouples reaches zero, or when the two temperature traces cross each other.
6. Steps 1 to 5 are repeated for new oven temperature or for a different size of the basket.

To cite an example of a practical study, Cuzzillo (1997) applied this method to study the auto-ignition behaviour of cooked and uncooked wood blocks, chips and sawdust, and found that the apparent kinetic data determined for these materials are in excellent agreement with the results obtained from another method – basket heating. Following the same lines, Nugroho and co-workers (2000) did the measurements with several types of Indonesian coals, with the typical results shown in Fig. 16.10. They also confirmed that the results coincide with those obtained from an alternative method, basket heating. However, using Chen's method the time spent for the tests was much shorter than that for runs with the basket-heating method.

As often stated in the literature (Chen and Chong 1995, 1998; Chong 1997; Cuzzillo 1997; Nugroho *et al.* 2000), the reliability of the results for the apparent kinetic data (A and E) depends on the accuracy in the determination of the centre temperature and the rate of temperature change. More fundamentally, and in the same way as the method of basket heating, the efficiency of this method depends on how well the equation describes the energy balance within the solid while the material is exposed to a hot environment. The method has the following restrictions:

- The sample is assumed to be non-porous, as the porosity of the material has not been taken into account in eqn 16.15.
- The chemical reactions occurring within the solid are simplified as a one-step steady-state reaction and the rate expression obeys the Arrehenius law, which is independent of the concentration of the reactants.
- Other physical processes, such as water evaporation, make only negligible contributions to the energy balance. In other words, Chen's method applies at temperatures higher than 100 °C.

A similar method was introduced by Jones and co-workers (Jones 1996; Jones *et al.* 1996a,b). This method is also based on simplification of the equation for the energy balance within a solid. It is argued that while a sample is heated in an oven at a preset temperature, heat transfer from the environment to the centre of the sample is through conduction. Once the temperature at the centre is close to the oven temperature, heat transfer from the environment is assumed to be negligible as compared to the rate of the enthalpy change and the heat generation term. Thus, the equation for energy balance for the sample can be simplified, when $T_0 = T_a$

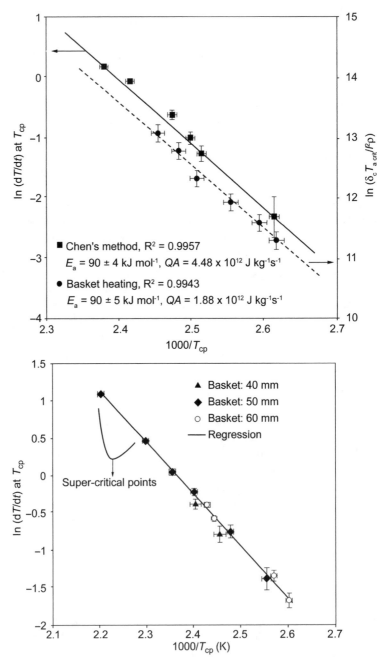

16.10 The determination of the apparent kinetic data using Chen's method and basket heating (Nugroho *et al.* 2000).

$$\rho c_p \frac{dT}{dt}\Big|_{T_0=T_a} = \rho Q A \exp\left(-\frac{E}{RT}\right)\Big|_{T_o=T_a} \qquad 16.16$$

Eliminating the density from both sides and taking the natural logarithm of the two sides yields

$$\ln \frac{dT}{dt}\Big|_{T_0=T_a} = \ln \frac{QA}{c_p} - \frac{E}{RT_a} \qquad 16.17$$

One measures the temperature of a sample undergoing external heating and calculates the rate of temperature change while the temperature at the sample centre approaches the oven temperature. This is necessarily repeated at various oven temperatures. By plotting $\ln(dT/dt)$ versus the reciprocal of the sample temperature at the crossing point, the apparent kinetic parameters for the exothermic reactions occurring during the self-heating of the solid are then determined. Using this approach, Jones and co-workers (Jones 1996; Jones et al. 1996a,b) have been successful in determining the kinetic data for a variety of materials, with the activation energies varying from 54 to 134 kJ/mol for the oxidation of two Scottish bituminous coals and one Irish peat.

The errors in the determination of the kinetic parameters come from two sources. One originates from neglecting the heat conduction term in the energy equation (eqn 16.16) at the time of the so-called crossing point; the other is a consequence of the simplification of the heat generation term caused by chemical reactions occurring during the self-heating process. The spontaneous combustion tests with the wood chips carried out by Cuzzillo (1997) indicate that, once the temperature at the sample centre reaches a steady value, the temperature at the sample centre is similar to but not the same as the oven temperature (Table 16.12).

Table 16.12 The T_p values obtained during the spontaneous combustion tests with Douglas fire chips using Chen's method (Cuzzillo 1997)

Sample status	Basket size/side of cube, mm	Oven temperature, °C	T_p, °C
Uncooked	50	250	246.32
Uncooked	100	200	197.98
Uncooked	200	125	115.59
Cooked	50	180	174.59
Cooked	50	220	227.74
Cooked	50	230	241.28
Cooked	50	240	273.69
Cooked	50	250	293.00
Cooked	100	180	173.21
Cooked	100	200	190.67

16.6   Heat-based methods: crossing-point temperature (CPT)

16.6.1 History of the development and experimental set-up

This method was first introduced at the beginning of the last century for the measurement of the so-called relative ignition temperature of a solid. Early investigators (Wheeler 1918; Parr and Coons 1925) put a powdered coal sample in a tube and placed it in an oil/sand bath and then introduced oxygen or air into the tube. By heating the oil bath gradually, the temperature rise was observed both in the sample and oil bath. The temperature was recorded once the sample reached the temperature of the bath. This temperature was named as the crossing-point temperature (CPT).

This method was further developed by Sherman and co-workers (1941). They used a glass tube to accommodate the sample. The furnace comprised a cylindrical copper block and an electric heater. The power supply for the heater was controlled by a variable transformer. The glass tube was placed in the furnace close to the centre. Air was introduced into the system and the flow rate was controlled using the capillary flow meter. Two thermocouples were used to measure the temperatures of the glass tube reactor and the furnace. During an experiment, the furnace temperature was raised at a steady heating rate, either 15 or 75 °C/min.

The system used by Nandy and co-workers (1972) is essentially identical to that reported by Sherman and co-workers. A sample was placed in a reaction vessel in the form of a U-tube, which was immersed in a glycerine bath. The bath temperature was able to rise at a steady rate using an electric heater and an auto-transformer. Two thermometers were used to monitor the temperatures of the sample centre and the bath. An air flow of 80 cm^3/min was saturated with water by passing through two consecutive water bubblers, before being introduced into the vessel. The experiment was started by increasing the bath temperature at the rate of 0.5 °C/min and was terminated once the crossing-point temperature was determined.

Nowadays, while carrying out a test with the CPT technique, one normally uses a programmable oven to provide a temperature enclosure where the rate of temperature rise can be accurately controlled. Another improvement is that thermocouples, in conjunction with an appropriate data logging system, are applied to measure the temperatures in the sample and in the oven. The experimental set-up utilised by Chakravorty and co-workers (1988) is shown in Fig. 16.11. It is seen that a typical CPT consists of an oven, a reactor to contain a sample, a gas line to introduce the air/oxygen into the reactor, and temperature sensors as well as a recording system. The experimental procedure is also quite simple and is very similar to that adopted by the early investigators. However, the experimental settings used by different researchers

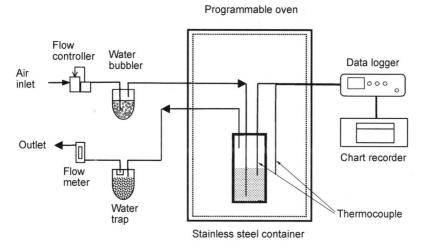

16.11 Schematic diagram of an experimental set-up for the CPT measurements, used by Chakravorty and co-investigators (1988).

often difffer in Table 16.13. The major differences in these measurements include the wide range in the heating rate and the oxidation medium used in these experiments.

Recently, a modified CPT method has been introduced in the literature (Ogunsola and Mikula 1991; Kadioglu and Varamaz 2003). As shown in Fig. 16.12, two reactors rather than one are used to obtain a CPT measurement, with the same amount of sample charged into each of them. The reactors are placed in a temperature-programmed oven. Air is introduced into one of the reactors, with the other receiving an inert gas such as nitrogen. During an experiment, the oven temperature is raised at a constant rate, and the oven temperature and the temperatures at the centre of the two samples are monitored continuously. An experiment is completed once the temperature of the sample in the reactor with flowing air reaches the oven temperature. The value of CPT is then determined and the heat evolution during the sample oxidation can be estimated by comparing the temperature curves obtained from the two reactors. This system is operated in a way similar to a DTA instrument (Ogunsola and Mikula 1991; Kadioglu and Varamaz 2003). Some workers also directly applied the DTA instrument to carry out the CPT tests. While carrying out the CPT tests using a DTA instrument, the crossing point is set at the time when the differential temperature between the sample and the inert material is zero. Details of the experimental procedure are reported in the articles of Tarafdar and Guha (1989) and Gouws and Wade (1989a,b).

Table 16.13 Typical parameters adopted in a CPT test

Author	Sample mass (g)	Particle size (mm)	Heating rate (°C/min)	Oxidation medium
Wheeler (1918)	40	150 × 150– 240 × 240 mesh	Not reported	Dried air
Parr and Coons (1925)	Not reported	60 mesh	10	Dried oxygen
Sherman *et al.* (1941)	31	40–60 mesh	15 or 75	Water saturated air or oxygen
Nandy *et al.* (1972)	20	−72 BSS mesh	0.5	Water saturated air at flow rate of 80 cm³/min
Chandra *et al.* (1983)	10	−72 BSS mesh	1	Oxygen flow at 300 cm³/min
Chakravorty *et al.* (1988); Ogunsola and Mikula (1991)	50	−60 + 200 mesh	0.5	Water saturated air at flow rate of 80 cm³/min
Unal *et al.* (1993)	∼0.7	<0.144	Not reported	Static air
Kadioglu and Varamaz (2003)	60	(1) −0.125; (2) 0.125–0.25; (3) 0.25–0.0.85.	1	Air flow at 50 cm³/min

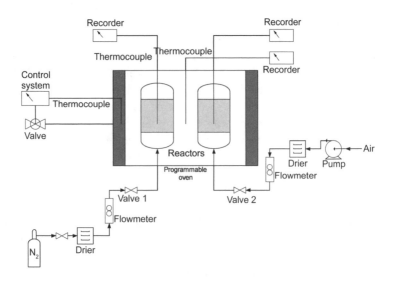

16.12 Schematic diagram of a CPT measurement with twin reactors (modified from Kadioglu and Varamaz 2003).

16.6.2   Indexes for evaluating self-heating behaviour and examples

The first index generated by this method is the crossing-point temperature itself. The CPT corresponds to the lowest temperature at which the exothermic reactions commence to govern the thermal behaviour of a specific system. It was often observed by previous workers (Wheeler 1918; Parr and Coons 1925; Sherman *et al.* 1941; Unal *et al.* 1993; Kadioglu and Varamaz 2003) that, the CPT value depends on experimental conditions, and often changes with the alteration in the experimental parameters, including the shape and dimension of the sample, particle size and the moisture content of the sample, thermal properties of the sample, partial pressure of oxygen in the environment, and heating rate set for raising the oven temperature. In this sense, the physical significance of the CPT is very contentious, as this temperature may be neither the temperature for the onset of self-heating of the solid nor the true temperature for auto-ignition. However, does CPT reflect the spontaneous combustion potential of a material as a combined result of exothermic reactions and the thermal conductivity of the material.

The significance of a temperature rise above the crossing point was first discussed by Sherman and co-workers (1941). They suggested that the rate of temperature rise is directly associated with the partial pressure of oxygen and the Arrhenius reaction rate, which is described by the following equation

$$\frac{dT}{dt}\Big|_{T_0=T_{CP}} = C_1 C_{O_2} \exp\left(-\frac{E}{RT}\right)\Big|_{T_0=T_{CP}} \qquad 16.18$$

where dT/dt represents the slope of the temperature history at the crossing point, C_{O_2} denotes the partial pressure of oxygen, E is the activation energy of the exothermic reactions, and C_1 is an empirical constant. Based on eqn 16.18, Sherman and co-workers (1941) developed a method to determine the apparent activation energy for the exothermic reactions occurring during the self-heating of the sample.

In 1973, Feng *et al.* (1973) proposed a new interpretation of the rate of temperature rise above the crossing point, suggesting that this term reflects the rate of heat generation at this stage. Feng *et al.* introduced an average heating rate between 110 and 230 °C (AHR) to quantify the rate of heat release by the exothermic reactions. This rate corresponds to the average rate of temperature rise or the slope of the temperature–time curve. They also introduced a composite index (FCC) to characterise a sample undergoing exothermic reactions during a CPT test, based on the relative auto-ignition temperature and the rate of heat generation (Feng *et al.* 1973). This index corresponds to the ratio of the rate of temperature rise at the crossing point (AHR) and the crossing point temperature (CPT), i.e.

$$FCC = \frac{AHR}{CPT} \times 1000 \qquad 16.19$$

Table 16.14 Values of FCC index of mountain coals (Feng *et al.* 1973)

Source	Location	Relative ignition temperature (°C)	Average heating rate (°C/min)	Composite index (min^{-1})	Relative self-heating tendency
A	Top 1'7"	164	0.69	4.2	Low
	1'7"–3'7"	168	0.96	5.7	Medium
	3'7"–6'7"	164	0.75	4.6	Low
	6'7"–9'7"	166	0.60	3.6	Low
	9'7"–12'7"	161	1.45	9.0	Medium
	12'7"–15'7"	161	0.54	3.5	Low
B	Top 3'	174	0.50	2.9	Low
	3'–6'	175	1.30	7.4	Medium
	6'–8'	179	1.20	6.7	Medium
C	Top 3'	188	1.34	7.1	Medium
	3'–6'	181	1.11	6.1	Medium
D	Top 3'	163	0.54	3.3	Low
	3'–6'	157	0.71	4.5	Low
	6'–9'	158	0.68	4.3	Low
	9'–12'	157	0.67	4.3	Low
	12'–15'	155	0.59	3.8	Low
E	Belt conv. (1)	162	0.84	5.2	Medium
	Belt conv. (2)	166	0.74	4.5	Low
F	Belt conv.	190	1.00	5.3	Medium
G	Area (1)	190	1.23	6.5	Medium
	Area (2)	168	0.70	4.2	Low
	Area (3)	162	0.90	5.5	Medium
	Area (4)	173	1.00	5.8	Medium
	Area (5)	172	0.77	4.5	Low

The unit of FCC is min^{-1}. This index is often called the FCC index. Feng and co-workers (1973) reported a series of results (Table 16.14) for the coal samples collected from different locations. The values within the same seam of the calculated composite index vary in a wider range, compared to those based on the CPT index.

An alternative method for the calculation of this index to replace the parameter AHR by a parameter, HR, which is defined as the rate of temperature rise at the crossing point (Ogunsola and Mikula 1991, Ogunsola 1996), i.e.

$$HR = \frac{T_{+10} - T_{-10}}{20}$$
16.20

where T_{-10} represents the sample temperature ten minutes after the crossing point, and T_{+10} denotes the sample temperature ten minutes ahead of the time when the sample temperature crosses the oven temperature (Ogunsola and Mikula 1991).

Another index, labelled I_a, is determined by calculating the incremental area between the temperature curves for samples exposed to the air and nitrogen atmospheres. This index reflects the heat evolution by the oxidative reactions (Ogunsola and Mikula 1991, Ogunsola 1996). Using this modified CPT method, Ogunsola and Mikula (1991) examined the capacity of spontaneous combustion of four Nigerian coals. The calculated values of HR, FCC and I_a for these coals are essentially identical (Fig. 16.13). This means that these three indexes, i.e. HR, FCC and I_a could provide more reliable evaluation of the spontaneous combustion capacity of these samples, compared to the index based solely on the CPT value.

Sarac (1993) determined the values of composite index FCC for 31 different coal samples, using the CPT technique. A multiple regression equation was then derived to correlate the liability index (FCC) with the intrinsic parameters of coal properties, including moisture content (MT), volatile matter (VM), fixed carbon (FC), total sulphur (TS) and ash (ASH), i.e.,

$$FCC = 1.2899(MT) - 0.1888(VM) + 0.1519(FC)$$
$$- 3.4199(TS) + 0.1204(ASH) \qquad 16.21$$

The above correlation suggests that the spontaneous combustion of a coal is not a simple phenomenon governed by a sole factor. As pointed out by Carras and Young (1994), 'self-heating behaviour depends in a complicated way on many factors, and no measurement of a single coal property can be used to predict the self-heating behaviour of coal stockpile'.

16.7 Non-heat-based methods: thermogravimetric analysis (TGA)

The self-heating and spontaneous combustion of a solid usually accompanies a change of mass of the material due to the interaction of the solid with oxygen and the generation of the oxidation products. As a result, the behaviour of self-heating and spontaneous combustion can be examined by monitoring the variation in mass of a sample exposed to the oxidation medium and heated within a temperature range. This is often done with a microbalance or a commercial TGA instrument. The operating principle of such a measurement is highlighted in Fig. 16.14. A specimen is loaded into a crucible and placed in a furnace. A gas medium containing oxygen is flowed into the furnace at a controlled flow rate. While heating the sample at a constant heating rate (dynamic mode) or retaining the sample at a constant temperature (static mode), mass change of the sample is recorded; this generates the so-called TG curve. A differential thermogravimetric plot (DTG) can be derived from the TG curve, illustrating the variation in the rate of mass loss with time or temperature (while an apparatus operates in the dynamic mode). The TG and DTG information can

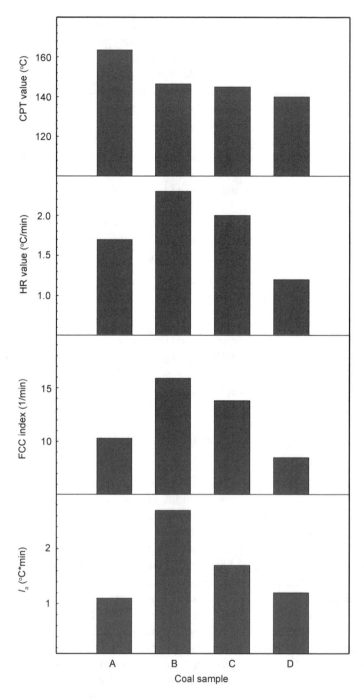

16.13 Values of the indexes for four Nigerian coals using the modified CPT method that contains two reactors (Ogunsola and Mikula 1991).

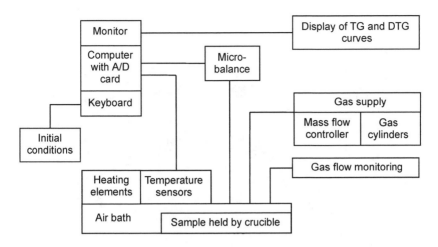

16.14 An illustration of major components and working mechanism of a TGA instrument.

be applied to evaluate the reactivity of a solid towards oxygen and to develop the kinetic models for oxidation of the solid.

This technique has often been used in the examination of oxygen uptake by inorganic materials, such as metals, as the oxidation of a metal results only in an increase in mass of the sample. Using a TG balance, Vernon *et al.* (1939) studied the oxidation of a pure zinc film with a surface area of $50 \times 50 \, mm^2$ in air at various temperatures between 25 and 400 °C. Prior to an experiment, samples were treated by annealing or etching. They found that the mass increase is not only time dependent, but also a function of temperature. The mass increase for a sample oxidised in air during a constant time period remains insignificant at temperatures below 225 °C, as shown in Fig. 16.15, while substantial oxidation of the pure zinc film occurs at higher temperatures with a transition point at ~225 °C. Using a commercial TGA instrument, the self-heating and spontaneous combustion of zinc dust was studied by Bylo and Lewandowski (1980). A sample of 2.0 g was oxidised in air by raising the temperature from 293 to 673 K at a heating rate of between 4.9 and 32.0 K/min. The TG curves illustrate that significant oxidation commences at 573 K, which is supposed to correspond to the onset temperature for spontaneous combustion of zinc dust.

The TG technique alone has rarely been used to investigate the low temperature oxidation of organic materials. This is due to the fact that the loss of the inherent water from an organic material, decomposition of volatiles and the generation of the oxidation products in both gas and solid phases may contribute to the mass change in distinct ways. As a result, the interpretation of the TG and DTG information presents a number of practical

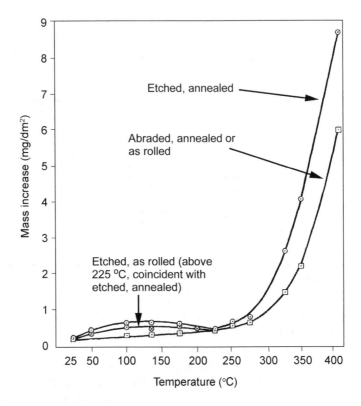

16.15 The amounts of oxygen consumed by the rectangular zinc samples of 50 × 50 × 7 mm in volume with purity of 99.9469% oxidised at various temperatures in air for a period of 200 h (Vernon *et al.* 1939).

difficulties. However, this technique has been successfully used to study the oxidation of coal, in conjunction with the DTA or DSC measurement. Examples can be seen in the articles of Marinov (1977) and Clemens and co-workers (1991).

It is clear that the TGA measurement is a quick and reliable method for determining the reactivity of a solid at different temperatures, provided that the interaction between the solid and oxygen can only form solid products and the reactions can be simplified to one step. In addition, it seems that the amount of oxygen adsorbed by a particular solid under specific conditions can be used as the index for assessing the self-heating behaviour of a solid. However, few attempts have been made to spply such information for evaluating the propensity of a solid toward self-heating and spontaneous combustion.

16.8   Non-heat-based methods: oxygen consumption and generation of gaseous oxidation products

16.8.1 Oxygen adsorption technique

The so-called oxygen adsorption technique examines the capacity of a solid material to consume oxygen and generate gaseous oxidation products under specified experimental conditions. In the description of the test method, the terms oxygen adsorption and oxygen consumption are used interchangeably. The test involves a simple apparatus that incorporates a glass flask, a few hundred millilitres in size, sealing solid samples between 10 and 50 g in mass under air or oxygen. The apparatus is then placed in a constant temperature oven for a period of several days to allow the sample to interact with the oxygen present in the void space of the reactor. As the test proceeds, one periodically records the reactor pressure with a manometer or a pressure transducer, as well as gas composition with an infra-red transducer or gas chromatograph. These measurements allow the determination of the partial pressure of oxygen and gaseous products of oxidation. The technique is very time consuming, as it necessitates long experimental times to establish the adsorption equilibrium, at temperatures near ambient, especially for materials characterised by vast internal surface areas, such as coal (Ponec *et al.* 1974), although the duration of experimental runs can be shortened for finely ground particles.

Although oxygen adsorption is not a heat-based test method, the release of a large amount of heat, in the order of 10 MJ/kg O_2, accompanies the consumption of oxygen. For this reason, the result of the test can be used as a proxy index for ranking the propensity of solids to self-heat. As early as 1915, Winmill (1915–16) studied the uptake of oxygen by various types of British coals and estimated the heat generated as a consequence of the consumption of oxygen. Subsequently, he suggested that coals that auto-ignite have the capacity to consume more than 300 cm^3 per 100 g coal in 96 h at 30 °C, whereas those with a consumption level below 200 cm^3/100 g coal do not tend to self-ignite. At present, Chinese coals are classified into three hazard categories depending on their capacity to adsorb oxygen, under prescribed conditions. Coals able to consume in excess of 0.8 cm^3/g coal are regarded to pose a high risk, those that consume between 0.4 and 0.8 cm^3/g coal as medium risk, and those below 0.4 cm^3/g coal as low risk (Cygankiewicz 2000; Xu 2000).

Another index generated from the 'oxygen adsorption' technique is the so-called Graham ratio or the CO index. First introduced by Graham (1920–21), this index is defined as the amount of CO generated in ppm divided by the amount of O_2 adsorbed in percent during a test. This index reflects both the capacity of a coal to adsorb oxygen and the ability of oxygen consumed to produce CO. Based on the examination of oxygen uptake by seven types of US coals, Kuchta and co-workers (1980) suggested that a value of CO index exceeding 180 appears to be characteristic of coals that are highly susceptible to self-heating. This index has also been used at the mining sites or storage yards for detecting the onset of the self-heating of coal, as it was found that for a specific coal, an increase in the value of Graham's ratio directly implies a temperature rise in the sample (Coward 1957; Burgess and Hayden 1976; Takakuwa et al. 1982; Smith et al. 1991; Cliff et al. 1996). The Graham ratio provides the same information as another index, CO make. Nevertheless, Graham's ratio is more reliable in practical applications, as this index takes into account two parameters, the amount of CO generated and the amount of O_2 adsorbed, and is essentially independent of the local oxygen concentration.

As reported by Chakravorty and co-workers (1988), a typical 'oxygen adsorption' test usually involves the following steps:

1. Around 50 g of coal is placed in a reactor of a known volume.
2. The moisture content of the sample is adjusted in situ by flowing humidified gas into the reactor.
3. The reactor is then sealed and placed in an oven at a preset constant temperature.
4. The pressure drop and temperature rise in the reactor are measured from time to time throughout the experiment.
5. Using a gas chromatograph or gas analyser, the concentration of O_2 and gaseous oxidation products is monitored during or at the end of the experiment.

Since there is no standard established for this technique, the experimental conditions used by different research groupsvary as summarised in Table 16.15. The major differences among these settings lie in the amount of sample utilised during a run, volume of the flask and time allowed for oxygen uptake by a sample. This makes it impossible to compare and utilise the results generated by different investigators. However, it appears that, recently various research groups started to adopt similar conditions for the operation of experimental apparatus to measure oxygen adsorption.

Typical data sets characterising oxygen consumed by coals of different ranks have been reported by Miron and coworkers (1990). A dried or undried sample was stored in a sealed flask of 500 cm^3 in volume for up to seven days. The concentrations of oxygen and some other carbon-containing species were measured, using a gas chromatograph. The amount of oxygen consumed and the CO index (Graham ratio) were then calculated, with the experimental results for the dried samples reported in Table 16.16. It can be seen that the amounts of O_2 consumed and the CO_2 and CO produced for each type of coal are essentially independent of the particle size, while CH_4 and C_2H_6 liberated are either present at very low levels (in the order of ppm) or display a sharp drop with a decrease in particle size. As stated in the literature (Smith et al. 1991; Wang et al. 2003b), hydrocarbons are not products of oxidation, but the residual gases formed during the coalification process.

Table 16.15 Experimental conditions and procedures used in the static 'oxygen adsorption' technique

Author	Reactor volume (cm^3)	Sample mass (g)	Particle size (mm)	Drying method	Operation temperature (°C)	Oxidation medium	Duration (h)	Measured parameters
Katz and Porter (1917)	~4600	100	80 to 100 mesh	Drying agents	25	Air saturated with water	2400	Pressure drop
Sevenster (1961b)	Not reported	~10	−60 BSS	Evacuation at 110°C	0, 32, 42, 55, 72, 92	Oxygen	~6	Pressure drop and concentrations of O_2 and oxidation products
Carpenter and Giddings (1964)	Not reported	~20	−72 + 100 BSS	Flow of dry N_2 at 115°C	75, 85, 95, 105, 115	Oxygen	5	Pressure drop
Kuchta et al. (1980)	250	50	10–20 mesh	Flow of dry N_2 at 70°C	25	Air	168	Pressure drop and concentrations of O_2 and oxidation products
Chakravorty et al. (1988)	2000	50	−60 + 200 mesh	Sample equilibrated at 60% R.H. over 48 h	23, 40, 60	Air saturated with water	96	Pressure drop and concentrations of O_2 and oxidation products
Procarione (1988)	563.6	10	Less than 80 mesh	Fresh sample added with glass wool and 1 ml of distilled water	30	Air	96	Concentration of O_2
Miron et al. (1990)	500	50	(1) 0.6–1.2; (2) 0.3–0.6; (3) 0.15–0.3; (4) 0.075–0.15	Flow of dry N_2 at 67°C	Not reported	Air	168	Pressure drop and concentrations of O_2 and oxidation products

Table 16.16 Average gas compositions in the sealed flasks containing dried coal samples after seven days (Miron *et al.* 1990)

Coal sample	Range of particle size (μm)	Gas composition					O_2 adsorbed (% vol)	CO index
		O_2 (% vol)	CO_2 (% vol)	CO (ppm)	CH_4 (ppm)	C_2H_6 (ppm)		
No. 80	1200 by 600	0.2	0.06	2800	12	0	20.8	135
	600 by 300	0.4	0.05	2703	11	0	20.6	131
	300 by 150	0.3	0.09	3153	15	0	20.6	153
	150 by 75	0.6	0.11	3237	11	0	20.4	159
F	1200 by 500	1.2	0.05	3483	17	3	19.7	177
	600 by 300	1.7	0.05	3727	14	1	19.2	194
	300 by 150	1.7	0.11	4543	15	0	19.3	235
	150 by 75	0.9	0.14	4660	12	0	20.1	232
No. 6	1200 by 500	6.9	0.06	1460	22	3	14.1	104
	600 by 300	7.3	0.07	1350	23	2	13.7	99
	300 by 150	3.9	0.13	1777	26	2	17.1	104
	150 by 75	5.0	0.13	1780	26	2	16.0	111
Pittsburgh	1200 by 500	13.1	0.07	785	10	1	7.8	101
	600 by 300	13.6	0.06	822	10	0	7.3	113
	300 by 150	13.9	0.07	752	20	1	7.1	106
	150 by 75	13.2	0.08	913	14	1	7.8	117
Mary Lee	1200 by 500	17.0	0.04	419	556	8	4.0	105
	600 by 300	17.1	0.04	475	28	6	3.9	122
	300 by 150	18.1	0.04	396	10	3	2.9	137
	150 by 75	18.3	0.04	397	10	0	2.7	147
Pocahontas 3	1200 by 500	18.2	0.05	239	3360	2184	2.8	85
	600 by 300	18.3	0.05	243	560	2160	2.6	93
	300 by 150	18.4	0.06	331	14	536	2.6	127
	150 by 75	18.6	0.06	332	19	42	2.4	138

As shown in Table 16.16, the capacity for oxygen depends on coal type, with the No. 80 and F coals exhibiting the amount of oxygen adsorbed and indicating that these two types of coals are more susceptible to self-heating. However, evaluation of the susceptibility of the coals based on Graham's ratio provides different results, as only the F coal reaches the highest value of between 177 and 232. This apparent inconsistency is a consequence of two distinct phenomena occurring during the interaction of coal with oxygen. The oxygen uptake by the coals includes both physical and chemical processes, while Graham's ratio reflects the capacity of a coal to convert the consumed oxygen into gaseous oxidation products.

16.8.2   Isothermal flow reactor

As opposed to the static 'oxygen adsorption' method, the present techniques involve gases continuously flowing through the reactor. The oxygen consumption and the oxidation products are monitored at the reactor exit. This technique quantifies accurately the time-dependent rates of oxygen consumption and the generation of oxidation products at a pre-set temperature. The isothermal flow reactor is particularly suitable for studying the reaction mechanism and the chemical kinetics of the exothermic reactions during the self-heating of a solid. For this reason, this technique has frequently been applied by various research groups to study the oxidation of coals and other materials (Kam *et al.* 1976a,b; Marinov 1977; Karsner and Perlmutter 1982; Liotta *et al.* 1983; Nelson 1989; Krishnaswamy *et al.* 1996; Wang *et al.* 2002a,b, 2003a,b).

A typical apparatus comprises a reactor body to accommodate the coal particles, gas lines for introducing the gaseous oxidation medium into the reactor, an oven to provide a constant temperature environment and diagnostic instrumentation for the measurement of oxygen consumption and the oxidation products, for example, a gas chromatograph. The experimental facility developed by Young and Nordon (Young 1978) is shown in Fig. 16.16. The cylindrical reactor is placed in a constant-temperature cabinet, and the gas mixture is preheated and then introduced into the reactor, with the water content altered using the water saturator. The gases at the reactor exit are dried and then analysed using a paramagnetic oxygen analyser, dispersive infrared analysers or a gas chromatograph allowing the oxidation products to be quantified accurately.

The sample size, the length of reactor and gas flow rate determine the contact time of coal particles with the oxidising medium. If the contact time is too short, the fresh oxidation medium may not have sufficient time to diffuse into the particles and interact with the active sites at the internal surface of the coal pores; otherwise, the adsorption of the oxidation products by the particles downstream may occur while the gas stream is flowing through the coal bed. Typical parameters of the reactors used by the previous investigators are listed in Table 16.17.

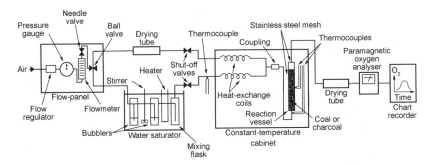

16.16 An isothermal flow reactor constructed by Young and Nordon (Young 1978).

As shown in Table 16.17, some research groups used very large coal samples. In such cases, the isothermal conditions may not be maintained throughout the coal bed, even for a test involving continuously flowing gas. This is because the rate of heat released by coal oxidation is significantly larger than the rate of heat withdrawn by the flowing gas (Smith *et al.* 1991). As the large-scale tests tend to simulate coal undergoing self-heating at mining sites, the test results reflect those expected in practical situations. However, such tests are very costly and time consuming. In addition, the individual variables influencing the self-heating of coal may not be easily controlled during large-scale tests.

Table 16.17 Typical parameters for operation of isothermal flow reactors adopted by various research groups

Author	Reactor shape	Reactor dimension (mm)	Capacity of coal bed (g)	Range of gas flowrate (ml/min)
Radspinner and Howard (1943)	Tube	Not reported	50–60	Not reported
Banerjee and Chakravorty (1967)	Tube	Not reported	Not reported	4–6
Kam et al. (1976a)	Tube	19 (i.d.), 350 (l)	44–45	45–180
Nordon et al. (1979)	Cylinder	22 (i.d.), 155 (l)	30–40	30
Kaji et al. (1985)	Tube	10 (i.d.)	~5	290
Krishnaswamy et al. (1996)	Tube	9.5 (i.d), 1 900 (l)	100	~22
Wang et al. (1999, 2002a)	Cylinder	38 (i.d), 296 (l)	140–180	45
Schmidt and Elder (1940)	Cylinder	660 (i.d.), 1 550 (l)	27 000–182 000	Not reported
Smith et al. (1991)	Cube	1 800 (w) × 1 800 (h) × 4 500 (l)	13 000 000	10 000–200 000

The isothermal flow reactor technique allowed important advances to be made in the study of low-temperature oxidation of coal. New reaction mechanisms and kinetic data have been summarised in a recent literature review written by the current authors (Wang *et al.* 2003b). So far, it seems that no attempts have been made to apply this technique to evaluate the potential of a solid toward self-heating, although the technique is rather straightforward and a relevant index could be easily established (e.g. O_2 consumed or the amount of CO index produced under defined conditions).

16.9   Non-heat-based methods: generation of solid oxidation products

A number of analytical methods have been used to identify the chemical species at solid surfaces, which are responsible for the interaction with oxygen. The techniques employed for this purpose include physical and chemical titrations, infra-red spectroscopy (IR), Fourier transform infra-red spectroscopy (FTIR), X-ray photoelectron spectroscopy (XPS), secondary ion mass spectrometry (SIMS) and ^{13}C nuclear magnetic resonance (^{13}C NMR), electron spin resonance (ESR) and electron paramagnetic resonance spectroscopy (EPR) (Jones and Townend 1949a,b; Chalishazar and Spooner 1957; Swann and Evans 1979; MacPhee and Nandi 1981; Dack *et al.* 1983, 1984; Liotta *et al.* 1983; Perry 1983; Gethner 1985, 1987; Martin *et al.* 1989; Nelson 1989; van Krevelen 1993; Carr *et al.* 1995; Kudynska and Buckmaster 1996; Wang *et al.* 2003b). A majority of the work done in this area is associated with various types of coals.

As early as 1945, Jones and Townend (1945) observed that the peroxygen (–O–O–) or the peroxide is an intermediate formed during the interaction of a coal with oxygen, playing an important role in the coal oxidation process. By separating the peroxygen species from the coal samples by using a titration method with ferrous thiocyanate solution, Jones and Townend (Jones and Townend 1945, 1949a,b; Jones *et al.* 1996b) found that this species starts to decompose at temperatures in excess of 80 °C. The amount of peroxygen formed after the coals oxidised in moist oxygen depends on temperature, with the amount reaching a maximum at about 70 °C. Following the same approach, Chakravorty (1960a,b) confirmed that peroxygen is an initial product of oxidation, and the amount of this species varies with coal type, oxidation temperature, and oxidation history.

In 1983, Chandra and co-workers (1983) measured the concentration of peroxygen species in various types of Indian coals. A sample of 0.2 g with particle size of less than (−72 mesh BSS) was treated with 25 ml (0.01 N) ferrous ammonium sulphate at room temperature, allowing the peroxygen to oxidise the Fe^{+2} into Fe^{+3}. The Fe^{+3} salts were titrated against standard mercurous nitrate solution using ammonium thiocyanate as an indicator. Their observations indicate that the amount of peroxygen in the coals essentially corresponds to the CPT value determined by CPT measurements. In other words, the concentration of peroxygen within a coal could be used as an index to evaluate the propensity of a coal to spontaneous combustion. The relative risk levels of the coal samples were then classified into three groups, in light of the amount of peroxygen in the samples (Table 16.18).

Recent observations suggest that the interaction of coal with oxygen is directly related to the free radicals present at solid surfaces. Using an ESR spectrometer, Dack and co-authors (1983, 1984) measured free radical concentration in Victorian Yallourn brown coal after the drying or oxidation process. The ESR spectra for the coal samples indicate that after vacuum drying, the free radical concentration of brown coal increases, and the oxidation of coal results in sharp fluctuations in the concentration of free radicals. Similar findings have been reported by later researchers using the same technique (Carr *et al.* 1995; Kudynska and Buckmaster 1996). It is clear that the reactivity of a coal towards oxygen is closely related to the quality of free radicals on the coal surface.

A significant amount of work has been performed to verify and quantify other species formed during coal oxidation. The observed species include hydroperoxide (–O–O–H), hydroxyl (phenolic –OH), ether (–C–O–C–), carbonyl (–CO) and carboxyl (–COOH) groups. Although methods have been established to measure the amount of these species formed during the oxidation process, so far no relationship has been established between the reactivity of a coal towards oxygen and the concentration of oxygenated compounds formed during coal oxidation.

There is no doubt that the reactivity of a solid toward oxygen is directly linked to the type and concentration of chemical species at the solid surface. From this perspective it should be possible to develop a new test method for ranking the propensity of coal toward oxidation. However, further effort needs to be made, including confirmation of the compounds responsible for oxygen uptake and the development of a detailed test procedure.

Table 16.18 The correspondence of the risk levels based on the peroxygen index to those based on the CPT values (Chandra *et al.* 1983)

Group No.	Peroxygen index at room temperature ($\times 10^{-5}$ g equivalents)	Crossing-point temperature, °C	Risk classification
I	< 2	> 145	Low
II	2–7	130–145	Moderate
III	> 7	< 130	High

16.10   Advantages and drawbacks of the various test methods

16.10.1   Features of the test methods

The existing methods evaluate the auto-ignition behaviour of a solid in a variety of ways. The suitability of a test method depends on its readiness of carrying out a measurement, the duration for completing a single run, and the reliability of the indexes.

It has been well recognised that adiabatic calorimetry provides the most realistic test methodology, as a run with a reasonable amount of sample can start from the ambient temperature and end after a thermal runaway occurs. Key environmental conditions can be simulated during a test by introducing a gaseous oxidation medium into the reactor at a controlled moisture content and flow rate. The method constitutes a heat-based measurement and directly examines the capacity of a solid towards self-heating. For this reason, this technique has often been used by earlier investigators, and a considerable volume of experimental results have been generated using this technique. It is clear, the reliability of the test results depends on the performance of the equipment, especially the maintenance of the adiabatic conditions in the furnace. However, the technique necessitates long experimental runs which are often a few days in duration.

Although CPT provides a means for rapid acquiring of meaningful measurements, such measurement are apparent in nature as they depend on the experimental conditions and physical properties of a solid, including the heating rate set for a run and the thermal conductivity of the solid. However, the composite index, FCC, which combines the CPT index with the AHR or HR index, provides an improved measure of the propensity of a material to self heat owing to the correction of the AHR or HR value by the CPT index.

Although promising, DTA/DSC techniques suffer from the restriction in the amount of sample used for an individual run. Specifically, DTA/DSC are unable to identify the onset of self-heating though they are suitable for determining the spontaneous ignition. For this reason, DTA/DSC techniques are only applied for determining the critical conditions for the commencement of spontaneous combustion of a solid, which can be characterised by the auto-ignition temperature of a solid.

The basket heating technique is based on rigorous theory, and is able to provide important information for determining critical dimensions for safe storage and transportation of pyrophoric materials. This method is also suitable for the determination of the apparent kinetic data for exothermic reactions occurring in a solid undergoing spontaneous heating. The related Chen method constitutes a more expedient technique for determining the apparent kinetic data for the exothermic reactions. Definitely, the accuracy of the results generated by these methods depends on how accurately the energy equation by itself can describe the phenomenon occurring during spontaneous heating.

Table 16.19 Comparison of the features of various testing methods

Method	Advantages	Major components	Typical operation duration	Indexes	Significance of the results
Adiabatic or isothermal calorimetry	Reflects realistic conditions	1. Temperature control system for maintaining isothermal or adiabatic conditions. 2. Temperature sensors in conjunction with the data logging system.	Varying from a few hours to a few days	1. Temperature for the onset of self-heating, R70. 2. Initial heating rate (IHR) and total temperature rise (TTR). 3. Minimum self-heating temperature (SHT).	1. Critical temperature for the onset of self-heating. 2. Rate of heat released by exothermic reactions during self-heating of a solid. 3. Apparent kinetic data for exothermic reactions.
Differential thermal analysis (DTA) and differential scanning calorimeter (DSC)	Yields fundamental properties	1. Temperature-programmed furnace. 2. Temperature sensors in conjunction with the data logging system.	A few hours for a single run	1. Temperature for the onset of auto-ignition. 2. Rate of heat release.	1. Critical temperature for spontaneous combustion. 2. Apparent kinetic parameters for exothermic reactions.
Basket heating	Provides data for scale-up calculations	1. Oven for setting constant temperature. 2. Temperature sensors in conjunction with the data logging system.	A few hours for a single run, but it is necessary to repeat a few times	1. Lowest temperature for spontaneous combustion. 2. Kinetic data for exothermic reactions.	1. Critical dimension for safe storage of the solid. 2. Capacity of the solid to liberate heat.

Method		Apparatus	Duration	Kinetic/measured data	Information obtained
Chen's and Jones's method	Ranks industrial materials	1. Oven for setting constant temperature. 2. Temperature sensors in conjunction with the data logging system.	A few hours for a single run	Kinetic data for exothermic reactions.	Capacity of the solid to generate heat.
Crossing-point temperature (CPT)	Ranks industrial materials	1. Temperature-programmed furnace. 2. Temperature sensors in conjunction with data logging system.	A few hours for a single run	1. Crossing-point temperature (CPT). 2. Rate of temperature rise at or after crossing point (AHR, HR). 3. Composite index, FCC.	1. Critical temperature for auto-ignition. 2. Rate of heat generation.
Thermogravimetric analysis (TGA)	Yields fundamental properties	1. Temperature-programmed furnace. 2. Micro-balance.	A few hours for a single run	1. Critical temperature for significant mass loss. 2. Kinetic data for oxygen uptake for inorganic materials.	1. Critical temperature for fast oxygen uptake. 2. Reactivity of the oxygen uptake.
'Oxygen adsorption' and isothermal flow reactor technique	Property test; close to reality in full-scale tests	1. Oven for setting constant temperature. 2. Gas analyser or gas chromatograph.	Varying in a vast range from several hours to up to 30 days	1. Amount of oxygen consumed. 2. Indexes based on oxidation products, such as CO make, Graham's ratio.	1. Capacity of the solid to adsorb oxygen. 2. Onset of the significant self-heating of solids based on the oxidation products.
Monitoring of the compounds at solid surface	Yields fundamental properties	Analytical techniques: chemical titration, electron spin resonance (ESR), etc.	A few hours for a test	The amount of active sites (free radicals) and the peroxygen observed at the solid surface.	The reactivity of the solid towards oxygen.

The 'oxygen adsorption' technique examines the capacity of a solid to adsorb oxygen. Although it is an indirect method for the determination of heat released by a sample, it provides reliable measurements of the reactivity of a solid toward oxygen, and yields rates of the consumption of oxygen and the liberation of oxidation products. However, the technique requires long experimental runs. The longest run reported in the literature lasted 100 days. The isothermal flow reactor technique also quantifies oxygen consumption and generation of oxidation products by a solid, but is a fast method to investigate the time-dependent oxidation process. Until the present, flow reactor technique has been applied for investigating the reaction mechanism and chemical kinetics of the interaction of a solid with oxygen. Unfortunately, it has not yet been well developed as a test method for industrial purposes.

The TGA technique is not suitable for the detailed examination of oxygen consumption by an organic material, such as coal, because of the complexity of the chemical reactions occurring during the oxidation process. On the other hand, it is a convenient technique for testing the capacity of an inorganic solid to adsorb oxygen under a range of experimental conditions. However, owing to small sample sizes applied during a run, the information on oxygen uptake by a sample may not be easily discerned at temperatures below 100 °C.

The techniques for the evolution of chemical species at solid surfaces (such as XPS, NMR or EPR) are still in development. This is a direction worthy of further investigation, as there is a potential to develop reliable indexes provided that the mechanism of the oxidation process is fully understood and the chemical species responsible for the exothermic reactions can be identified. Table 16.19 summarises salient features of our experimental methods introduced in this chapter. The reliability of the established indexes is discussed separately in the next section.

16.10.2   Reliability of the established indexes

Much work has been dedicated to evaluating the performance of the existing methods used as tools for assessing the behaviour of spontaneous combustion of a solid. As argued by previous investigators, a generic problem with these methods is the difficulty in comparing the results from different methods. The obverse indexes generated by these methods can be conveniently summarised in four groups:

- critical temperature for the self-heating and auto-ignition of a solid, including SHT established from adiabatic reactor technique; auto-ignition temperature determined by DTA/DSC techniques; crossing-point temperature of CPT method; minimum temperature for auto-ignition from basket heating
- the reactivity of a solid toward oxygen, which may be determined by TGA measurement, oxygen adsorption, isothermal flow reactor, or the measure-

ment of the concentration of chemical species which are responsible for the interaction with oxygen

- the capacity of a solid to generate oxidation products in the gas or solid phases, adiabatic heating, 'oxygen adsorption' and isothermal flow reactor techniques, as well as identification and quantification of chemical species in the solid
- kinetic data for characterising the reactivity of the solid while interacting with oxygen or the exothermic reactions occurring during spontaneous combustion of a solid: 'oxygen adsorption' and isothermal flow reactor techniques, measurement of the oxidation products in solid phase, and some heat-based methods, such as DTA/DSC, basket heating, and Chen's method.

In the past, efforts have been made by previous investigators to correlate the results generated from different test methods. Miron and co-workers (1990) examined the auto-ignition behaviour of six types of bituminous coals by determining the minimum SHT values in an adiabatic oven and the amounts of oxygen consumed in the sealed flask tests. The following relationship was found to correlate the minimum SHT value and the amount of oxygen consumed during the sealed flask tests

$$SHT = 116.6 - 3.9\Delta_{O_2} \hspace{4cm} 16.22$$

where Δ_{O_2} denotes the oxygen consumed (% vol.). This correlation is independent of particle size. It is clear that a higher capacity of coal to adsorb oxygen may correspond to a lower minimum SHT value. The important information contained by eqn 16.22 is that the results from the SHT index are in agreement with those obtained from the static 'oxygen adsorption' measurements.

The correlation between the temperature rise in an adiabatic reactor and the CO index observed during the self-heating of a solid has been confirmed by the work of Smith and Lazzara (1987). As shown in Fig. 16.6, in the adiabatic system, the concentration of CO increases directly with an increase in the temperature of the sample. The significant temperature rise is related to the swift oxygen consumption of the sample.

Conflicting results have been found in the application of different test methods to characterise the spontaneous combustion potential of the same solid. This was highlighted by the work of Chandra et al. (1983). These authors found that two coals characterised by the same CPT value may differ in their propensity to spontaneous combustion. As shown in Table 16.20, the samples denoted as E and F have the same peroxygen index at room temperature but different crossing-point temperatures, while three different coals have the same crossing-point temperature but different peroxygen index values.

Using the CPT and 'oxygen adsorption' techniques, Chakravorty and co-investigators (1988) examined the potential for spontaneous combustion of several western Canadian coals. The results are illustrated in Table 16.21. The

Table 16.20 Results of the indexes generated from two independent testing methods (Chandra *et al.* 1983)

Sample no.	Peroxygen index at room temperature ($\times 10^{-5}$ g equivalents)	Crossing-point temperature, °C
E	7.14	118
F	7.14	141
G	3.06	163
J	1.70	163
U	0.34	163

samples covered all ranks from lignite to the low-volatility bituminous coals. It was found that the risk levels determined by the HR index are identical to that of the FCC index. The risk levels of the coals based on the amount of O_2 consumed by fresh coal are similar to those determined from the CO index. The pattern based on the amount of O_2 consumed by the dried coals is in agreement with that determined by the HR and FCC indexes. Interestingly, the pattern determined by the CPT index is in disagreement with all the other indexes.

An attempt was made by Ogunsola (1996) to establish a universal index that provides an accurate indication of the liability of a material to spontaneous combustion. The autogenous combustion behaviour of 13 coals was studied, using CPT and oxygen adsorption techniques. By analysing the values of CPT, HR, FCC and I_a generated from the CPT method and the amount of oxygen consumed by the 'oxygen adsorption' test, Ogunsola (1996) found that all these indexes can be well correlated by two property parameters of the coals, i.e., equilibrium moisture content and concentration of oxygen-containing species in

Table 16.21 Comparison of the results from two different test methods (Chakravorty *et al.* 1988)

Sample no.	Crossing-point temperature (°C)	Heating rate (°C/min)	FCC index (1/min)	Oxygen consumed after 96 h (cm^3/100 g)		CO concentration after 96 h (ppm)
				As received (60% R.H.)	Dry ash free basis	
A1	203	1.95	9.61	165.24	283.86	1717.02
A2	205	1.92	9.37	135.10	265.47	1069.96
B1	185	1.30	7.03	114.02	166.19	1335.86
B2	180	1.30	7.22	100.50	148.78	1220.42
C1	168	1.38	8.21	207.88	265.35	1922.86
C2	170	1.17	6.88	76.58	214.37	1290.80
D1	203	0.70	3.45	1.80	1.98	62.42
D2	197	0.82	4.16	2.10	2.17	70.06

the samples. Unfortunately, Ogunsola did not provide a universal index for determining the propensity of coals to combust spontaneously.

Work is still in progress to identify the relationships among these indexes and then forecasting accurately the likelihood of spontaneous combustion of a solid. An ideal universal index should take into account at least two or three indexes generated from two or three independent test methods. The risk posed by a solid could then be classified, based on the results computed from such a universal index.

16.11   Conclusions

Several test methods have been developed to evaluate the propensity of a solid toward self-heating and spontaneous combustion. These techniques were designed to evaluate one of the phenomena occurring during the spontaneous combustion process: oxygen consumption by a solid, formation of the oxidation products in gas or solid phase, evolution of heat, and mass change in the solid. The classical methods include adiabatic calorimetry, crossing-point temperature and 'oxygen adsorption' techniques.

The indexes established for ranking the susceptibility of a solid to self-heating or auto-ignition can be classified into four categories: (i) critical temperature for the onset of self-heating or auto-ignition; (ii) reactivity of the solid toward oxygen; (iii) capacity of the solid to liberate oxidation products; (iv) apparent kinetic data for the exothermic reactions occurring during the auto-ignition process. However, no concensus exists in the literature on the most reliable index for assessing the behaviour of the spontaneous combustion of a solid.

A universal index, which could unite the indexes generated from two or three independent test methods is urgently needed. Such an index needs to preserve the advantages of the selected methods but needs to eliminate their drawbacks. As a result, this index should provide more reliable assessment for ranking the propensity of the materials to spontaneous combustion. One possibility is to combine the minimum self heating temperaturte (SHT) with the determination of the amount of oxygen consumed by a solid. In fact, some preliminary work wrongly supports this type of inquiry. Further efforts are urgently required to accelerate progress.

Except for the measurement of the relative auto-ignition temperature using a DTA/DSC instrument, no widely recognised standards have been developed. This makes it difficult to compare the results generated by different research groups even those using the same technique. Attempts have been made by previous investigators to standardise the measurement procedure for the 'oxygen adsorption' technique. As a result, a standard can be readily established for the operation with the 'oxygen adsorption' technique. Adiabatic calorimetry is one of the most popular methods widely used in the evaluation of the self-heating of a solid, and commercial instruments are available. A standardised test procedure for this technique would make it possible for researchers to share and compare the results collected for different materials in separate laboratories.

16.12 Nomenclature

A	pre-exponential factor for an Arrhenius type of reaction (1/s)
AHR	slope of the temperature-time curve between 110 and 230 °C (°C/min)
C_1	empirical constant
C_0	constant
C_{O_2}	oxygen concentration in the gas stream ($kmol/m^3$)
CPT	crossing-point temperature (°C)
c_p	thermal capacity of a solid (J/(kg K) or J/(mol K))
E	activation energy (J/mol)
FCC	composite index based on a CPT test (1/min)
HR	rate of temperature rise at the crossing point (°C/min)
h	convective heat transfer coefficient ($W/(m^2 K)$)
I_a	incremental area between the temperature curves obtained from the samples in the air and nitrogen flow using the modified CPT technique with twin reactors (°C min)
j	geometric factor
k	thermal conductivity (W/(m K))
Q	amount of heat released per unit volume (W/m^3)
q_1	heat liberated within the reactor in unit time (W)
q_2	heat dissipated through the wall to the environment (W)
R	universal gas constant (J/(mol K))
R_{70}	gradient of the linear sections in the plot of temperature rise from 40 to 70 °C for a solid obtained from an adiabatic calorimetric measurement
r	radial distance (m)
SHT	minimum self-heating temperature of a solid determined by a modified adiabatic calorimeter (°C)
T	temperature (°C)
T_0	temperature at the sample centre (°C)
T_a	environmental temperature/oven temperature (°C)
T_{cr}	critical temperature when the temperature at the sample centre reaches the temperature at the point adjacent to the sample centre during a measurement using the Chen method (°C)
T_{ib}	temperature at the starting point of the steepest slope on a thermogram determined by DTA or DSC (°C)
t	time (min or h)
x	distance (m)
z	dimensionless distance, x/r

Greek symbols

δ	Frank-Kamenetskii parameter
δ_c	critical Frank-Kamenetskii parameter
Δ_{O_2}	oxygen consumed during an 'oxygen adsorption' test (% vol)
θ	non-dimensional temperature
θ_0	non-dimensional temperature at the sample centre
ρ	density of the solid (kg/m^3)

Subscripts

0	at sample centre
a	in the environment
CP	at crossing point during the CPT measurement
O_2	oxygen
s	on the reactor wall surface

16.13   References

ASTM E537-02 (2002) Standard test method for the thermal stability of chemicals by differential scanning calorimetry. West Conshohocken, PA, USA, ASTM International.

ASTM E698-05 (2005) Standard test method for Arrhenius kinetic constants for thermally unstable materials using differential scanning calorimetry and the Flynn/Wall/Ozawa methods. West Conshohocken, PA, USA, ASTM International.

Banerjee, S. C., and Chakravorty, R. N. (1967). 'Use of DTA in the study of spontaneous combustion of coal.' *J. Mines, Metals Fuels*: 1–5.

Beamish, B. B., Barakat, M. A., and George, J. D. (2001). 'Spontaneous-combustion propensity of New Zealand coals under adiabatic conditions.' *Inter. J. Coal Geol.* **45**: 217–24.

Bhattacharyya, K. K., Hodges, D. J., and Hinsley, F. B. (1969). 'The influence of humidity on the initial stages of the spontaneous heating.' *The Mining Engineer*: 274–84.

Bowes, P. C. (1984). *Self-heating: evaluating and controlling the hazards*. New York, Elsevier.

Bowes, P. C., and Cameron, A. (1971). 'Self-heating and ignition of chemically activated carbon.' *J. Appl. Chem. Biotechnol.* **21**: 244–50.

Burgess, D., and Hayden, H. H. (1976). 'Carbon monoxide index monitoring system in an underground coal mine.' *Trans. Soc. Mining Engineers, AIME* **260**: 312–17.

Bylo, Z., and Lewandowski, P. (1980). 'Investigation of reaction mechanism of self-oxidation of zinc dust.' *Archiwum Hutnictwa* **25**(4): 723–44.

Carpenter, D. L. and Giddings, D.G. (1964). 'The initial stages of the oxidation of coal with molecular oxygen. I. Effect of time, temperature and coal rank on rate of oxygen consumption.' *Fuel* **43**: 247–66.

Carr, R. M., Kumagai, H., Peake, B. M., Robinson, B. H., Clemens, A.H., and Matheson, T. W. (1995). 'Formation of free radical during drying and oxidation of a lignite and a bituminous coal.' *Fuel* **74**: 389–94.

Carras, J. N., and Young, B. C. (1994). 'Self-heating of coal and related materials: models, application and test methods.' *Prog. Energy Combust. Sci.*: 1–15.

Chakravorty, S. L. (1960a). 'Auto-oxidation of Indian coals. Part I – Influence of variables on the formation of peroxidic bodies.' *J. Mines, Metals, Fuels*: 1–4.

Chakravorty, S. L. (1960b). 'Auto-oxidation of Indian coals. Part II – Mechanism of oxidation.' *J. Mines, Metals, Fuels*: 10-5.

Chakravorty, R. N., Kar, K., and Mansour, N. (1988). Categorization of western Canadian coals with respect to their susceptibility to spontaneous combustion, CANMET.

Chalishazar, B. H., and Spooner, C. E. (1957). 'A rapid method for the determination of peroxide groups on coal.' *Fuel* **36**: 127–28.

Chandra, D., Bhattacharya, S. K., Ghosh, R., and Dasgupta, N. (1983). 'On evaluation and classification of coal with respect to proneness to spontaneous combustion.' *Quart. J. Geol. Min. Metall. Soc. India* **55**(3): 130–6.

Chen, X. D., and Chong, L. V. (1995). 'Some characteristics of transient self-heating inside an exothermically reactive porous solid slab.' *Trans. IChemE* **73 (B)**: 101–7.

Chen, X. D., and Chong, L. V. (1998). 'Several important issues related to the crossing-point-temperature (CPT) method for measuring self-ignition kinetics of combustible solids.' *Trans IChemE* **76 (B)**: 90–3.

Chen, Z., Ge, X., and Gu, X. (1990). *Techniques for measurements of heat and thermal properties of materials*. Hefei, Anhui, China, Press of China University of Science and Technology.

Chong, L. V. (1997). *Thermal ignition of dairy powders*. Auckland, New Zealand, The University of Auckland.

Clemens, A. H., Matheson, T. W., and Rogers, D. E. (1991). 'Low temperature oxidation studies of dried New Zealand coals.' *Fuel* **70**: 215–21.

Cliff, D., Rowlands, D., and Sleeman, J. (1996). *Spontaneous combustion in Australian underground coal mines*. Queensland, SIMTARS.

Coward, H. F. (1957). *Research on spontaneous combustion of coal in mines – A review*, Safety in Mines Research Establishment, Great Britain.

Cudmore, J. F. (1964). 'Determination of self-heating rates of coal.' *Chem. Ind. (London)* **41**: 1720–1.

Cuzzillo, B. R. (1997). *Pyrophoria*. Berkeley, CA, USA, The University of California.

Cygankiewicz, J. (2000). 'Determination of susceptibility of coals to spontaneous combustion by using an adiabatic test method.' *Archives Min. Sci.* **45**(2): 247–73.

Dack, S. W., Hobday, M. D., Smith, T. D., and Pilbrow, J. R. (1983). 'Free radical involvemnet in the oxidation of Victorian brown coal.' *Fuel* **62**: 1510–2.

Dack, S. W., Hobday, M. D., Smith, T. D., and Pilbrow, J. R. (1984). 'Free-radical involvement in the drying and oxidation of Victorian brown coal.' *Fuel* **63**: 39-42.

Davies, D., and Horrocks, A. R. (1983). 'Ignition studies on cotton cellulose by DTA.' *Thermchim. Acta* **63**: 351–62.

Davis, J. D., and Byrne, J. F. (1924). 'An adiabatic method for studying spontaneous heating of coal.' *J. Amer. Ceram. Soc.* **7**: 809–16.

Feng, K. K., Chakravorty, R. N., and Cochrane, T. S. (1973). 'Spontaneous combustion. Coal mining hazard.' *Canadian Mining Metall.* **66**: 75–84.

Garcia, P., Hall, P. J., and Mondragon, F. (1999). 'The use of differential scanning calorimetry to identify coals susceptible to spontaneous combustion.' *Thermochim. Acta* **336**: 41–6.

Gethner, J. S. (1985). 'Thermal and oxidation chemistry of coal at low temperatures.' *Fuel* **64**: 1443–6.

Gethner, J. S. (1987). 'The mechanism of the low temperature oxidation of coal by O_2: Observation and separation of simultaneous reactions using *in situ* FT-IR difference spectroscopy.' *Applied Spectroscopy* **41**: 50–63.

Gouws, M. J., and Knoetze, T. P. (1995). 'Coal self-heating and explosibility.' *J. S. Afr. Inst. Min. Metall.*: 37–43.

Gouws, M. J., and Wade, L. (1989a). 'The self-heating liability of coal: Predictions based on simple indexes.' *Mining Sci. Technol.* **9**(1): 75–80.

Gouws, M. J., and Wade, L. (1989b). 'The self-heating liability of coal: Predictions based on composite indexes.' *Mining Sci. Technol.* **9**(1): 81–5.

Graham, J. I. (1920–21). 'The normal production of carbon monoxide in coal-mine.' *Trans. Inst. Mining Eng., London* **60**: 222–34.

Gray, B. and Macaskill, C. (2004). 'The role of self-heating in the estimation of kinetic constants for thermally unstable materials using differential scanning calorimetry (DSC).' Proc. Interflam 2004, Edinburgh, UK, 961–972, Interscience Communications.

Guney, M., and Hodges, D. J. (1968). 'An adiabatic apparatus designed to study the self-heating rates of coal.' *Chem. Ind.*: 1429–33.

Hill, S. M., and Quintiere, J. G. (2000). 'Investigating materials from fires using a test method for spontaneous ignition.' *Fire Mater.* **24**: 61–6.

Huang, C., and Wu, T. (1994). 'A simple method for estimating the auto-ignition temperature of solid energetic materials with a single non-isothermal DSC or DTA curve.' *Thermochim. Acta* **239**: 105–14.

Humphreys, D., Rowlands, D., and Cudmore, J. F. (1981). *Spontaneous combustion of some Queensland coals.* Proc. Ignitions, Explosions and Fires in Coal Mines Symp., Aus. I.M.M. Illawara Branch.

Jones, J. C. (1996). 'Steady behaviour of long duration in the spontaneous heating of a bituminous coal.' *J. Fire Sci.* **14**: 159–66.

Jones, J. C., and Raj, S. C. (1988). 'The self-heating and ignition of vegetation debris.' *Fuel* **67**: 1208–10.

Jones, R. E., and Townend, D. T. A. (1945). 'Mechanism of the oxidation of coal.' *Nature* **155**: 424–5.

Jones, R. E., and Townend, D. T. A. (1949a). 'The oxidation of coal.' *J. Soc. Chem. Ind.* **68**: 197–201.

Jones, R. E., and Townend, D. T. A. (1949b). 'The role of oxygen complexes in the oxidation of carbonaceous materials.' *Trans. Far. Soc.* **42**: 297–305.

Jones, J. C., Chiz, P. S., Koh, R., and Matthew, J. (1996a). 'Continuity of kinetics between sub- and supercritical regimes in the oxidation of a high-volatile solid substrate.' *Fuel* **75**(15): 1733–6.

Jones, J. C., Chiz, P. S., Koh, R., and Matthew, J. (1996b). 'Kinetic parameters of oxidation of bituminous coals from heat-release rate measurements.' *Fuel* **75**: 1755–7.

Kadioglu, Y., and Varamaz, M. (2003). 'The effect of moisture cotent and air-drying on spontaneous combustion characteristics of two Turkish lignites.' *Fuel* **82**: 1985–93.

Kaji, R., Hishinuma, Y. and Nakamura, Y. (1985). 'Low temperature oxidation of coals. Effects of pore structure and coal composition.' *Fuel* **64**: 297–302.

Kam, A. Y., Hixson, A. N., and Perlmutter, D. D. (1976a). 'The oxidation of bituminous coal - I. Development of a mathematical model.' *Chem. Eng. Sci.* **31**: 815–19.

Kam, A. Y., Hixson, A. N., and Perlmutter, D. D. (1976b). 'The oxidation of bituminous coal - II. Exerimental kinetics and interpretation.' *Chem. Eng. Sci.* **31**: 821–34.

Karsner, G. G., Perlmutter, D. D. (1982). 'Model for coal oxidation kinetics. 1. Reaction under chemical control.' *Fuel* **61**: 29–34.

Katz, S. H., and Porter, H. C. (1917). *Effects of moisture on the spontaneous heating of stored coal*, US Bureau of Mines.

Kim, A. G. (1977). *Laboratory studies on spontaneous heating of coal. A summary of information in the literature*, US Bureau of Mines.

Krishnaswamy, S., Gunn, R. D., and Agarwal, P. K. (1996). 'Low-tempertaure oxidation of coal. 2. An experimental and modelling investigation using a fixed-bed isothermal flow reactor.' *Fuel* **75**: 344–52.

Kuchta, J. M., Rowe, V. R., and Burgess, D.S. (1980). *Spontaneous combustion susceptibility of U.S. coals*, US Bureau of Mines.

Kudynska, J., and Buckmaster, H. A. (1996). 'Low-temperature oxidation kinetics of high-volatile bituminous coal studied by dynamic *in situ* 9 GHz c.w. e.p.r. spectroscopy.' *Fuel* **75**: 872–8.

Liotta, R., Brons, G., and Isaacs, J. (1983). 'Oxidative weathering of Illinois No. 6 coal.' *Fuel* **62**: 781–91.

MacPhee, J. A., and Nandi, B. N. (1981). '^{13}C n.m.r. as a probe for the characterization of the low-temperature oxdiation of coal.' *Fuel* **60**: 169–70.

Marinov, V. N. (1977). 'Self-ignition and mechanisms of interaction of coal with oxygen at low temperatures. 2. Changes in weight and thermal effects on gradual heating of coal in air in the range 20–300 °C.' *Fuel* **56**: 158–64.

Martin, R. R., MacPhee, J. A., Workinton, M., and Lindsay, E. (1989). 'Measurement of the activation energy of the low temperature oxidation of coal using secondary ion mass spectrometry.' *Fuel* **68**: 1077–79.

Miron, Y., Smith, A. C., and Lazzara, C. P. (1990). *Sealed flask test for evaluating the self-heating tendencies of coals*, US Bureau of Mines.

Nandy, D. K., Banerjee, D. D., and Chakravorty, R. N. (1972). 'Application of crossing point temperature for determining the spontaneous heating characteristics of coals.' *J. Mines, Metals Fuels* **20**(2): 41–8.

Nelson, C. R. (1989). 'Coal weathering: Chemical processes and pathways.' *Chemistry of Coal Weathering*. N.C.R. Amsterdam, Elsevier.

Nordon, P., Young, B. C., and Bainbridge, N. W. (1979). 'The rate of oxidation of char and coal in relation to their tendency to self-heating.' *Fuel* **58**: 443–9.

Nugroho, Y. S., McIntosh, A. C., and Gibbs, B. M. (2000). 'Low-temperature oxidation of single and blended coals.' *Fuel* **79**: 1951–61.

Ogunsola, O. I. (1996). *Autogenous combustion propensity of coal: in search of a universal index and an accurate method of determination*. Proceedings: NOBCChE'96.

Ogunsola, O. I., and Mikula, R. J. (1991). 'A study of spontaneous combustion characteristics of Nigerian coals.' *Fuel* **70**: 258–61.

Parr, S. W., and Coons, C. C. (1925). 'Carbon dioxide as an index of the critical oxidation temperature for coal in storage.' *J. Ind. Eng. Chem.* **17**: 118–20.

Perry, D. L. and Grint, A. (1983). 'Application of XPS to coal characterization.' *Fuel* **62**: 1024–33.

Ponec, V., Knor, Z. and Cerny, S. (1974). *Adsorption on solids*. London, Butterworths.

Pope, M. I., and Judd, M. D. (1980). *Differential thermal analysis. A guide to the technique and its applications.* London, Heyden.

Procarione, J. A. (1988). 'Spontaneous combustion tests applied to abandoned coal mine refuse.' *Mining Sci. Technol.* **6**(2): 147–52.

Radspinner, J. A. and Howard, H. C. (1943). 'Determination of surface oxidation of bituminous coal.' *Ind. Eng. Chem.: Analy. Ed.* **15**: 566–70.

Ren, T. X., Edwards, J. S., and Clarke, D. (1999). 'Adiabatic oxidation study on the propensity of pulverised coals to spontaneous combustion.' *Fuel* **78**: 1611–20.

Sarac, S. (1993). 'A statistical approach to spontaneous combustion tendency of coal.' *J. Mines, Metals Fuels* **41**(6–7): 146–9.

Schmidt, L. D., and Elder, J. L. (1940). 'Atmospheric oxidation of coal at moderate temperatures: Rates of the oxidation reaction for representative coking coals.' *Ind. Eng. Chem.* **32**: 249–56.

Sevenster, P. G. (1961a). 'Studies on the interaction of oxygen with coal in the temperature range 0 to 90 °C, Part I.' *Fuel* **40**: 7–17.

Sevenster, P. G. (1961b). 'Studies on the interaction of oxygen with coal in the temperature range 0 to 90 °C, Part II – Consideration of the kinetics of the process.' *Fuel* **40**: 18–32.

Sherman, R. A., Pilcher, J. M., and Ostborg, H. N. (1941). 'A laboratory test for the ignitibility of coal.' *Fuel* **20**(8): 194–202.

Singh, R. N., and Demirbilek, S. (1987). 'Statistical appraisal of intrinsic factors affecting spontaneous combustion of coal.' *Mining Sci. Technol.* **4**(2): 155–65.

Singh, R. V. K., Sharma, A., and Jhanwar, J. C. (1993). 'Different techniques for study of spontaneous heating susceptibility of coal with special reference to Indian coals.' *J. Mines Metals Fuels* **41**(2–3): 60–4.

Smith, A. C., and Lazzara, C. P. (1987). *Spontaneous combustion studies of U.S. coals*, US Bureau of Mines.

Smith, A. C., Miron, Y. and Lazzara, C. P. (1991). *Large-scale studies of spontaneous combustion of coal*, US Bureau of Mines.

Swann, P. D., and Evans, D. G. (1979). 'Low-temperature oxidation of brown coal. 3. Reaction with molecular oxygen at temperatures close to ambient.' *Fuel* **58**: 276–80.

Takakuwa, I., Kikuchi, A., and Haneda, H. (1982). 'A judgment method of ceasing tendency of coal spontaneous combustion by gas analysis.' *Saiko to Hoan* **28**(5): 69–75.

Tarafdar, M. N., and Guha, D. (1989). 'Application of wet oxidation processes for the assessment of the spontaneous heating of coal.' *Fuel* **68**: 315–17.

Unal, S., Piskin, S., and Dincer, S. (1993). 'Autoignition tendencies of some Turkish lignites.' *Fuel* **72**: 1357–9.

van Krevelen, D. W. (1993). *Coal: Typology – chemistry – physics – constitution.* London, Elsevier.

Vance, W. E., Chen, X. D., and Scott, S. C. (1996). 'The rate of temperature rise of a subbituminous coal during spontaneous combustion in an adiabatic device: The effect of moisture content and drying methods.' *Combust. Flame* **106**: 261–70.

Vernon, H. J., Akeroyd, E. I., and Stroud, E. G. (1939). 'The direct oxidation of zinc.' *J. Inst. Metals* **65**: 301–43.

Wang, H., Dlugogorski, B. Z., and Kennedy, E. M. (1999). 'Experimental study on low-temperature oxidation of an Australian coal.' *Energy Fuel* **13**: 1173–9.

Wang, H., Dlugogorski, B. Z., and Kennedy, E. M. (2002a). 'Examination of CO_2, CO and H_2O formation during low-temperature oxidation of a bituminous coal.' *Energy Fuels* **16**: 586–92.

Wang, H., Dlugogorski, B. Z., and Kennedy, E. M. (2002b). 'Kinetic modeling of low-temperature oxidation of coal.' *Combust. Flame* **131**: 452–69.

Wang, H., Dlugogorski, B. Z., and Kennedy, E. M. (2003a). 'Analysis of the mechanism of the low-temperature oxidation of coal.' *Combust. Flame* **134**: 107–17.

Wang, H., Dlugogorski, B. Z., and Kennedy, E. M. (2003b). 'Coal oxidation at low temperatures: Oxygen consumption, oxidation products, reaction mechanism and kinetic modelling.' *Prog. Energy Combust. Sci.* **29**: 487–513.

Wheeler, R. V. (1918). 'Oxidation and ignition of coal.' *J. Chem. Soc., Abstracts* **113**: 945–55.

Winmill, T. F. (1915–16). 'The absorption of oxygen by coal. IX. Comparison of rates of adsorption of oxygen by different varieties of coal.' *Trans. Inst. Mining Eng., London* **51**: 510–47.

Xu, B. (2000). 'Research on risk assessment index of coal spontaneous combustion.' *Shandong Keji Daxue Xuebao, Ziran Kexueban* **19**(4): 97–101.

Young, B. C. and Nordon, P. (1978). 'Method for determining the rate of oxygen sorption by coals and chars at low temperatures.' *Fuel* **57**(9): 574–5.